D0122702

Inventions
and
Inventors

Inventions and Inventors

Volume 1

Abortion pill — Laminated glass

1 – 458

edited by
Roger Smith

SALEM PRESS, INC.
Pasadena, California   Hackensack, New Jersey

Copyright © 2002, by SALEM PRESS, INC.

All rights in this book are reserved. No part of this work may be
used or reproduced in any manner whatsoever or transmitted
in any form or by any means, electronic or mechanical, including photocopy, recording, or any information storage and re-
trieval system, without written permission from the copyright
owner except in the case of brief quotations embodied in critical
articles and reviews. For information address the publisher, Salem Press, Inc., P.O. Box 50062, Pasadena, California 91115.

Essays originally appeared in *Twentieth Century: Great Events*
(1992, 1996), *Twentieth Century: Great Scientific Achievements* (1994),
and *Great Events from History II: Business and Commerce Series*
(1994). New material has been added.

∞ The paper used in these volumes conforms to the American
National Standard for Permanence of Paper for Printed Library
Materials, Z39.48-1992 (R1997).

Library of Congress Cataloging-in-Publication Data
Inventions and inventors / edited by Roger Smith
    p.cm. — (Magill's choice)
    Includes bibliographical reference and index
    ISBN 1-58765-016-9 (set : alk. paper) — ISBN 1-58765-017-7
(vol 1 : alk. paper) — ISBN 1-58765-018-5 (vol 2. : alk. paper)
1. Inventions—History—20th century—Encyclopedias. 2. Inventors—Biography—Encyclopedias. I. Smith, Roger, 1953-   .
II. Series.

T20 .I59 2001
609—dc21   2001049412

TABLE OF CONTENTS

Publisher's Note

To many people, the word "invention" brings to mind cleverly contrived gadgets and devices, such as safety pins, zippers, typewriters, and telephones—all of which have fascinating stories of invention behind them. However, the word actually has a much broader meaning, one that goes back to the Latin word *invenire*, for "to come upon." In its broad sense, an invention can be *any* tangible device or contrivance, or even a process, that is brought into being by human imagination. It is in this broad sense that the term is used in *Inventions and Inventors*, the latest contribution to the Magill's Choice reference books.

This two-volume set contains articles on 195 twentieth century inventions, which span the full range of human imagination—from simple gadgets, such as disposable razors, to unimaginably complex medical breakthroughs, such as genetically engineered insulin. This set is not an encyclopedic catalog of the past century's greatest inventions but rather a selective survey of noteworthy breakthroughs in the widest possible variety of fields.

A combination of several features sets *Inventions and Inventors* apart from other reference works on this subject: the diversity of its subject matter, the depth of its individual articles, and its emphasis on the people behind the inventions. The range of subjects covered here is unusually wide. In addition to articles on what might be considered "classic" inventions—such as airplanes, television, and satellites—the set has articles on inventions in fields as diverse as agriculture, biology, chemistry, computer science, consumer products, drugs and vaccines, energy, engineering, food science, genetic engineering, medical procedures, music, photography, physics, synthetics, transportation, and weapons technology.

Most of this set's essays appeared earlier in *Twentieth Century: Great Events* (1992, 1996) and *Twentieth Century: Great Scientific Achievements* (1994). Its longest essays are taken from *Great Events from History II: Business and Commerce Series* (1994). Information in the articles has been updated, and completely new bibliographical notes have been added to all of them. Half the essays also have original sidebars on people behind the inventions.

At least one thousand words in length, each essay opens with a brief summary of the invention and its significance, followed by an annotated list of important personages behind it—including scientists, engineers, technicians, and entrepreneurs. The essay then examines the background to the invention, its process of discovery and innovation, and its impact on the world. Half the articles have entirely new sidebars on individuals who played important roles in the inventions' development and promotion.

Users can find topics by using any of several different methods. Articles are alphabetically arranged under their titles, which use the names of the inventions themselves, such as "Abortion pill," "Airplane," "Alkaline storage battery," "Ammonia," and "Amniocentesis." Many inventions are known by more than one name, however, and users may find what they are looking for in the general index, which lists topics under multiple terms.

Several systems of cross-referencing direct users to articles of interest. Appended to every essay is a list of articles on related or similar inventions. Further help in can be found in appendices at the end of volume two. The first, a Time Line, lists essay topics chronologically, by the years in which the inventions were first made. The second, Topics by Category list, organizes essay topics under broader headings, with most topics appearing under at least two category headings. Allowing for the many topics counted more than once, these categories include Consumer products (36 essays), Electronics (28), Communications (27), Medicine (25), Measurement and detection (24), Computer science (23), Home products (20), Materials (18), Medical procedures (17), Synthetics (17), Photography (16), Energy (16), Engineering (16), Physics (13), Food science (13), Drugs and vaccines (13), Transportation (11), Weapons technology (11), Genetic engineering (11), Aviation and space (10), Biology (9), Chemistry (9), Exploration (8), Music (7), Earth science (6), Manufacturing (6), and Agriculture (5).

More than one hundred scholars wrote the original articles used in these volumes. Because their names did not appear with their articles in the *Twentieth Century* sets, we cannot, unfortunately, list them here. However, we extend our thanks for their contributions. We also are indebted to Roger Smith for his help in assembling the topic list and in writing all the biographical sidebars.

Editor's Foreword

The articles in *Inventions and Inventors* recount the birth and growth of important components in the technology of the twentieth centuries. They concern inventions ranging from processes, methods, sensors, and tests to appliances, tools, machinery, vehicles, electronics, and materials. To explain these various inventions, the essays deal with principles of physics, chemistry, engineering, biology, and computers—all intended for general readers. From complex devices, such as electron microscopes, and phenomena difficult to define, such as the Internet, to things so familiar that they are seldom thought of as having individual histories at all, such as Pyrex glass and Velcro, all the inventions described here increased the richness of technological life. Some of these inventions, such as the rotary-dial telephone, have passed out of common use, at least in the United States and Europe, while others, such as the computer, are now so heavily relied upon that mass technological culture could scarcely exist without them. Each article, then, is at the same time a historical sketch and technical explanation of an invention, written to inform and, I hope, intrigue.

Brief biographical sidebars accompany half the articles. The sidebars outline the lives of people who are in some way responsible for the inventions discussed: the original inventor, a person who makes important refinements, an entrepreneur, or even a social crusader who fostered acceptance for a controversial invention, as Margaret Sanger did for the birth control pill. These little biographies, although offering only basic information, call forth the personal struggles behind inventions. And that is a facet to inventions that needs emphasizing, because it shows that technology, which can seem bewilderingly impersonal and complex, is always rooted in human need and desire.

Roger Smith
Portland, Oregon

Inventions
and
Inventors

ABORTION PILL

THE INVENTION: RU-486 was the first commercially available drug that prevented fertilized eggs from implanting themselves in the walls of women's uteruses.

THE PEOPLE BEHIND THE INVENTION:
Étienne-Émile Baulieu (1926-   ), a French biochemist and endocrinologist
Georges Teutsch, a French chemist
Alain Bélanger, a French chemist
Daniel Philibert, a French physicist and pharmacologist

DEVELOPING AND TESTING

In 1980, Alain Bélanger, a research chemist, was working with Georges Teutsch at Roussel Uclaf, a French pharmaceutical company. Teutsch and Bélanger were interested in understanding how changes in steroids affect the chemicals' ability to bind to their steroid receptors. (Receptors are molecules on cells that can bind with certain chemical substances such as hormones. Receptors therefore act as connecting links to promote or prevent specific bodily activities or processes.) Bélanger synthesized several steroids that bonded to steroid receptors. Among these steroids was a compound that came to be called "RU-486."

Another member of the research project, Daniel Philibert, found that RU-486 blocked the activities of progesterone by binding tightly to the progesterone receptor. Progesterone is a naturally occurring steroid hormone that prepares the wall of the uterus to accept a fertilized egg. Once this is done, the egg can become implanted and can begin to develop. The hormone also prevents the muscles of the uterus from contracting, which might cause the uterus to reject the egg. Therefore RU-486, by acting as a kind of shield between hormone and receptor, essentially stopped the progesterone from doing its job.

At the time, Teutsch's group did not consider that RU-486 might be useful for deliberately interrupting human pregnancy. It was

Étienne-Émile Baulieu, a biochemist and endocrinologist and a consultant for Roussel Uclaf, who made this connection. He persuaded the company to test RU-486 for its effects on fertility control.

Many tests were performed on rabbits, rats, and monkeys; they showed that, even in the presence of progesterone, RU-486 could prevent secretory tissue from forming in the uterus, could change the timing of the menstrual cycle, and could terminate a pregnancy—that is, cause an abortion. The compound also seemed to be nontoxic, even in high doses.

In October of 1981, Baulieu began testing the drug with human volunteers. By 1985, major tests of RU-486 were being done in

---

ÉTIENNE-EMILE BAULIEU

Étienne-Émile Baulieu was born in Strasbourg, France, in 1926. He moved to Paris for his advanced studies at the Faculty of Medicine and Faculty of Science of Pasteur College. He was an Intern of Paris from 1951 until he received a medical degree in 1955. He passed examinations qualifying him to become a teacher at state schools in 1958 and during the 1961-1962 academic year was a visiting scientist in Columbia University's Department of Biochemistry.

In 1963 Baulieu was made a Doctor of Science and appointed director of a research unit at France's National Institute of Health and Medical Science, a position he held until he retired in 1997. He also served as Head of Service of Hormonal Biochemistry of the Hospital of Bicêtre (1970-1997), professor of biochemistry at University of Paris-South (1970-1993), and consultant for Roussel Uclaf (1963-1997).

Among his many honors are the Gregory Pincus Memorial Award (1978), awards from the National Academy of Medicine, the Christopher Columbus Discovery Award in Biomedical Research (1992), the Joseph Bolivar DeLee Humanitarian Award (1994), and Commander of the Legion of Honor (1990). Although busy with research and teaching duties, Baulieu was on the editorial board of several French and international newspapers, a member of scientific councils, and a participant in the Special Program in Human Reproduction of the World Health Organization.

France, Great Britain, The Netherlands, Sweden, and China. When a relatively low dose of RU-486 was given orally, there was an 85 percent success rate in ending pregnancy; the woman's body expelled the embryo and all the endometrial surface. Researchers found that if a low dose of a prostaglandin (a hormonelike substance that causes the smooth muscles of the uterus to contract, thereby expelling the embryo) was given two days later, the success rate rose to 96 percent. There were few side effects, and the low doses of RU-486 did not interfere with the actions of other steroid hormones that are necessary to keep the body working.

In the March, 1990, issue of *The New England Journal of Medicine*, Baulieu and his coworkers reported that with one dose of RU-486, followed in thirty-six to forty-eight hours with a low dose of prostaglandin, 96 percent of the 2,040 women they studied had a complete abortion with few side effects. The women were monitored after receiving the prostaglandin to watch for side effects, which included nausea, vomiting, abdominal pain, and diarrhea. When they returned for a later checkup, fewer than 2 percent of the women complained of side effects. The researchers used two different prostaglandins; they found that one caused a quicker abortion but also brought about more pain and a longer period of bleeding.

Using the Drug

In September, 1988, the French government approved the distribution of RU-486 for use in government-controlled clinics. The next month, however, Roussel Uclaf stopped selling the drug because people opposed to abortion did not want RU-486 to be available and were threatening to boycott the company.

Then, however, there were threats and pressure from the other side. For example, members of the World Congress of Obstetrics and Gynecology announced that they might boycott Roussel Uclaf if it did not make RU-486 available. The French government, which controlled a 36 percent interest in Roussel Uclaf, ordered the company to start distributing the drug once more.

By the fall of 1989, more than one-fourth of all early abortions in France were being done with RU-486 and a prostaglandin. The French government began helping to pay the cost of using RU-486 in 1990.

Testing for approval of RU-486 was completed in Great Britain and The Netherlands, but Roussel Uclaf's parent company, Hoechst AG, did not try to market the drug there or in any other country outside France. (In the United States, government regulations did not allow RU-486 to be tested using government funds.)

Medical researchers believe that RU-486 may be useful not only for abortions but also in other ways. For example, it may help in treating certain breast cancers and other tumors. RU-486 is also being investigated as a possible treatment for glaucoma—to lower pressure in the eye that may be caused by a high level of steroid hormone. It may be useful in promoting the healing of skin wounds and softening the cervix at birth, easing delivery. Researchers hope as well that some form of RU-486 may prove useful as a contraceptive—that is, not to prevent a fertilized egg from implanting itself in the mother's uterus but to prevent ovulation in the first place.

IMPACT

Groups opposed to abortion rights have spoken out against RU-486, while those who favor the right to abortion have urged its acceptance. The drug has been approved for use in China as well as in France. In the United States, however, the government has avoided giving its approval to the drug. Officials of the World Health Organization (WHO) have argued that RU-486 could prevent the deaths of women who undergo botched abortions. Under international law, WHO has the right to take control of the drug and make it available in poor countries at low cost. Because of the controversy surrounding the drug, however, WHO called for more testing to ensure that RU-486 is quite safe for women.

See also Amniocentesis; Antibacterial drugs; Artificial hormone; Birth control pill; Salvarsan.

FURTHER READING

Baulieu, Etienne-Emile, and Mort Rosenblum. *The "Abortion Pill":
    RU-486, a Woman's Choice.* New York: Simon & Schuster, 1991.
Butler, John Douglas, and David F. Walbert. *Abortion, Medicine, and
    the Law.* 4th ed. New York: Facts on File, 1992.

Lyall, Sarah. "Britain Allows Over-the-Counter Sales of Morning-After Pill." *New York Times* (January 15, 2001).

McCuen, Gary E. *RU 486: The Abortion Pill Controversy.* Hudson, Wis.: GEM Publications, 1992.

Nemecek, Sasha. "The Second Abortion Pill." *Scientific American* 283, no. 6 (December, 2000).

Zimmerman, Rachel. "Ads for Controversial Abortion Pill Set to Appear in National Magazines." *Wall Street Journal* (May 23, 2001).

Airplane

THE INVENTION: The first heavier-than-air craft to fly, the airplane revolutionized transportation and symbolized the technological advances of the twentieth century.

THE PEOPLE BEHIND THE INVENTION:
Wilbur Wright (1867-1912), an American inventor
Orville Wright (1871-1948), an American inventor
Octave Chanute (1832-1910), a French-born American civil engineer

A CAREFUL SEARCH

Although people have dreamed about flying since the time of the ancient Greeks, it was not until the late eighteenth century that hot-air balloons and gliders made human flight possible. It was not until the late nineteenth century that enough experiments had been done with kites and gliders that people could begin to think seriously about powered, heavier-than-air flight. Two of these people were Wilbur and Orville Wright.

The Wright brothers making their first successful powered flight, at Kitty Hawk, North Carolina. (Library of Congress)

The Wright brothers were more than just tinkerers who accidentally found out how to build a flying machine. In 1899, Wilbur wrote the Smithsonian Institution for a list of books to help them learn about flying. They used the research of people such as George Cayley, Octave Chanute, Samuel Langley, and Otto Lilienthal to help them plan their own experiments with birds, kites, and gliders. They even built their own wind tunnel. They never fully trusted the results of other people's research, so they repeated the experiments of others and drew their own conclusions. They shared these results with Octave Chanute, who was able to offer them lots of good advice. They were continuing a tradition of excellence in engineering that began with careful research and avoided dangerous trial and error.

SLOW SUCCESS

Before the brothers had set their minds to flying, they had built and repaired bicycles. This was a great help to them when they put their research into practice and actually built an airplane. From building bicycles, they knew how to work with wood and metal to make a lightweight but sturdy machine. Just as important, from riding bicycles, they got ideas about how an airplane needed to work.

They could see that both bicycles and airplanes needed to be fast and light. They could also see that airplanes, like bicycles, needed to be kept under constant control to stay balanced, and that this control would probably take practice. This was a unique idea. Instead of building something solid that was controlled by levers and wheels like a car, the Wright brothers built a flexible airplane that was controlled partly by the movement of the pilot, like a bicycle.

The result was the 1903 *Wright Flyer*. The Flyer had two sets of wings, one above the other, which were about 12 meters from tip to tip. They made their own 12-horsepower engine, as well as the two propellers the engine spun. The craft had skids instead of wheels. On December 14, 1903, the Wright brothers took the *Wright Flyer* to the shores of Kitty Hawk, North Carolina, where Wilbur Wright made the first attempt to fly the airplane.

The first thing Wilbur found was that flying an airplane was not as easy as riding a bicycle. One wrong move sent him tumbling into

THE WRIGHT BROTHERS

Orville and his older brother Wilbur first got interested in aircraft when their father gave them a toy helicopter in 1878. Theirs was a large, supportive family. Their father, a minister, and their mother, a college graduate and inventor of household gadgets, encouraged all five of the children to be creative. Although Wilbur, born in 1867, was four years older than Orville, they were close as children. While in high school, they put out a weekly newspaper together, *West Side News*, and they opened their bicycle shop in 1892. Orville was the mechanically adept member of the team, the tinkerer; Wilbur was the deliberative one, the planner and designer.

Since the bicycle business was seasonal, they had time to pursue their interest in aircraft, puzzling out the technical problems and studying the successes and failures of others. They started with gliders, flying their first, which had a five-foot wing span, in 1899. They developed their own technique to control the gliders, the "wing-warping technique," after watching how birds fly. They attached wires to the trailing edges of the wings and pulled the wires to deform the wings' shape. They built a sixteen-foot glider in 1900 and spent a vacation in North Carolina gaining flying experience. Further designs and many more tests followed, including more than two hundred shapes of wing studied in their home-built wind tunnel, before their first successful engine-powered flight in 1903.

Neither man ever married. After Wilbur died of typhoid in 1912, Orville was stricken by the loss of his brother but continued to run their business until 1915. He last piloted an airplane himself in 1918 and died thirty years later.

Their first powered airplane, the *Wright Flyer*, lives on at the National Air and Space Museum in Washington, D.C. Small parts from the aircraft were taken to the Moon by Neil Armstrong and Edwin Aldrin when they made the first landing there in 1969.

the sand only moments after takeoff. Wilbur was not seriously hurt, but a few more days were needed to repair the *Wright Flyer*.

On December 17, 1903, at 10:35 A.M., after eight years of research and planning, Orville Wright took to the air for a historic twelve seconds. He covered 37 meters of ground and 152 meters of air space. Both brothers took two flights that morning. On the fourth flight, Wilbur flew for fifty-nine seconds over 260 meters of ground and through more than 800 meters of air space. After he had landed, a sudden gust of wind struck the plane, damaging it beyond repair. Yet no one was able to beat their record for three years.

IMPACT

Those first flights in 1903 got little publicity. Only a few people, such as Octave Chanute, understood the significance of the Wright brothers' achievement. For the next two years, they continued to work on their design, and by 1905 they had built the *Wright Flyer III*. Although Chanute tried to get them to enter flying contests, the brothers decided to be cautious and try to get their machine patented first, so that no one would be able to steal their ideas.

News of their success spread slowly through the United States and Europe, giving hope to others who were working on airplanes of their own. When the Wright brothers finally went public with the *Wright Flyer III*, they inspired many new advances. By 1910, when the brothers started flying in air shows and contests, their feats were matched by another American, Glen Hammond Curtiss. The age of the airplane had arrived.

Later in the decade, the Wright brothers began to think of military uses for their airplanes. They signed a contract with the U.S. Army Signal Corps and agreed to train military pilots.

Aside from these achievements, the brothers from Dayton, Ohio, set the standard for careful research and practical experimentation. They taught the world not only how to fly but also how to design airplanes. Indeed, their methods of purposeful, meaningful, and highly organized research had an impact not only on airplane design but also on the field of aviation science in general.

See also Bullet train; Cruise missile; Dirigible; Gas-electric car; Propeller-coordinated machine gun; Rocket; Stealth aircraft; Supersonic passenger plane; Turbojet; V-2 rocket.

FURTHER READING

Brady, Tim. *The American Aviation Experience: A History.* Carbondale: Southern Illinois University Press, 2000.

Chanute, Octave, Marvin Wilks, Orville Wright, and Wilbur Wright. *The Papers of Wilbur and Orville Wright: Including the Chanute-Wright Letters and Other Papers of Octave Chanute.* New York: McGraw-Hill, 2000.

Culik, Fred, and Spencer Dunmore. *On Great White Wings: The Wright Brothers and the Race for Flight.* Toronto: McArthur, 2001.

Howard, Fred. *Wilbur and Orville: A Biography of the Wright Brothers.* Mineola, N.Y.: Dover Publications, 1998.

ALKALINE STORAGE BATTERY

THE INVENTION: The nickel-iron alkaline battery was a lightweight, inexpensive portable power source for vehicles with electric motors.

THE PEOPLE BEHIND THE INVENTION:

Thomas Alva Edison (1847-1931), American chemist, inventor, and industrialist

Henry Ford (1863-1947), American inventor and industrialist

Charles F. Kettering (1876-1958), American engineer and inventor

A THREE-WAY RACE

The earliest automobiles were little more than pairs of bicycles harnessed together within a rigid frame, and there was little agreement at first regarding the best power source for such contraptions. The steam engine, which was well established for railroad and ship transportation, required an external combustion area and a boiler. Internal combustion engines required hand cranking, which could cause injury if the motor backfired. Electric motors were attractive because they did not require the burning of fuel, but they required batteries that could store a considerable amount of energy and could be repeatedly recharged. Ninety percent of the motorcabs in use in New York City in 1899 were electrically powered.

The first practical storage battery, which was invented by the French physicist Gaston Planté in 1859, employed electrodes (conductors that bring electricity into and out of a conducting medium) of lead and lead oxide and a sulfuric acid electrolyte (a solution that conducts electricity). In somewhat improved form, this remained the only practical rechargeable battery at the beginning of the twentieth century. Edison considered the lead acid cell (battery) unsuitable as a power source for electric vehicles because using lead, one of the densest metals known, resulted in a heavy battery that added substantially to the power requirements of a motorcar. In addition, the use of an acid electrolyte required that the battery container be either nonmetallic or coated with a non-metal and thus less dependable than a steel container.

THE EDISON BATTERY

In 1900, Edison began experiments aimed at developing a rechargeable battery with inexpensive and lightweight metal electrodes and an alkaline electrolyte so that a metal container could be used. He had already been involved in manufacturing the nonrechargeable battery known as the Lalande cell, which had zinc and copper oxide electrodes and a highly alkaline sodium hydroxide electrolyte. Zinc electrodes could not be used in a rechargeable cell because the zinc would dissolve in the electrolyte. The copper electrode also turned out to be unsatisfactory. After much further experimentation, Edison settled on the nickel-iron system for his new storage battery. In this system, the power-producing reaction involved the conversion of nickel oxide to nickel hydroxide together with the oxidation of iron metal to iron oxide, with both materials in contact with a potassium hydroxide solution. When the battery was recharged, the nickel hydroxide was converted into oxide and the iron oxide was converted back to the pure metal.

Although the basic ingredients of the Edison cell were inexpensive, they could not readily be obtained in adequate purity for battery use. Edison set up a new chemical works to prepare the needed materials. He purchased impure nickel alloy, which was then dissolved in acid, purified, and converted to the hydroxide. He prepared pure iron powder by using a multiple-step process. For use in the battery, the reactant powders had to be packed in pockets made of nickel-plated steel that had been perforated to al-

Thomas A. Edison. (Library of Congress)

low the iron and nickel powders to come into contact with the electrolyte. Because the nickel compounds were poor electrical conductors, a flaky type of graphite was mixed with the nickel hydroxide at this stage.

Sales of the new Edison storage battery began in 1904, but within six months it became apparent that the battery was subject to losses in power and a variety of other defects. Edison took the battery off

---

THOMAS ALVA EDISON

Thomas Alva Edison (1847-1931) was America's most famous and prolific inventor. His astonishing success story, rising from a home-schooled child who worked as a newsboy to a leader in American industry, was celebrated in children's books, biographies, and movies. Corporations still bear his name, and his inventions and improvements of others' inventions—such as the light bulb, phonograph, and motion picture—shaped the way Americans live, work, and entertain themselves. The U.S. Patent Office issued Edison 1,093 patents during his lifetime, the most granted to one person.

Hailed as a genius, Edison himself emphasized the value of plain determination. Genius is one percent inspiration and 99 percent perspiration, he insisted. He also understood the value of working with others. In fact, one of his greatest contributions to American technology involved organized research. At age twenty-three he sold the rights to his first major invention, an improved ticker-tape machine for Wall Street brokers, for $40,000. He invested the money in building an industrial research laboratory, the first ever. It led to his large facilities at Menlo Park, New Jersey, and, later, labs in other locations. At times as many as one hundred people worked for him, some of whom, such as Nikola Tesla and Reginald Fessenden, became celebrated inventors in their own right.

At his labs Edison not only developed electrical items, such as the light bulb and storage battery; he also produced an efficient mimeograph and worked on innovations in metallurgy, organic chemistry, photography and motion pictures, and phonography. The phonograph, he once said, was his favorite invention. Edison never stopped working. He was still receiving patents the year he died.

the market in 1905 and offered full-price refunds for the defective batteries. Not a man to abandon an invention, however, he spent the next five years examining the failed batteries and refining his design. He discovered that the repeated charging and discharging of the battery caused a shift in the distribution of the graphite in the nickel hydroxide electrode. By using a different type of graphite, he was able to eliminate this problem and produce a very dependable power source.

The Ford Motor Company, founded by Henry Ford, a former Edison employee, began the large-scale production of gasoline-powered automobiles in 1903 and introduced the inexpensive, easy-to-drive Model T in 1908. The introduction of the improved Edison battery in 1910 gave a boost to electric car manufacturers, but their new position in the market would be short-lived. In 1911, Charles Kettering invented an electric starter for gasoline-powered vehicles that eliminated the need for troublesome and risky hand cranking. By 1915, this device was available on all gasoline-powered automobiles, and public interest in electrically powered cars rapidly diminished. Although the Kettering starter required a battery, it required much less capacity than an electric motor would have and was almost ideally suited to the six-volt lead-acid battery.

IMPACT

Edison lost the race to produce an electrical power source that would meet the needs of automotive transportation. Instead, the internal combustion engine developed by Henry Ford became the standard. Interest in electrically powered transportation diminished as immense reserves of crude oil, from which gasoline could be obtained, were discovered first in the southwestern United States and then on the Arabian peninsula. Nevertheless, the Edison cell found a variety of uses and has been manufactured continuously throughout most of the twentieth century much as Edison designed it.

Electrically powered trucks proved to be well suited for local deliveries, and some department stores maintained fleets of such trucks into the mid-1920's. Electrical power is still preferable to internal combustion for indoor use, where exhaust fumes are a significant problem, so forklifts in factories and passenger transport vehicles at airports still make use of the Edison-type power source. The Edison battery also continues to be used in mines, in railway signals, in some communications equipment, and as a highly reliable source of standby emergency power.

See also Compressed-air-accumulating power plant; Internal combustion engine; Photoelectric cell; Photovoltaic cell.

FURTHER READING

Baldwin, Neil. *Edison: Inventing the Century.* Chicago: University of Chicago Press, 2001.

Boyd, Thomas Alvin. *Professional Amateur: The Biography of Charles Franklin Kettering.* New York: Arno Press, 1972.

Bryan, Ford R. *Beyond the Model T: The Other Ventures of Henry Ford.* Rev. ed. Detroit: Wayne State University Press, 1997.

Cramer, Carol. *Thomas Edison.* San Diego, Calif.: Greenhaven Press, 2001.

Israel, Paul. *Edison: A Life of Invention.* New York: Wiley, 2000.

AMMONIA

THE INVENTION: The first successful method for converting nitrogen from the atmosphere and combining it with hydrogen to synthesize ammonia, a valuable compound used as a fertilizer.

THE PERSON BEHIND THE INVENTION:
Fritz Haber (1868-1934), a German chemist who won the 1918 Nobel Prize in Chemistry

THE NEED FOR NITROGEN

The nitrogen content of the soil, essential to plant growth, is maintained normally by the deposition and decay of old vegetation and by nitrates in rainfall. If, however, the soil is used extensively for agricultural purposes, more intensive methods must be used to maintain soil nutrients such as nitrogen. One such method is crop rotation, in which successive divisions of a farm are planted in rotation with clover, corn, or wheat, for example, or allowed to lie fallow for a year or so. The clover is able to absorb nitrogen from the air and deposit it in the soil through its roots. As population has increased, however, farming has become more intensive, and the use of artificial fertilizers—some containing nitrogen—has become almost universal.

Nitrogen-bearing compounds, such as potassium nitrate and ammonium chloride, have been used for many years as artificial fertilizers. Much of the nitrate used, mainly potassium nitrate, came from Chilean saltpeter, of which a yearly amount of half a million tons was imported at the beginning of the twentieth century into Europe and the United States for use in agriculture. Ammonia was produced by dry distillation of bituminous coal and other low-grade fuel materials. Originally, coke ovens discharged this valuable material into the atmosphere, but more economical methods were found later to collect and condense these ammonia-bearing vapors.

At the beginning of the twentieth century, Germany had practically no source of fertilizer-grade nitrogen; almost all of its supply came from the deserts of northern Chile. As demand for nitrates increased, it became apparent that the supply from these vast deposits would not be enough. Other sources needed to be found, and the almost unlimited supply of nitrogen in the atmosphere (80 percent nitrogen) was an obvious source.

Temperature and Pressure

When Fritz Haber and coworkers began his experiments on ammonia production in 1904, Haber decided to repeat the experiments of the British chemist Sir William Ramsay and Sydney Young, who in 1884 had studied the decomposition of ammonia at about 800 degrees Celsius. They had found that a certain amount of ammonia was always left undecomposed. In other words, the reaction between ammonia and its constituent elements—nitrogen and hydrogen—had reached a state of equilibrium.

Haber decided to determine the point at which this equilibrium took place at temperatures near 1,000 degrees Celsius. He tried several approaches, reacting pure hydrogen with pure nitrogen, and starting with pure ammonia gas and using iron filings as a catalyst. (Catalytic agents speed up a reaction without affecting it otherwise).

Having determined the point of equilibrium, he next tried different catalysts and found nickel to be as effective as iron, and calcium and manganese even better. At 1,000 degrees Celsius, the rate of reaction was enough to produce practical amounts of ammonia continuously.

Further work by Haber showed that increasing the pressure also increased the percentage of ammonia at equilibrium. For example, at 300 degrees Celsius, the percentage of ammonia at equilibrium at 1 atmosphere of pressure was very small, but at 200 atmospheres, the percentage of ammonia at equilibrium was far greater. A pilot plant was constructed and was successful enough to impress a chemical company, Badische Anilin-und Soda-Fabrik (BASF). BASF agreed to study Haber's process and to investigate different catalysts on a large scale. Soon thereafter, the process became a commercial success.

FRITZ HABER

Fritz Haber's career is a warning to inventors: Beware of what you create, even if your intentions are honorable.

Considered a leading chemist of his age, Haber was born in Breslau (now Wrocław, Poland) in 1868. A brilliant student, he earned a doctorate quickly, specializing in organic chemistry, and briefly worked as an industrial chemist. Although he soon took an academic job, throughout his career Haber believed that science must benefit society—new theoretical discoveries must find practical applications.

(Nobel Foundation)

Beginning in 1904, he applied new chemical techniques to fix atmospheric nitrogen in the form of ammonia. Nitrogen in the form of nitrates was urgently sought because nitrates were necessary to fertilize crops and natural sources were becoming rare. Only artificial nitrates could sustain the amount of agriculture needed to feed expanding populations. In 1908 Haber succeeded in finding an efficient, cheap process to make ammonia and convert it to nitrates, and by 1910 German manufacturers had built large plants to exploit his techniques. He was lauded as a great benefactor to humanity.

However, his efforts to help Germany during World War I, even though he hated war, turned his life into a nightmare. His wife committed suicide because of his chlorine gas research, which also poisoned his international reputation and tainted his 1918 Nobel Prize in Chemistry. After the war he redirected his energies to helping Germany rebuild its economy. Eight years of experiments in extracting gold from seawater ended in failure, but he did raise the Kaiser Wilhelm Institute for Physical Chemistry, which he directed, to international prominence. Nonetheless, Haber had to flee Adolf Hitler's Nazi regime in 1933 and died a year later, better known for his war research than for his fundamental service to agriculture and industry.

IMPACT

With the beginning of World War I, nitrates were needed more urgently for use in explosives than in agriculture. After the fall of Antwerp, 50,000 tons of Chilean saltpeter were discovered in the harbor and fell into German hands. Because the ammonia from Haber's process could be converted readily into nitrates, it became an important war resource. Haber's other contribution to the German war effort was his development of poison gas, which was used for the chlorine gas attack on Allied troops at Ypres in 1915. He also directed research on gas masks and other protective devices.

At the end of the war, the 1918 Nobel Prize in Chemistry was awarded to Haber for his development of the process for making synthetic ammonia. Because the war was still fresh in everyone's memory, it became one of the most controversial Nobel awards ever made. A headline in *The New York Times* for January 26, 1920, stated: "French Attack Swedes for Nobel Prize Award: Chemistry Honor Given to Dr. Haber, Inventor of German Asphyxiating Gas." In a letter to the *Times* on January 28, 1920, the Swedish legation in Washington, D.C., defended the award.

Haber left Germany in 1933 under duress from the anti-Semitic policies of the Nazi authorities. He was invited to accept a position with the University of Cambridge, England, and died on a trip to Basel, Switzerland, a few months later, a great man whose spirit had been crushed by the actions of an evil regime.

See also Fuel cell; Refrigerant gas; Silicones; Thermal cracking process.

FURTHER READING

Goran, Morris Herbert. *The Story of Fritz Haber.* Norman: University of Oklahoma Press, 1967.

Jansen, Sarah. "Chemical-Warfare Techniques for Insect Control: Insect 'Pests' in Germany Before and After World War I." *Endeavour* 24, no. 1 (March, 2000).

Smil, Vaclav. *Enriching the Earth: Fritz Haber, Carl Bosch, and the Transformation of World Food Production.* Cambridge, Mass.: MIT Press, 2001.

Amniocentesis

THE INVENTION: A technique for removing amniotic fluid from pregnant women, amniocentesis became a life-saving tool for diagnosing fetal maturity, health, and genetic defects.

THE PEOPLE BEHIND THE INVENTION:
Douglas Bevis, an English physician
Aubrey Milunsky (1936-   ), an American pediatrician

How Babies Grow

For thousands of years, the inability to see or touch a fetus in the uterus was a staggering problem in obstetric care and in the diagnosis of the future mental and physical health of human offspring. A beginning to the solution of this problem occurred on February 23, 1952, when *The Lancet* published a study called "The Antenatal Prediction of a Hemolytic Disease of the Newborn." This study, carried out by physician Douglas Bevis, described the use of amniocentesis to assess the risk factors found in the fetuses of Rh-negative women impregnated by Rh-positive men. The article is viewed by many as a landmark in medicine that led to the wide use of amniocentesis as a tool for diagnosing fetal maturity, fetal health, and fetal genetic deects.

At the beginning of a human pregnancy (conception) an egg and a sperm unite to produce the fertilized egg that will become a new human being. After conception, the fertilized egg passes from the oviduct into the uterus, while dividing and becoming an organized cluster of cells capable of carrying out different tasks in the nine-month-long series of events leading up to birth.

About a week after conception, the cluster of cells, now a "vesicle" (a fluid-filled sac containing the new human cells), attaches to the uterine lining, penetrates it, and becomes intimately intertwined with uterine tissues. In time, the merger between the vesicle and the uterus results in formation of a placenta that connects the mother and the embryo, and an amniotic sac filled with the amniotic fluid in which the embryo floats.

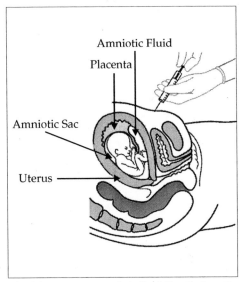

Amniotic Fluid

Placenta

Amniotic Sac

Uterus

Physicians extract amniotic fluid directly from the womb and examine it to determine the health of the fetus.

Eight weeks after conception, the embryo (now a fetus) is about 2.5 centimeters long and possesses all the anatomic elements it will have when it is born. At this time, about two and one-half months after her last menstruation, the expectant mother typically visits a physician and finds out she is pregnant. Also at this time, expecting mothers often begin to worry about possible birth defects in the babies they carry. Diabetic mothers and mothers older than thirty-five years have higher than usual chances of delivering babies who have birth defects.

Many other factors inferred from the medical history an expecting mother provides to her physician can indicate the possible appearance of birth defects. In some cases, knowledge of possible physical problems in a fetus may allow their treatment in the uterus and save the newborn from problems that could persist throughout life or lead to death in early childhood. Information is obtained through the examination of the amniotic fluid in which the fetus is suspended throughout pregnancy. The process of obtaining this fluid is called "amniocentesis."

DIAGNOSING DISEASES BEFORE BIRTH

Amniocentesis is carried out in several steps. First, the placenta and the fetus are located by the use of ultrasound techniques. Next, the expecting mother may be given a local anesthetic; a long needle is then inserted carefully into the amniotic sac. As soon as amniotic fluid is seen, a small sample (about four teaspoons) is drawn into a hypodermic syringe and the syringe is removed. Amniocentesis is nearly painless, and most patients feel only a little abdominal pressure during the procedure.

The amniotic fluid of early pregnancy resembles blood serum. As pregnancy continues, its content of substances from fetal urine and other fetal secretions increases. The fluid also contains fetal cells from skin and from the gastrointestinal, reproductive, and respiratory tracts. Therefore, it is of great diagnostic use. Immediately after the fluid is removed from the fetus, the fetal cells are separated out. Then, the cells are used for genetic analysis and the amniotic fluid is examined by means of various biochemical techniques.

One important use of the amniotic fluid from amniocentesis is the determination of its lecithin and sphingomyelin content. Lecithins and sphingomyelins are two types of body lipids (fatty molecules) that are useful diagnostic tools. Lecithins are important because they are essential components of the so-called pulmonary surfactant of mature lungs. The pulmonary surfactant acts at lung surfaces to prevent the collapse of the lung air sacs (alveoli) when a person exhales.

Subnormal lecithin production in a fetus indicates that it most likely will exhibit respiratory distress syndrome or a disease called "hyaline membrane disease" after birth. Both diseases can be fatal, so it is valuable to determine whether fetal lecithin levels are adequate for appropriate lung function in the newborn baby. This is particularly important in fetuses being carried by diabetic mothers, who frequently produce newborns with such problems. Often, when the risk of respiratory distress syndrome is identified through amniocentesis, the fetus in question is injected with hormones that help it produce mature lungs. This effect is then confirmed by the repeated use of amniocentesis. Many other problems can also be identified by the use of amniocentesis and corrected before the baby is born.

Consequences

In the years that have followed Bevis's original observation, many improvements in the methodology of amniocentesis and in the techniques used in gathering and analyzing the genetic and biochemical information obtained have led to good results. Hundreds of debilitating hereditary diseases can be diagnosed and some ameliorated—by the examination of amniotic fluid and fetal cells isolated by amniocentesis. For many parents who have had a child afflicted by some hereditary disease, the use of the technique has become a major consideration in family planning. Furthermore, many physicians recommend strongly that all mothers over the age of thirty-four be tested by amniocentesis to assist in the diagnosis of Down syndrome, a congenital but nonhereditary form of mental deficiency.

There remains the question of whether such solutions are morally appropriate, but parents—and society—now have a choice resulting from the techniques that have developed since Bevis's 1952 observation. It is also hoped that these techniques will lead to means for correcting and preventing diseases and preclude the need for considering the therapeutic termination of any pregnancy.

See also Abortion pill; Birth control pill; CAT scanner; Electrocardiogram; Electroencephalogram; Mammography; Nuclear magnetic resonance; Pap test; Ultrasound; X-ray image intensifier.

FURTHER READING

Milunsky, Aubrey. *Genetic Disorders and the Fetus: Diagnosis, Prevention, and Treatment*. 3d ed. Baltimore: Johns Hopkins University Press, 1992.

Rapp, Rayna. *Testing Women, Testing the Fetus: The Social Impact of Amniocentesis in America*. New York: Routledge, 1999.

Rothenberg, Karen H., and Elizabeth Jean Thomson. *Women and Prenatal Testing: Facing the Challenges of Genetic Technology*. Columbus: Ohio State University Press, 1994.

Rothman, Barbara Katz. *The Tentative Pregnancy: How Amniocentesis Changes the Experience of Motherhood*. New York: Norton, 1993.

ANTIBACTERIAL DRUGS

THE INVENTION: Sulfonamides and other drugs that have proved effective in combating many previously untreatable bacterial diseases.

THE PEOPLE BEHIND THE INVENTION:
Gerhard Domagk (1895-1964), a German physician who was awarded the 1939 Nobel Prize in Physiology or Medicine
Paul Ehrlich (1854-1915), a German chemist and bacteriologist who was the cowinner of the 1908 Nobel Prize in Physiology or Medicine

THE SEARCH FOR MAGIC BULLETS

Although quinine had been used to treat malaria long before the twentieth century, Paul Ehrlich, who discovered a large number of useful drugs, is usually considered the father of modern chemotherapy. Ehrlich was familiar with the technique of using dyes to stain microorganisms in order to make them visible under a microscope, and he suspected that some of these dyes might be used to poison the microorganisms responsible for certain diseases without hurting the patient. Ehrlich thus began to search for dyes that could act as "magic bullets" that would destroy microorganisms and cure diseases. From 1906 to 1910, Ehrlich tested numerous compounds that had been developed by the German dye industry. He eventually found that a number of complex trypan dyes would inhibit the protozoans that caused African sleeping sickness.

Ehrlich and his coworkers also synthesized hundreds of organic compounds that contained arsenic. In 1910, he found that one of these compounds, salvarsan, was useful in curing syphilis, a sexually transmitted disease caused by the bacterium *Treponema*. This was an important discovery, because syphilis killed thousands of people each year. Salvarsan, however, was often toxic to patients, because it had to be taken in large doses for as long as two years to effect a cure. Ehrlich thus searched for and found a less toxic arsenic compound, neosalvarsan, which replaced salvarsan in 1912.

In 1915, tartar emetic (a compound containing the metal antimony) was found to be useful in treating kala-azar, which was caused by a protozoan. Kala-azar affected millions of people in Africa, India, and Asia, causing much suffering and many deaths each year. Two years later, it was discovered that injection of tartar emetic into the blood of persons suffering from bilharziasis killed the flatworms infecting the bladder, liver, and spleen. In 1920, suramin, a colorless compound developed from trypan red, was introduced to treat African sleeping sickness. It was much less toxic to the patient than any of the drugs Ehrlich had developed, and a single dose would give protection for more than a month. From the dye methylene blue, chemists made mepacrine, a drug that was effective against the protozoans that cause malaria. This chemical was introduced in 1933 and used during World War II; its principal drawback was that it could cause a patient's skin to become yellow.

Well Worth the Effort

Gerhard Domagk had been trained in medicine, but he turned to research in an attempt to discover chemicals that would inhibit or kill microorganisms. In 1927, he became director of experimental pathology and bacteriology at the Elberfeld laboratories of the German chemical firm I. G. Farbenindustrie. Ehrlich's discovery that trypan dyes selectively poisoned microorganisms suggested to Domagk that he look for antimicrobials in a new group of chemicals known as azo dyes. A number of these dyes were synthesized from sulfonamides and purified by Fritz Mietzsch and Josef Klarer. Domagk found that many of these dyes protected mice infected with the bacteria *Streptococcus pyogenes*. In 1932, he discovered that one of these dyes was much more effective than any tested previously. This red azo dye containing a sulfonamide was named prontosil rubrum.

From 1932 to 1935, Domagk began a rigorous testing program to determine the effectiveness and dangers of prontosil use at different doses in animals. Since all chemicals injected into animals or humans are potentially dangerous, Domagk determined the doses that harmed or killed. In addition, he worked out the lowest doses that would eliminate the pathogen. The firm supplied samples of the drug to physicians to carry out clinical trials on humans. (Animal experimentation can give only an indication of which chemicals might be useful in humans and which doses are required.)

Domagk thus learned which doses were effective and safe. This knowledge saved his daughter's life. One day while knitting, Domagk's daughter punctured her finger with a needle and was infected with a virulent bacteria, which quickly multiplied and spread from the wound into neighboring tissues. In an attempt to alleviate the swelling, the infected area was lanced and allowed to drain, but this did not stop the infection from spreading. The child became critically ill with developing septicemia, or blood poisoning.

In those days, more than 75 percent of those who acquired blood infections died. Domagk realized that the chances for his daughter's survival were poor. In desperation, he obtained some of the powdered prontosil that had worked so well on infected animals. He extrapolated from his animal experiments how much to give his daughter so that the bacteria would be killed but his daughter would not be poisoned. Within hours of the first treatment, her fever dropped, and she recovered completely after repeated doses of prontosil.

Impact

Directly and indirectly, Ehrlich's and Domagk's work served to usher in a new medical age. Prior to the discovery that prontosil could be use to treat bacterial infection and the subsequent development of a series of sulfonamides, or "sulfa drugs," there was no chemical defense against this type of disease; as a result, illnesses such as streptococcal infection, gonorrhea, and pneumonia held terrors of which they have largely been shorn. A small injury could easily lead to death.

By following the clues presented by the synthetic sulfa drugs and how they worked to destroy bacteria, other scientists were able to develop an even more powerful type of drug, the antibiotic. When the American bacteriologist Rene Dubos discovered that natural organisms could also be used to fight bacteria, interest was renewed in an earlier discovery by the Scottish bacteriologist Sir Alexander: the development of penicillin.

Antibiotics such as penicillin and streptomycin have become some of the most important tools in fighting disease. Antibiotics have replaced sulfa drugs for most uses, in part because they cause fewer side effects, but sulfa drugs are still used for a handful of purposes. Together, sulfonamides and antibiotics have offered the possibility of a cure to millions of people who previously would have had little chance of survival.

See also Penicillin; Polio vaccine (Sabin); Polio vaccine (Salk); Salvarsan; Tuberculosis vaccine; Typhus vaccine; Yellow fever vaccine.

FURTHER READING

Alstaedter, Rosemarie. *From Germanin to Acylureidopenicillin: Research That Made History: Documentation of a Scientific Revolution: Dedicated to Gerhardt Domagk on the Eighty-fifth Anniversary of His Birth.* Leverkausen, West Germany: Bayer AG, 1980.
Baumler, Ernst. *Paul Ehrlich: Scientist for Life.* New York: Holmes and Meier, 1984.
Galdston, Iago. Behind the Sulfa Drugs, a Short History of Chemotherapy. New York: D. Appleton-Century, 1943.
Physiology or Medicine, 1922-1941. River Edge, N.J.: World Scientific, 1999.

Apple II computer

THE INVENTION: The first commercially available, preassembled personal computer, the Apple II helped move computers out of the workplace and into the home.

THE PEOPLE BEHIND THE INVENTION:

Stephen Wozniak (1950-   ), cofounder of Apple and designer of the Apple II computer

Steven Jobs (1955-   ), cofounder of Apple

Regis McKenna (1939-   ), owner of the Silicon Valley public relations and advertising company that handled the Apple account

Chris Espinosa (1961-   ), the high school student who wrote the BASIC program shipped with the Apple II

Randy Wigginton (1960-   ), a high school student and Apple software programmer

Inventing the Apple

As late as the 1960's, not many people in the computer industry believed that a small computer could be useful to the average person. It was through the effort of two friends from the Silicon Valley—the high-technology area between San Francisco and San Jose—that the personal computer revolution was started.

Both Steven Jobs and Stephen Wozniak had attended Homestead High School in Los Altos, California, and both developed early interests in technology, especially computers. In 1971, Wozniak built his first computer from spare parts. Shortly after this, he was introduced to Jobs. Jobs had already developed an interest in electronics (he once telephoned William Hewlett, cofounder of Hewlett-Packard, to ask for parts), and he and Wozniak became friends. Their first business together was the construction and sale of "blue boxes," illegal devices that allowed the user to make long-distance telephone calls for free.

After attending college, the two took jobs within the electronics industry. Wozniak began working at Hewlett-Packard, where he studied calculator design, and Jobs took a job at Atari, the video company. The friendship paid off again when Wozniak, at Jobs's request, designed the game "Breakout" for Atari, and the pair was paid seven hundred dollars.

In 1975, the Altair computer, a personal computer in kit form, was introduced by Micro Instrumentation and Telemetry Systems (MITS). Shortly thereafter, the first personal computer club, the Homebrew Computer Club, began meeting in Menlo Park, near Stanford University. Wozniak and Jobs began attending the meeting regularly. Wozniak eagerly examined the Altairs that others brought. He thought that the design could be improved. In only a few more weeks, he produced a circuit board and interfaces that connected it to a keyboard and a video monitor. He showed the machine at a Homebrew meeting and distributed photocopies of the design.

In this new machine, which he named an "Apple," Jobs saw a big opportunity. He talked Wozniak into forming a partnership to develop personal computers. Jobs sold his car, and Wozniak sold his two Hewlett-Packard calculators; with the money, they ordered printed circuit boards made. Their break came when Paul Terrell, a retailer, was so impressed that he ordered fifty fully assembled Apples. Within thirty days, the computers were completed, and they sold for a fairly high price: $666.66.

During the summer of 1976, Wozniak kept improving the Apple. The new computer would come with a keyboard, an internal power supply, a built-in computer language called the *Beginner's All-Purpose Symbolic Instruction Code"* (BASIC), hookups for adding printers and other devices, and color graphics, all enclosed in a plastic case. The output would be seen on a television screen. The machine would sell for twelve hundred dollars.

SELLING THE APPLE

Regis McKenna was the head of the Regis McKenna Public Relations agency, the best of the public relations firms that served the high-technology industries of the valley, which Jobs wanted to handle the Apple account. At first, McKenna rejected the offer, but Jobs's constant pleading finally convinced him. The agency's first

STEVEN JOBS

While IBM and other corporations were devoting massive resources and talent to designing a small computer in 1975, Steven Paul Jobs and Stephen Wozniak, members of the tiny Homebrew Computer Club, put together the first truly user-friendly personal computer in Wozniak's home. Jobs admitted later that "Woz" was the engineering brains. Jobs himself was the brains of design and marketing. Both had to scrape together money for the project from their small salaries as low-level electronics workers. Within eight years, Jobs headed the most progressive company in the new personal computer industry and was worth an estimated $210 million.

Little in his background foretold such fast, large material success. Jobs was born in 1955 and became an orphan. Adopted by Paul and Clara Jobs, he grew up in California towns near the area that became known as Silicon Valley. He did not like school much and was considered a loner, albeit one who always had a distinctive way of thinking about things. Still in high school, he impressed William Hewlett, founder of Hewlett-Packard in Palo Alto, and won a summer job at the company, as well as some free equipment for one of his school projects.

However, he dropped out of Reed College after one semester and became a hippie. He studied philosophy and Chinese and Indian mysticism. He became a vegetarian and practiced meditation. He even shaved his head and traveled to India on a spiritual pilgrimage. When he returned to America, however, he also returned to his interest in electronics and computers. Through various jobs at his original company, Apple, and elsewhere, he stayed there.

contributions to Apple were the colorful striped Apple logo and a color ad in *Playboy* magazine.

In February, 1977, the first Apple Computer office was opened in Cupertino, California. By this time, two of Wozniak's friends from Homebrew, Randy Wigginton and Chris Espinosa—both high school students—had joined the company. Their specialty was writing software. Espinosa worked through his Christmas vacation so that BASIC (the built-in computer language) could ship with the computer.

The team pushed ahead to complete the new Apple in time to display it at the First West Coast Computer Faire in April, 1977. At this time, the name "Apple II" was chosen for the new model. The Apple II computer debuted at the convention and included many innovations. The "motherboard" was far simpler and more elegantly designed than that of any previous computer, and the ease of connecting the Apple II to a television screen made it that much more attractive to consumers.

CONSEQUENCES

The introduction of the Apple II computer launched what was to be a wave of new computers aimed at the home and small-business markets. Within a few months of the Apple II's introduction, Commodore introduced its PET computer and Tandy Corporation/Radio Shack brought out its TRS-80. Apple continued to increase the types of things that its computers could do and worked out a distribution deal with the new ComputerLand chain of stores.

In December, 1977, Wozniak began work on creating a floppy disk system for the Apple II. (A floppy disk is a small, flexible plastic disk coated with magnetic material. The magnetized surface enables computer data to be stored on the disk.) The cassette tape storage on which all personal computers then depended was slow and unreliable. Floppy disks, which had been introduced for larger computers by the International Business Machines (IBM) Corporation in 1970, were fast and reliable. As he did with everything that interested him, Wozniak spent almost all of his time learning about and designing a floppy disk drive. When the final drive shipped in June, 1978, it made possible development of more powerful software for the computer.

By 1980, Apple had sold 130,000 Apple II's. That year, the company went public, and Jobs and Wozniak, among others, became wealthy. Three years later, Apple became the youngest company to make the Fortune 500 list of the largest industrial companies. By then, IBM had entered the personal computer field and had begun to dominate it, but the Apple II's earlier success ensured that personal computers would not be a market fad. By the end of the 1980's, 35 million personal computers would be in use.

See also BINAC computer; Colossus computer; ENIAC computer; Floppy disk; Hard disk; IBM Model 1401 computer; Personal computer; UNIVAC computer.

FURTHER READING

Carlton, Jim. *Apple: The Inside Story of Intrigue, Egomania, and Business Blunders*. Rev. ed. London: Random House, 1999.

Gold, Rebecca. *Steve Wozniak: A Wizard Called Woz*. Minneapolis: Lerner, 1994.

Linzmayer, Owen W. *Apple Confidential: The Real Story of Apple Computer, Inc*. San Francisco: No Starch Press, 1999.

Moritz, Michael. *The Little Kingdom: The Private Story of Apple Computer*. New York: Morrow, 1984.

Rose, Frank. *West of Eden: The End of Innocence at Apple Computer*. New York: Viking, 1989.

Aqualung

The invention: A device that allows divers to descend hundreds of meters below the surface of the ocean by enabling them to carry the oxygen they breathe with them.

The people behind the invention:
  Jacques-Yves Cousteau (1910-1997), a French navy officer, undersea explorer, inventor, and author
  Émile Gagnan, a French engineer who invented an automatic air-regulating device

The Limitations of Early Diving

Undersea dives have been made since ancient times for the purposes of spying, recovering lost treasures from wrecks, and obtaining natural treasures (such as pearls). Many attempts have been made since then to prolong the amount of time divers could remain underwater. The first device, described by the Greek philosopher Aristotle in 335 B.C.E., was probably the ancestor of the modern snorkel. It was a bent reed placed in the mouth, with one end above the water.

In addition to depth limitations set by the length of the reed, pressure considerations also presented a problem. The pressure on a diver's body increases by about one-half pound per square centimeter for every meter ventured below the surface. After descending about 0.9 meter, inhaling surface air through a snorkel becomes difficult because the human chest muscles are no longer strong enough to inflate the chest. In order to breathe at or below this depth, a diver must breathe air that has been pressurized; moreover, that pressure must be able to vary as the diver descends or ascends.

Few changes were possible in the technology of diving until air compressors were invented during the early nineteenth century. Fresh, pressurized air could then be supplied to divers. At first, the divers who used this method had to wear diving suits, complete with fishbowl-like helmets. This "tethered" diving made divers relatively immobile but allowed them to search for sunken treasure or do other complex jobs at great depths.

THE DEVELOPMENT OF SCUBA DIVING

The invention of scuba gear gave divers more freedom to move about and made them less dependent on heavy equipment. ("Scuba" stands for self-contained underwater breathing apparatus.) Its development occurred in several stages. In 1880, Henry Fleuss of England developed an outfit that used a belt containing pure oxygen. Belt and diver were connected, and the diver breathed the oxygen over and over. A version of this system was used by the U.S. Navy in World War II spying efforts. Nevertheless, it had serious drawbacks: Pure oxygen was toxic to divers at depths greater than 9 meters, and divers could carry only enough oxygen for relatively short dives. It did have an advantage for spies, namely, that the oxygen—breathed over and over in a closed system—did not reach the surface in the form of telltale bubbles.

The next stage of scuba development occurred with the design of metal tanks that were able to hold highly compressed air. This enabled divers to use air rather than the potentially toxic pure oxygen. More important, being hooked up to a greater supply of air meant that divers could stay under water longer. Initially, the main problem with the system was that the air flowed continuously through a mask that covered the diver's entire face. This process wasted air, and the scuba divers expelled a continual stream of air bubbles that made spying difficult. The solution, according to Axel Madsen's *Cousteau* (1986), was "a valve that would allow inhaling and exhaling through the same mouthpiece."

Jacques-Yves Cousteau's father was an executive for Air Liquide—France's main producer of industrial gases. He was able to direct Cousteau to Émile Gagnan, an engineer at thecompany's Paris laboratory who had been developing an automatic gas shutoff valve for Air Liquide. This valve became the Cousteau-Gagnan regulator, a breathing device that fed air to the diver at just the right pressure whenever he or she inhaled.

With this valve—and funding from Air Liquide—Cousteau and Gagnan set out to design what would become the Aqualung. The first Aqualungs could be used at depths of up to 68.5 meters. During testing, however, the dangers of Aqualung diving became apparent. For example, unless divers ascended and descended in slow stages,

JACQUES-YVES COUSTEAU

The son of a businessman who liked to travel, Jacques-Yves Cousteau acquired the same wanderlust. Born in 1910 in Saint-André-de-Cubzac, France, he was a sickly child, but he learned to love swimming and the ocean. He also took an interest in movies, producing his first film when he was thirteen.

Cousteau graduated from France's naval academy, but his career as an officer ended with a nearly fatal car accident in 1936. He went to Toulon, where he returned to his interests in the sea and photography, a period that culminated in his invention of the aqualung with Émile Gagnan in 1944. During World War II he also won a Légion d'honneur for his photographic espionage. The French Navy established the Underwater Research Group for Cousteau in 1944, and after the war the venture evolved into the freewheeling, worldwide voyages that Cousteau became famous for. Aboard the *Calypso*, a converted U.S. minesweeper, he and his crew conducted research and pioneered underwater photography. His 1957 documentary *The Silent World* (based on a 1953 book) won an Oscar and the Palm d'Or of the Cannes film festival. Subsequent movies and *The Undersea World of Jacques Cousteau*, a television series, established Cousteau as a leading environmentalist and science educator. His Cousteau Society, dedicated to exploring and protecting the oceans, attracted millions of members worldwide. Through it he launched another innovative technology, "Turbosails," towering non-rotating cylinders that act as sails to reduce ships' dependency on oil-fueled engines. A new ship propelled by them, the *Alcyone*, eventually replaced the *Calypso*.

(Library of Congress)

Cousteau inspired legions of oceanographers and environmentalists while calling attention to pressing problems in the world's oceans. Although his later years where marked by family tragedies and controversy, he was revered throughout the world and had received many honors when he died in 1997.

it was likely that they would get "the bends" (decompression sickness), the feared disease of earlier, tethered deep-sea divers. Another problem was that, below 42.6 meters, divers encountered nitrogen narcosis. (This can lead to impaired judgment that may cause fatal actions, including removing a mouthpiece or developing an overpowering desire to continue diving downward, to dangerous depths.)

Cousteau believed that the Aqualung had tremendous military potential. During World War II, he traveled to London soon after the Normandy invasion, hoping to persuade the Allied Powers of its usefulness. He was not successful. So Cousteau returned to Paris and convinced France's new government to use Aqualungs to locate and neutralize underwater mines laid along the French coast by the German navy. Cousteau was commissioned to combine mine-sweeping with the study of the physiology of scuba diving. Further research revealed that the use of helium-oxygen mixtures increased to 76 meters the depth to which a scuba diver could go without suffering nitrogen narcosis.

IMPACT

One way to describe the effects of the development of the Aqualung is to summarize Cousteau's continued efforts to the present. In 1946, he and Philippe Tailliez established the Undersea Research Group of Toulon to study diving techniques and various aspects of life in the oceans. They studied marine life in the Red Sea from 1951 to 1952. From 1952 to 1956, they engaged in an expedition supported by the National Geographic Society. By that time, the Research Group had developed many techniques that enabled them to identify life-forms and conditions at great depths.

Throughout their undersea studies, Cousteau and his coworkers continued to develop better techniques for scuba diving, for recording observations by means of still and television photography, and for collecting plant and animal specimens. In addition, Cousteau participated (with Swiss physicist Auguste Piccard) in the construction of the deep-submergence research vehicle, or bathyscaphe. In the 1960's, he directed a program called Conshelf, which tested a human's ability to live in a specially built underwater habitat. He also wrote and produced films on underwater exploration that attracted, entertained, and educated millions of people.

Cousteau has won numerous medals and scientific distinctions. These include the Gold Medal of the National Geographic Society

(1963), the United Nations International Environment Prize (1977), membership in the American and Indian academies of science (1968 and 1978, respectively), and honorary doctor of science degrees from the University of California, Berkeley (1970), Harvard University (1979), and Rensselaer Polytechnic Institute (1979).

See also Bathyscaphe; Bathysphere.

FURTHER READING

Cousteau, Jacques Yves. *The Silent World*. New York: Harper & Brothers, 1952.
_____. "Lord of The Depths. *Time* 153, no. 12 (March 29, 1999).
_____, and James Dugan. *The Living Sea*. London: Elm Tree, 1988.
Madsen, Axel. *Cousteau: An Unauthorized Biography*. New York: Beaufort Books, 1986.
Munson, Richard. *Cousteau: The Captain and His World*. New York: Paragon House, 1991.
Zanelli, Leo, and George T. Skuse. *Sub-Aqua Illustrated Dictionary*. New York: Oxford University Press, 1976.

Artificial blood

The invention: A perfluorocarbon emulsion that serves as a blood plasma substitute in the treatment of human patients.

The person behind the invention:
Ryoichi Naito (1906-1982), a Japanese physician

Blood Substitutes

The use of blood and blood products in humans is a very complicated issue. Substances present in blood serve no specific purpose and can be dangerous or deadly, especially when blood or blood products are taken from one person and given to another. This fact, combined with the necessity for long-term blood storage, a shortage of donors, and some patients' refusal to use blood for religious reasons, brought about an intense search for a universal bloodlike substance.

The life-sustaining properties of blood (for example, oxygen transport) can be entirely replaced by a synthetic mixture of known chemicals. Fluorocarbons are compounds that consist of molecules containing only fluorine and carbon atoms. These compounds are interesting to physiologists because they are chemically and pharmacologically inert and because they dissolve oxygen and other gases.

Studies of fluorocarbons as blood substitutes began in 1966, when it was shown that a mouse breathing a fluorocarbon liquid treated with oxygen could survive. Subsequent research involved the use of fluorocarbons to play the role of red blood cells in transporting oxygen. Encouraging results led to the total replacement of blood in a rat, and the success of this experiment led in turn to trials in other mammals, culminating in 1979 with the use of fluorocarbons in humans.

Clinical Studies

The chemical selected for the clinical studies was Fluosol-DA, produced by the Japanese Green Cross Corporation. Fluosol-DA consists of a 20 percent emulsion of two perfluorocarbons (per-fluorodecalin and perfluorotripopylamine), emulsifiers, and salts that are included to give the chemical some of the properties of blood plasma. Fluosol-DA had been tested in monkeys, and it had shown a rapid reversible uptake and release of oxygen, a reasonably rapid excretion, no carcinogenicity or irreversible changes in the animals' systems, and the recovery of blood components to normal ranges within three weeks of administration.

The clinical studies were divided into three phases. The first phase consisted of the administration of Fluosol-DA to normal human volunteers. Twelve healthy volunteers were administered the chemical, and the emulsion's effects on blood pressure and composition and on heart, liver, and kidney functions were monitored. No adverse effects were found in any case. The first phase ended in March, 1979, and based on its positive results, the second and third phases were begun in April, 1979.

Twenty-four Japanese medical institutions were involved in the next two phases. The reasons for the use of Fluosol-DA instead of blood in the patients involved were various, and they included refusal of transfusion for religious reasons, lack of compatible blood, "bloodless" surgery for protection from risk of hepatitis, and treatment of carbon monoxide intoxication.

Among the effects noticed by the patients were the following: a small increase in blood pressure, with no corresponding effects on respiration and body temperature; an increase in blood oxygen content; bodily elimination of half the chemical within six to nineteen hours, depending on the initial dose administered; no change in red-cell count or hemoglobin content of blood; no change in whole-blood coagulation time; and no significant blood-chemistry changes. These results made the clinical trials a success and opened the door for other, more extensive ones.

IMPACT

Perfluorocarbon emulsions were initially proposed as oxygen-carrying resuscitation fluids, or blood substitutes, and the results of the pioneering studies show their success as such. Their success in this area, however, led to advanced studies and expanded use of these compounds in many areas of clinical medicine and biomedical research.

Perfluorocarbon emulsions are useful in cancer therapy, because they increase the oxygenation of tumor cells and therefore sensitize them to the effects of radiation or chemotherapy. Perfluorocarbons can also be used as "contrasting agents" to facilitate magnetic resonance imaging studies of various tissues; for example, the uptake of particles of the emulsion by the cells of malignant tissues makes it possible to locate tumors. Perfluorocarbons also have a high nitrogen solubility and therefore can be used to alleviate the potentially fatal effects of decompression sickness by "mopping up" nitrogen gas bubbles from the circulation system. They can also be used to preserve isolated organs and amputated extremities until they can be reimplanted or reattached. In addition, the emulsions are used in cell cultures to regulate gas supply and to improve cell growth and productivity.

The biomedical applications of perfluorocarbon emulsions are multidisciplinary, involving areas as diverse as tissue imaging, organ preservation, cancer therapy, and cell culture. The successful clinical trials opened the door for new applications of these compounds, which rank among the most versatile compounds exploited by humankind.

See also Artificial heart; Artificial hormone; Artificial kidney; Blood transfusion; Coronary artery bypass surgery; Electrocardiogram; Heart-lung machine.

FURTHER READING

"Artificial Blood Product May Debut in Two Years." *Health Care Strategic Management* 18, no. 8 (August, 2000).
"The Business of Blood: Ryoichi Naito and Fluosol-DA Artificial Blood." *Forbes* 131 (January 17, 1983).
Glanz, James. "Pulse Quickens in Search for Blood Substitute." *Research & Development* 34, no. 10 (September, 1992).
Tsuchida, E. *Artificial Red Cells: Materials, Performances, and Clinical Study as Blood Substitutes*. New York: Wiley, 1997.

ARTIFICIAL CHROMOSOME

THE INVENTION: Originally developed for use in the study of natural chromosome behavior, the artificial chromosome proved to be a valuable tool for recombinant DNA technology.

THE PEOPLE BEHIND THE INVENTION:
Jack W. Szostak (1952-   ), a British-born Canadian professor at Harvard Medical School
Andrew W. Murray (1956-   ), a graduate student

THE VALUE OF ARTIFICIAL CHROMOSOMES

The artificial chromosome gives biologists insight into the fundamental mechanisms by which cells replicate and plays an important role as a tool in genetic engineering technology. Soon after its invention in 1983 by Andrew W. Murray and Jack W. Szostak, the artificial chromosome was judged by scientists to be important and its value in the field of medicine was exploited.

Chromosomes are essentially carriers of genetic information; that is, they possess the genetic code that is the blueprint for life. In higher organisms, the number and type of chromosomes that a cell contains in its nucleus are characteristic of the species. For example, each human cell has forty-six chromosomes, while the garden pea has fourteen and the guinea pig has sixty-four. The chromosome's job in a dividing cell is to replicate and then distribute one copy of itself into each new "daughter" cell. This process, which is referred to as "mitosis" or "meiosis," depending upon the actual mechanism by which the process occurs, is of supreme importance to the continuation of life.

In 1953, when biophysicists James D. Watson and Francis Crick discovered the structure of deoxyribonucleic acid (DNA), an achievement for which they won the 1962 Nobel Prize in Physiology or Medicine, it was immediately apparent to them how the double-helical form of DNA (which looks something like a twisted ladder) might explain the mechanism behind cell division. During DNA replication, the chromosome unwinds to expose the thin threads of

DNA. The two strands of the double helix separate, and each acts as a template for the formation of a new complementary strand, thus forming two complete and identical chromosomes that can be distributed to each new cell. This distribution process, which is referred to as "segregation," relies on the chromosomes being pulled along a microtubule framework in the cell called the "mitotic spindle."

CREATING ARTIFICIAL CHROMOSOMES

An artificial chromosome is a laboratory-designed chromosome that possesses only those functional elements its creators choose. In order to be a true working chromosome, however, it must, at minimum, maintain the machinery necessary for replication and segregation.

By the early 1980's, Murray and Szostak had recognized the possible advantages of using a simple, controlled model to study chromosome behavior, since there are several difficulties associated with studying chromosomes in their natural state. Since natural chromosomes are large and have poorly defined structures, it is almost impossible to sift out for study those elements that are essential for replication and segregation. Previous methods of altering a natural chromosome and observing the effects were difficult to use because the cells containing that altered chromosome usually died. Furthermore, even if the cell survived, analysis was complicated by the extensive amount of genetic information carried by the chromosome. Artificial chromosomes are simple and have known components, although the functions of those components may be poorly understood. In addition, since artificial chromosomes are extra chromosomes that are carried by the cell, their alteration does not kill the cell.

Prior to the synthesis of the first artificial chromosome, the essential functional chromosomal elements of replication and segregation had to be identified and harvested. One of the three chromosomal elements thought to be required is the origin of replication, the site at which the synthesis of new DNA begins. The relatively weak interaction between DNA strands at this site facilitates their separation, making possible—with the help of appropriate enzymes—the subsequent replication of the strands into "sister chromatids."

The second essential element is the "centromere," a thinner segment of the chromosome that serves as the attachment site for the mitotic spindle. Sister chromatids are pulled into diametric ends of the dividing cell by the spindle apparatus, thus forming two identical daughter cells. The final functional elements are repetitive sequences of DNA called "telomeres," which are located at both ends of the chromosome. The telomeres are needed to protect the terminal genes from degradation.

With all the functional elements at their disposal, Murray and Szostak proceeded to construct their first artificial chromosome. Once made, this chromosome would be inserted into yeast cells to replicate, since yeast cells are relatively simple and well characterized but otherwise resemble cells of higher organisms. Construction begins with a commonly used "bacterial plasmid," a small, circular, autonomously replicating section of DNA. Enzymes are then called upon to create a gap in this "cloning vector" into which the three chromosomal elements are spliced. In addition, genes that confer some distinct trait, such as color, to yeast cells are also inserted, thus making it possible to determine which cells have actually taken up the new chromosome. Although their first attempt resulted in a chromosome that failed to segregate properly, by September, 1983, Murray and Szostak had announced in the prestigious British journal *Nature* their success in creating the first artificial chromosome.

CONSEQUENCES

One of the most exciting aspects of the artificial chromosome is its application to recombinant DNA technology, which involves creating novel genetic materials by combining segments of DNA from various sources. For example, the artificial yeast chromosome can be used as a cloning vector. In this process, a segment of DNA containing some desired gene is inserted into an artificial chromosome and is then allowed to replicate in yeast until large amounts of the gene are produced. David T. Burke, Georges F. Carle, and Maynard Victor Olson at Washington University in St. Louis have pioneered the technique of combining human genes with artificial yeast chromosomes and have succeeded in cloning large segments of human DNA.

Although amplifying DNA in this manner has been done before, using bacterial plasmids as cloning vectors, the artificial yeast chromosome has the advantage of being able to hold much larger segments of DNA, thus allowing scientists to clone very large genes. This is of great importance, since the genes that cause diseases such as hemophilia and Duchenne's muscular dystrophy are enormous. The most ambitious project for which the artificial yeast chromosome is being used is the national project whose intent is to clone the entire human genome.

See also Artificial blood; Artificial hormone; Genetic "fingerprinting"; Genetically engineered insulin; In vitro plant culture; Synthetic DNA; Synthetic RNA.

FURTHER READING

"Evolving RNA with Enzyme-Like Action." *Science News* 144 (August 14, 1993).
Freedman, David H. "Playing God: The Handmade Cell." *Discover* 13, no. 8 (August, 1992).
Varshavsky, Alexander. "The 2000 Genetics Society of America Medal: Jack W. Szostak." *Genetics* 157, no. 2 (February, 2001).

Artificial Heart

THE INVENTION: The first successful artificial heart, the Jarvik-7, has helped to keep patients suffering from otherwise terminal heart disease alive while they await human heart transplants.

THE PEOPLE BEHIND THE INVENTION:
Robert Jarvik (1946-   ), the main inventor of the Jarvik-7
William Castle DeVries (1943-   ), a surgeon at the University of Utah in Salt Lake City
Barney Clark (1921-1983), a Seattle dentist, the first recipient of the Jarvik-7

EARLY SUCCESS

The Jarvik-7 artificial heart was designed and produced by researchers at the University of Utah in Salt Lake City; it is named for the leader of the research team, Robert Jarvik. An air-driven pump made of plastic and titanium, it is the size of a human heart. It is made up of two hollow chambers of polyurethane and aluminum, each containing a flexible plastic membrane. The heart is implanted in a human being but must remain connected to an external air pump by means of two plastic hoses. The hoses carry compressed air to the heart, which then pumps the oxygenated blood through the pulmonary artery to the lungs and through the aorta to the rest of the body. The device is expensive, and initially the large, clumsy air compressor had to be wheeled from room to room along with the patient.

The device was new in 1982, and that same year Barney Clark, a dentist from Seattle, was diagnosed as having only hours to live. His doctor, cardiac specialist William Castle DeVries, proposed surgically implanting the Jarvik-7 heart, and Clark and his wife agreed. The Food and Drug Administration (FDA), which regulates the use of medical devices, had already given DeVries and his coworkers permission to implant up to seven Jarvik-7 hearts for permanent use.

The operation was performed on Clark, and at first it seemed quite successful. Newspapers, radio, and television reported this medical breakthrough: the first time a severely damaged heart had been re-

---

WILLIAM C. DEVRIES

William Castle DeVries did not invent the artificial heart himself; however, he did develop the procedure to implant it. The first attempt took him seven and a half hours, and he needed fourteen assistants. A success, the surgery made DeVries one of the most talked-about doctors in the world.

DeVries was born in Brooklyn, New York, in 1943. His father, a Navy physician, was killed in action a few months later, and his mother, a nurse, moved with her son to Utah. As a child DeVries showed both considerable mechanical aptitude and athletic prowess. He won an athletic scholarship to the University of Utah, graduating with honors in 1966. He entered the state medical school and there met Willem Kolff, a pioneer in designing and testing artificial organs. Under Kolff's guidance, DeVries began performing experimental surgeries on animals to test prototype mechanical hearts. He finished medical school in 1970 and from 1971 until 1979 was an intern and then a resident in surgery at the Duke University Medical Center in North Carolina.

DeVries returned to the University of Utah as an assistant professor of cardiovascular and thoracic surgery. In the meantime, Robert K. Jarvik had devised the Jarvik-7 artificial heart. DeVries experimented, implanting it in animals and cadavers until, following approval from the Federal Drug Administration, Barney Clark agreed to be the first test patient. He died 115 days after the surgery, having never left the hospital. Although controversy arose over the ethics and cost of the procedure, more artificial heart implantations followed, many by DeVries.

Long administrative delays getting patients approved for surgery at Utah frustrated DeVries, so he moved to Humana Hospital-Audubon in Louisville, Kentucky, in 1984 and then took a professorship at the University of Louisville. In 1988 he left experimentation for a traditional clinical practice. The FDA withdrew its approval for the Jarvik-7 in 1990.

In 1999 DeVries retired from practice, but not from medicine. The next year he joined the Army Reserve and began teaching surgery at the Walter Reed Army Medical Center.

---

placed by a totally artificial heart. It seemed DeVries had proved that an artificial heart could be almost as good as a human heart.

Soon after Clark's surgery, DeVries went on to implant the device in several other patients with serious heart disease. For a time, all of them survived the surgery. As a result, DeVries was offered a position at Humana Hospital in Louisville, Kentucky. Humana offered to pay for the first one hundred implant operations.

THE CONTROVERSY BEGINS

In the three years after DeVries's operation on Barney Clark, however, doubts and criticism arose. Of the people who by then had received the plastic and metal device as a permanent replacement for their own diseased hearts, three had died (including Clark) and four had suffered serious strokes. The FDA asked Humana Hospital and Symbion (the company that manufactured the Jarvik-7) for complete, detailed histories of the artificial-heart recipients.

It was determined that each of the patients who had died or been disabled had suffered from infection. Life-threatening infection, or "foreign-body response," is a danger with the use of any artificial organ. The Jarvik-7, with its metal valves, plastic body, and Velcro attachments, seemed to draw bacteria like a magnet—and these bacteria proved resistant to even the most powerful antibiotics.

By 1988, researchers had come to realize that severe infection was almost inevitable if a patient used the Jarvik-7 for a long period of time. As a result, experts recommended that the device be used for no longer than thirty days.

Questions of values and morality also became part of the controversy surrounding the artificial heart. Some people thought that it was wrong to offer patients a device that would extend their lives but leave them burdened with hardship and pain. At times DeVries claimed that it was worth the price for patients to be able live another year; at other times, he admitted that if he thought a patient would have to spend the rest of his or her life in a hospital, he would think twice before performing the implant.

There were also questions about "informed consent"—the patient's understanding that a medical procedure has a high risk of failure and may leave the patient in misery even if it succeeds. Getting truly informed consent from a dying patient is tricky, because, understandably, the patient is probably willing to try anything. The Jarvik-7 raised several questions in this regard: Was the ordeal worth the risk? Was the patient's suffering justifiable? Who should make the decision for or against the surgery: the patient, the researchers, or a government agency?

Also there was the issue of cost. Should money be poured into expensive, high-technology devices such as the Jarvik heart, or should it be reserved for programs to help prevent heart disease in the first place? Expenses for each of DeVries's patients had amounted to about one million dollars.

Humana's and DeVries's earnings were criticized in particular. Once the first one hundred free Jarvik-7 implantations had been performed, Humana Hospital could expect to make large amounts of money on the surgery. By that time, Humana would have so much expertise in the field that, though the surgical techniques could not be patented, it was expected to have a practical monopoly. DeVries himself owned thousands of shares of stock in Symbion. Many people wondered whether this was ethical.

CONSEQUENCES

Given all the controversies, in December of 1985 a panel of experts recommended that the FDA allow the experiment to continue, but only with careful monitoring. Meanwhile, cardiac transplantation was becoming easier and more common. By the end of 1985, almost twenty-six hundred patients in various countries had received human heart transplants, and 76 percent of these patients had survived for at least four years. When the demand for donor hearts exceeded the supply, physicians turned to the Jarvik device and other artificial hearts to help see patients through the waiting period.

Experience with the Jarvik-7 made the world keenly aware of how far medical science still is from making the implantable permanent mechanical heart a reality. Nevertheless, the device was a breakthrough in the relatively new field of artificial organs. Since then, other artificial body parts have included heart valves, blood vessels, and inner ears that help restore hearing to the deaf.

See also Artificial blood; Artificial kidney; Blood transfusion; Coronary artery bypass surgery; Electrocardiogram; Heart-lung machine; Pacemaker; Velcro.

FURTHER READING

Fox, Renee C., and Judith P. Swazy. *Spare Parts: Organ Replacement in American Society.* New York: Oxford University Press, 1992.

Kunin, Calvin M., Joanne J. Debbins, and Julio C. Melo. "Infectious Complications in Four Long-Term Recipients of the Jarvik-7 Artificial Heart." *JAMA* 259 (February 12, 1988).

Kunzig, Robert. "The Beat Goes On." *Discover* 21, no. 1 (January, 2000).

Lawrie, Gerald M. "Permanent Implantation of the Jarvik-7 Total Artificial Heart: A Clinical Perspective." *JAMA* 259 (February 12, 1988).

Artificial hormone

The invention: Synthesized oxytocin, a small polypeptide hormone from the pituitary gland that has shown how complex polypeptides and proteins may be synthesized and used in medicine.

The people behind the invention:
Vincent du Vigneaud (1901-1978), an American biochemist and winner of the 1955 Nobel Prize in Chemistry
Oliver Kamm (1888-1965), an American biochemist
Sir Edward Albert Sharpey-Schafer (1850-1935), an English physiologist
Sir Henry Hallett Dale (1875-1968), an English physiologist and winner of the 1936 Nobel Prize in Physiology or Medicine
John Jacob Abel (1857-1938), an American pharmacologist and biochemist

Body-Function Special Effects

In England in 1895, physician George Oliver and physiologist Edward Albert Sharpey-Schafer reported that a hormonal extract from the pituitary gland of a cow produced a rise in blood pressure (a pressor effect) when it was injected into animals. In 1901, Rudolph Magnus and Sharpey-Schafer discovered that extracts from the pituitary also could restrict the flow of urine (an antidiuretic effect). This observation was related to the fact that when a certain section of the pituitary was removed surgically from an animal, the animal excreted an abnormally large amount of urine.

In addition to the pressor and antidiuretic activities in the pituitary, two other effects were found in 1909. Sir Henry Hallett Dale, an English physiologist, was able to show that the extracts could cause the uterine muscle to contract (an oxytocic effect), and Isaac Ott and John C. Scott found that when lactating (milk-producing) animals were injected with the extracts, milk was released from the mammary gland.

Following the discovery of these various effects, attempts were made to concentrate and isolate the substance or substances that were responsible. John Jacob Abel was able to concentrate the pressor activity at The Johns Hopkins University using heavy metal salts and extraction with organic solvents. The results of the early work, however, were varied. Some investigators came to the conclusion that only one substance was responsible for all the activities, while others concluded that two or more substances were likely to be involved.

In 1928, Oliver Kamm and his coworkers at the drug firm of Parke, Davis and Company in Detroit reported a method for the separation of the four activities into two chemical fractions with high potency. One portion contained most of the pressor and antidiuretic activities, while the other contained the uterine-contracting and milk-releasing activities. Over the years, several names have been used for the two substances responsible for the effects. The generic name "vasopressin" generally has become the accepted term for the substance causing the pressor and antidiuretic effects, while the name "oxytocin" has been used for the other two effects. The two fractions that Kamm and his group had prepared were pure enough for the pharmaceutical firm to make them available for medical research related to obstetrics, surgical shock, and diabetes insipidus.

A COMPLICATED SYNTHESIS

The problem of these hormones and their nature interested Vincent du Vigneaud at the George Washington University School of Medicine. Working with Kamm, he was able to show that the sulfur content of both the oxytocin and the vasopressin fractions was a result of the amino acid cystine. This helped to strengthen the concept that these hormones were polypeptide, or proteinlike, substances. Du Vigneaud and his coworkers next tried to find a way of purifying oxytocin and vasopressin. This required not only the separation of the hormones themselves but also the separation from other impurities present in the preparations.

During World War II (1939-1945) and shortly thereafter, other techniques were developed that would give du Vigneaud the tools he needed to complete the job of purifying and characterizing the two hormonal factors. One of the most important was the countercurrent distribution method of chemist Lyman C. Craig at the Rockefeller Institute. Craig had developed an apparatus that could do multiple extractions, making possible separations of substances with similar properties. Du Vigneaud had used this technique in purifying his synthetic penicillin, and when he returned to the study of oxytocin and vasopressin in 1946, he used it on his purest preparations. The procedure worked well, and milligram quantities of pure oxytocin were available in 1949 for chemical characterization.

Using the available techniques, Vigneaud and his coworkers were able to determine the structure of oxytocin. It was du Vigneaud's goal to make synthetic oxytocin by duplicating the structure his group had worked out. Eventually, du Vigneaud's synthetic oxytocin was obtained and the method published in the *Journal of the American Chemical Society* in 1953.

Du Vigneaud's oxytocin was next tested against naturally occurring oxytocin, and the two forms were found to act identically in every respect. In the final test, the synthetic form was found to induce labor when given intravenously to women about to give birth. Also, when microgram quantities of oxytocin were given intravenously to women who had recently given birth, milk was released from the mammary gland in less than a minute.

CONSEQUENCES

The work of du Vigneaud and his associates demonstrated for the first time that it was possible to synthesize peptides that have properties identical to the natural ones and that these can be useful in certain medical conditions. Oxytocin has been used in the last stages of labor during childbirth. Vasopressin has been used in the treatment of diabetes insipidus, when an individual has an insufficiency in the natural hormone, much as insulin is used by persons having diabetes mellitus.

After receiving the Nobel Prize in Chemistry in 1955, du Vigneaud continued his work on synthesizing chemical variations of the two hormones. By making peptides that differed from oxytocin and vasopressin by one or more amino acids, it was possible to study how the structure of the peptide was related to its physiological activity.

After the structure of insulin and some of the smaller proteins were determined, they, too, were synthesized, although with greater difficulty. Other methods of carrying out the synthesis of peptides and proteins have been developed and are used today. The production of biologically active proteins, such as insulin and growth hormone, has been made possible by efficient methods of biotechnology. The genes for these proteins can be put inside microorganisms, which then make them in addition to their own proteins. The microorganisms are then harvested and the useful protein hormones isolated and purified.

See also Abortion pill; Artificial blood; Birth control pill; Genetically engineered insulin; Pap test.

FURTHER READING

Basa, Channa, and G. M. Anantharamaiah. *Peptides: Design, Synthesis, and Biological Activity*. Boston: Birkauser, 1994.

Bodanszky, Miklos. "Vincent du Vigneaud, 1901-1978." *Nature* 279, no. 5710 (1979).

Vigneud, Vincent du. "A Trail of Sulfur Research from Insulin to Oxytocin" [Nobel lecture]. In *Chemistry, 1942-1962*. River Edge, N.J.: World Scientific, 1999.

Artificial Insemination

THE INVENTION: Practical techniques for the artificial insemination of farm animals that have revolutionized livestock breeding practices throughout the world.

THE PEOPLE BEHIND THE INVENTION:
Lazzaro Spallanzani (1729-1799), an Italian physiologist
Ilya Ivanovich Ivanov (1870-1932), a Soviet biologist
R. W. Kunitsky, a Soviet veterinarian

Reproduction Without Sex

The tale is told of a fourteenth-century Arabian chieftain who sought to improve his mediocre breed of horses. Sneaking into the territory of a neighboring hostile tribe, he stimulated a prize stallion to ejaculate into a piece of cotton. Quickly returning home, he inserted this cotton into the vagina of his own mare, who subsequently gave birth to a high-quality horse. This may have been the first case of "artificial insemination," the technique by which semen is introduced into the female reproductive tract without sexual contact.

The first scientific record of artificial insemination comes from Italy in the 1770's. Lazzaro Spallanzani was one of the foremost physiologists of his time, well known for having disproved the theory of spontaneous generation, which states that living organisms can spring "spontaneously" from lifeless matter. There was some disagreement at that time about the basic requirements for reproduction in animals. It was unclear if the sex act was necessary for an embryo to develop, or if it was sufficient that the sperm and eggs come into contact. Spallanzani began by studying animals in which union of the sperm and egg normally takes place outside the body of the female. He stimulated males and females to release their sperm and eggs, then mixed these sex cells in a glass dish. In this way, he produced young frogs, toads, salamanders, and silkworms.

Next, Spallanzani asked whether the sex act was also unnecessary for reproduction in those species in which fertilization normally takes place inside the body of the female. He collected semen that had been ejaculated by a male spaniel and, using a syringe, injected the semen into the vagina of a female spaniel in heat. Two months later, she delivered a litter of three pups, which bore some resemblance to both the mother and the male that had provided the sperm.

It was in animal breeding that Spallanzani's techniques were to have their most dramatic application. In the 1880's, an English dog breeder, Sir Everett Millais, conducted several experiments on artificial insemination. He was interested mainly in obtaining offspring from dogs that would not normally mate with one another because of difference in size. He followed Spallanzani's methods to produce a cross between a short, low, basset hound and the much larger bloodhound.

Long-Distance Reproduction

Ilya Ivanovich Ivanov was a Soviet biologist who was commissioned by his government to investigate the use of artificial insemination on horses. Unlike previous workers who had used artificial insemination to get around certain anatomical barriers to fertilization, Ivanov began the use of artificial insemination to reproduce thoroughbred horses more effectively. His assistant in this work was the veterinarian R. W. Kunitsky.

In 1901, Ivanov founded the Experimental Station for the Artificial Insemination of Horses. As its director, he embarked on a series of experiments to devise the most efficient techniques for breeding these animals. Not content with the demonstration that the technique was scientifically feasible, he wished to ensure further that it could be practiced by Soviet farmers.

If sperm from a male were to be used to impregnate females in another location, potency would have to be maintained for a long time. Ivanov first showed that the secretions from the sex glands were not required for successful insemination; only the sperm itself was necessary. He demonstrated further that if a testicle were removed from a bull and kept cold, the sperm would remain alive. More useful than preservation of testicles would be preservation of the ejaculated sperm. By adding certain salts to the sperm-containing fluids, and by keeping these at cold temperatures, Ivanov was able to preserve sperm for long periods.

Ivanov also developed instruments to inject the sperm, to hold the vagina open during insemination, and to hold the horse in place during the procedure. In 1910, Ivanov wrote a practical textbook with technical instructions for the artificial insemination of horses. He also trained some three hundred veterinary technicians in the use of artificial insemination, and the knowledge he developed quickly spread throughout the Soviet Union. Artificial insemination became the major means of breeding horses.

Until his death in 1932, Ivanov was active in researching many aspects of the reproductive biology of animals. He developed methods to treat reproductive diseases of farm animals and refined methods of obtaining, evaluating, diluting, preserving, and disinfecting sperm. He also began to produce hybrids between wild and domestic animals in the hope of producing new breeds that would be able to withstand extreme weather conditions better and that would be more resistant to disease. His crosses included hybrids of ordinary cows with aurochs, bison, and yaks, as well as some more exotic crosses of zebras with horses.

Ivanov also hoped to use artificial insemination to help preserve species that were in danger of becoming extinct. In 1926, he led an expedition to West Africa to experiment with the hybridization of different species of anthropoid apes.

IMPACT

The greatest beneficiaries of artificial insemination have been dairy farmers. Some bulls are able to sire genetically superior cows that produce exceptionally large volumes of milk. Under natural conditions, such a bull could father at most a few hundred offspring in its lifetime. Using artificial insemination, a prize bull can inseminate ten to fifteen thousand cows each year. Since frozen sperm may be purchased through the mail, this also means that dairy farmers no longer need to keep dangerous bulls on the farm. Artificial insemination has become the main method of reproduction of dairy cows, with about 150 million cows (as of 1992) produced this way throughout the world.

In the 1980's, artificial insemination gained added importance as a method of breeding rare animals. Animals kept in zoo cages, animals that are unable to take part in normal mating, may still produce sperm that can be used to inseminate a female artificially. Some species require specific conditions of housing or diet for normal breeding to occur, conditions not available in all zoos. Such animals can still reproduce using artificial insemination.

See also Abortion pill; Amniocentesis; Artificial chromosome; Birth control pill; Cloning; Genetic "fingerprinting"; Genetically engineered insulin; In vitro plant culture; Rice and wheat strains; Synthetic DNA.

FURTHER READING

Bearden, Henry Joe, and John W. Fuquay. *Applied Animal Reproduction.* 5th ed. Upper Saddle River, N.J.: Prentice Hall, 2000.

Foote, Robert H. *Artificial Insemination to Cloning: Tracing Fifty Years of Research.* Ithaca, N.Y.: Cornell University Press, 1998.

Hafez, Elsayed Saad Eldin. *Reproduction in Farm Animals.* 6th ed. Philadelphia: Lea and Febiger, 1993.

Herman, Harry August. *Improving Cattle by the Millions: NAAB and the Development and Worldwide Application of Artificial Insemination.* Columbia: University of Missouri Press, 1981.

Artificial kidney

The invention: A machine that removes waste end-products and poisons out of the blood when human kidneys are not working properly.

The people behind the invention:

John Jacob Abel (1857-1938), a pharmacologist and biochemist known as the "father of American pharmacology"

Willem Johan Kolff (1911-   ), a Dutch American clinician who pioneered the artificial kidney and the artificial heart

Cleansing the Blood

In the human body, the kidneys are the dual organs that remove waste matter from the bloodstream and send it out of the system as urine. If the kidneys fail to work properly, this cleansing process must be done artifically—such as by a machine.

John Jacob Abel was the first professor of pharmacology at Johns Hopkins University School of Medicine. Around 1912, he began to study the by-products of metabolism that are carried in the blood. This work was difficult, he realized, because it was nearly impossible to detect even the tiny amounts of the many substances in blood. Moreover, no one had yet developed a method or machine for taking these substances out of the blood.

In devising a blood filtering system, Abel understood that he needed a saline solution and a membrane that would let some substances pass through but not others. Working with Leonard Rowntree and Benjamin B. Turner, he spent nearly two years figuring out how to build a machine that would perform dialysis—that is, remove metabolic by-products from blood. Finally their efforts succeeded.

The first experiments were performed on rabbits and dogs. In operating the machine, the blood leaving the patient was sent flowing through a celloidin tube that had been wound loosely around a drum. An anticlotting substance (hirudin, taken out of leeches) was added to blood as the blood flowed through the tube. The drum, which was immersed in a saline and dextrose solution, rotated slowly. As blood flowed through the immersed tubing, the pressure of osmosis removed urea and other substances, but not the plasma or cells, from the blood. The celloidin membranes allowed oxygen to pass from the saline and dextrose solution into the blood, so that purified, oxygenated blood then flowed back into the arteries.

Abel studied the substances that his machine had removed from the blood, and he found that they included not only urea but also free amino acids. He quickly realized that his machine could be useful for taking care of people whose kidneys were not working properly. Reporting on his research, he wrote, "In the hope of providing a substitute in such emergencies, which might tide over a dangerous crisis . . . a method has been devised by which the blood of a living animal may be submitted to dialysis outside the body, and again returned to the natural circulation." Abel's machine removed large quantities of urea and other poisonous substances fairly quickly, so that the process, which he called "vividiffusion," could serve as an artificial kidney during cases of kidney failure.

For his physiological research, Abel found it necessary to remove, study, and then replace large amounts of blood from living animals, all without dissolving the red blood cells, which carry oxygen to the body's various parts. He realized that this process, which he called "plasmaphaeresis," would make possible blood banks, where blood could be stored for emergency use.

In 1914, Abel published these two discoveries in a series of three articles in the *Journal of Pharmacology and Applied Therapeutics*, and he demonstrated his techniques in London, England, and Groningen, The Netherlands. Though he had suggested that his techniques could be used for medical purposes, he himself was interested mostly in continuing his biochemical research. So he turned to other projects in pharmacology, such as the crystallization of insulin, and never returned to studying vividiffusion.

REFINING THE TECHNIQUE

Georg Haas, a German biochemist working in Giessen, West Germany, was also interested in dialysis; in 1915, he began to experiment with "blood washing." After reading Abel's 1914 writings, Haas tried substituting collodium for the celloidin that Abel had used as a filtering membrane and using commercially prepared heparin instead of the homemade hirudin Abel had used to prevent blood clotting. He then used this machine on a patient and found that it showed promise, but he knew that many technical problems had to be worked out before the procedure could be used on many patients.

In 1937, Willem Johan Kolff was a young physician at Groningen. He felt sad to see patients die from kidney failure, and he wanted to find a way to cure others. Having heard his colleagues talk about the possibility of using dialysis on human patients, he decided to build a dialysis machine.

Kolff knew that cellophane was an excellent membrane for dialyzing, and that heparin was a good anticoagulant, but he also realized that his machine would need to be able to treat larger volumes of blood than Abel's and Haas's had. During World War II (1939-

JOHN JACOB ABEL

Born in 1857, John Jacob Abel grew up in Cleveland, Ohio, and then attended the University of Michigan. He graduated in 1883 and studied for six years in Germany, which boasted the finest medical researchers of the times. He received a medical degree in 1888 in Strasbourg, transferred to Vienna, Austria, for more clinical experience, and then returned to the United States in 1891 to teach pharmacology at the University of Michigan. He had to organize his own laboratory, journal, and course of instruction. His efforts attracted the notice of Johns Hopkins University, which then had the nation's most progressive medical school. In 1893 Abel moved there and became the first American to hold the title of professor of pharmacology. He remained at Johns Hopkins until his retirement in 1932.

His biochemical research illuminated the complex interaction in the endocrine system. He isolated epinephrine (adrenaline), used his artificial kidney apparatus to demonstrate the presence of amino acids in the blood, and investigated pituitary gland hormones and insulin.

Abel died in 1938, but his influence did not. His many students took Abel's interest in the biochemical basis of pharmacology to other universities and commercial laboratories, modernizing American drug research.

1945), with the help of the director of a nearby enamel factory, Kolff built an artificial kidney that was first tried on a patient on March 17, 1943. Between March, 1943, and July 21, 1944, Kolff used his secretly constructed dialysis machines on fifteen patients, of whom only one survived. He published the results of his research in *Acta Medica Scandinavica*. Even though most of his patients had not survived, he had collected information and developed the technique until he was sure dialysis would eventually work.

Kolff brought machines to Amsterdam and The Hague and encouraged other physicians to try them; meanwhile, he continued to study blood dialysis and to improve his machines. In 1947, he brought improved machines to London and the United States. By the time he reached Boston, however, he had given away all of his machines. He did, however, explain the technique to John P. Merrill, a physician at the Harvard Medical School, who soon became the leading American developer of kidney dialysis and kidney-transplant surgery.

Kolff himself moved to the United States, where he became an expert not only in artificial kidneys but also in artificial hearts. He helped develop the Jarvik-7 artificial heart (named for its chief inventor, Robert Jarvik), which was implanted in a patient in 1982.

Impact

Abel's work showed that the blood carried some substances that had not been previously known and led to the development of the first dialysis machine for humans. It also encouraged interest in the possibility of organ transplants.

After World War II, surgeons had tried to transplant kidneys from one animal to another, but after a few days the recipient began to reject the kidney and die. In spite of these failures, researchers in Europe and America transplanted kidneys in several patients, and they used artificial kidneys to take care of the patients who were waiting for transplants. In 1954, Merrill—to whom Kolff had demonstrated an artificial kidney—successfully transplanted kidneys in identical twins.

After immunosuppressant drugs (used to prevent the body from rejecting newly transplanted tissue) were discovered in 1962, transplantation surgery became much more practical. After kidney transplants became common, the artificial kidney became simply a way of keeping a person alive until a kidney donor could be found.

See also Artificial blood; Artificial heart; Blood transfusion; Genetically engineered insulin; Reserpine.

FURTHER READING

Cogan, Martin G., Patricia Schoenfeld, and Frank A. Gotch. *Introduction to Dialysis*. 2d ed. New York: Churchill Livingstone, 1991.
DeJauregui, Ruth. *One Hundred Medical Milestones That Shaped World History*. San Mateo, Calif.: Bluewood Books, 1998.
Noordwijk, Jacob van. *Dialysing for Life: The Development of the Artificial Kidney*. Boston: Kluwer Academic Publishers, 2001.

Artificial satellite

The invention: *Sputnik I*, the first object put into orbit around the earth, which began the exploration of space.

The people behind the invention:
Sergei P. Korolev (1907-1966), a Soviet rocket scientist
Konstantin Tsiolkovsky (1857-1935), a Soviet schoolteacher and the founder of rocketry in the Soviet Union
Robert H. Goddard (1882-1945), an American scientist and the founder of rocketry in the United States
Wernher von Braun (1912-1977), a German who worked on rocket projects
Arthur C. Clarke (1917-   ), the author of more than fifty books and the visionary behind telecommunications satellites

A Shocking Launch

In Russian, *sputnik* means "satellite" or "fellow traveler." On October 4, 1957, *Sputnik 1*, the first artificial satellite to orbit Earth, was placed into successful orbit by the Soviet Union. The launch of this small aluminum sphere, 0.58 meter in diameter and weighing 83.6 kilograms, opened the doors to the frontiers of space.

Orbiting Earth every 96 minutes, at 28,962 kilometers per hour, *Sputnik 1* came within 215 kilometers of Earth at its closest point and 939 kilometers away at its farthest point. It carried equipment to measure the atmosphere and to experiment with the transmission of electromagnetic waves from space. Equipped with two radio transmitters (at different frequencies) that broadcast for twenty-one days, *Sputnik 1* was in orbit for ninety-two days, until January 4, 1958, when it disintegrated in the atmosphere.

Sputnik 1 was launched using a Soviet intercontinental ballistic missile (ICBM) modified by Soviet rocket expert Sergei P. Korolev. After the launch of Sputnik 2, less than a month later, Chester Bowles, a former United States ambassador to India and Nepal, wrote: "Armed with a nuclear warhead, the rocket which launched

Sergei P. Korolev

Sergei P. Korolev's rocket launched the Space Age: *Sputnik I* climbed into outer space aboard one of his R-7 missiles. Widely considered the Soviet Union's premiere rocket scientist, he almost died in Joseph Stalin's infamous Siberian prison camps before he could build the launchers that made his country a military superpower and pioneer of space exploration.

Born in 1907, Korolev studied aeronautical engineering at the Kiev Polytechnic Institute. Upon graduation he helped found the Group for Investigation of Reactive Motion, which in the early 1930's tested liquid-fuel rockets. His success attracted the military's attention. It created the Reaction Propulsion Scientific Research Institute for him, and he was on the verge of testing a rocket-propelled airplane when he was arrested during a political purge in 1937 and sent as a prison laborer to the Kolyma gold mines. After Germany attacked Russia in World War II, Korolev was transferred to a prison research institute to help develop advanced aircraft.

After World War II, rehabilitated in the eyes of the Soviet authorities, Korolev was placed in charge of long-range ballistic missile research. In 1953 he began to build the R-7 intercontinental ballistic missile (ICBM). While other design bureaus concentrated on developing the ICBM into a Cold War weapon, Korolev built rockets that explored the Moon with probes. His goal was to send cosmonauts there too. With his designs and guidance, the Soviet space program proved that human space flight was possible in 1961, and so in 1962 he began development of the N-1, a booster that like the American Saturn V was powerful enough to send a crewed vehicle to the Moon. Tragically, Korolev died following minor surgery in 1966. The N-1 project was cancelled in 1971, along with Russian dreams of settling its citizens on the Moon.

Sputnik 1 could destroy New York, Chicago, or Detroit 18 minutes after the button was pushed in Moscow."

Although the launch of Sputnik 1 came as a shock to the general public, it came as no surprise to those who followed rocketry. In June, 1957, the United States Air Force had issued a nonclassified memo stating that there was "every reason to believe that the Russian satellite shot would be made on the hundredth anniversary" of Konstantin Tsiolkovsky's birth.

THOUSANDS OF LAUNCHES

Rockets have been used since at least the twelfth century, when Europeans and the Chinese were using black powder devices. In 1659, the Polish engineer Kazimir Semenovich published his *Roketten für Luft und Wasser* (rockets for air and water), which had a drawing of a three-stage rocket. Rockets were used and perfected for warfare during the nineteenth and twentieth centuries. Nazi Germany's V-2 rocket (thousands of which were launched by Germany against England during the closing years of World War II) was the model for American and Soviet rocket designers between 1945 and 1957. In the Soviet Union, Tsiolkovsky had been thinking about and writing about space flight since the last decade of the nineteenth century, and in the United States, Robert H. Goddard had been thinking about and experimenting with rockets since the first decade of the twentieth century.

Wernher von Braun had worked on rocket projects for Nazi Germany during World War II, and, as the war was ending in May, 1945, von Braun and several hundred other people involved in German rocket projects surrendered to American troops in Europe. Hundreds of other German rocket experts ended up in the Soviet Union to continue with their research. Tom Bower pointed out in his book *The Paperclip Conspiracy: The Hunt for the Nazi Scientists* (1987)—so named because American "recruiting officers had identified [Nazi] scientists to be offered contracts by slipping an ordinary paperclip onto their files"—that American rocketry research was helped tremendously by Nazi scientists who switched sides after World War II.

The successful launch of *Sputnik 1* convinced people that space travel was no longer simply science fiction. The successful launch of *Sputnik 2* on November 3, 1957, carrying the first space traveler, a dog named Laika (who was euthanized in orbit because there were no plans to retrieve her), showed that the launch of *Sputnik 1* was only the beginning of greater things to come.

CONSEQUENCES

After October 4, 1957, the Soviet Union and other nations launched more experimental satellites. On January 31, 1958, the United States sent up Explorer 1, after failing to launch a Vanguard satellite on December 6, 1957.

Arthur C. Clarke, most famous for his many books of science fiction, published a technical paper in 1945 entitled "Extra-Terrestrial Relays: Can Rocket Stations Give World-Wide Radio Coverage?" In that paper, he pointed out that a satellite placed in orbit at the correct height and speed above the equator would be able to hover over the same spot on Earth. The placement of three such "geostationary" satellites would allow radio signals to be transmitted around the world. By the 1990's, communications satellites were numerous.

In the first twenty-five years after *Sputnik 1* was launched, from 1957 to 1982, more than two thousand objects were placed into various Earth orbits by more than twenty-four nations. On the average, something was launched into space every 3.82 days for this twenty-five-year period, all beginning with *Sputnik 1*.

See also Communications satellite; Cruise missile; Rocket; V-2 rocket; Weather satellite.

FURTHER READING

Dickson, Paul. *Sputnik: The Shock of the Century.* New York: Walker, 2001.

Heppenheimer, T. A. *Countdown: A History of Space Flight.* New York: John Wiley & Sons, 1997.

Logsdon, John M., Roger D. Launius, and Robert W. Smith. *Reconsidering Sputnik: Forty Years Since the Soviet Satellite.* Australia: Harwood Academic, 2000.

Aspartame

THE INVENTION: An artificial sweetener with a comparatively natural taste widely used in carbonated beverages.

THE PEOPLE BEHIND THE INVENTION:
Arthur H. Hayes, Jr. (1933-   ), a physician and commissioner of the U.S. Food and Drug Administration (FDA)
James M. Schlatter (1942-   ), an American chemist
Michael Sveda (1912-   ), an American chemist and inventor
Ludwig Frederick Audrieth (1901-   ), an American chemist and educator
Ira Remsen (1846-1927), an American chemist and educator
Constantin Fahlberg (1850-1910), a German chemist

SWEETNESS WITHOUT CALORIES

People have sweetened food and beverages since before recorded history. The most widely used sweetener is sugar, or sucrose. The only real drawback to the use of sucrose is that it is a nutritive sweetener: In addition to adding a sweet taste, it adds calories. Because sucrose is readily absorbed by the body, an excessive amount can be life-threatening to diabetics. This fact alone would make the development of nonsucrose sweeteners attractive.

There are three common nonsucrose sweeteners in use around the world: saccharin, cyclamates, and aspartame. Saccharin was the first of this group to be discovered, in 1879. Constantin Fahlberg synthesized saccharin based on the previous experimental work of Ira Remsen using toluene (derived from petroleum). This product was found to be three hundred to five hundred times as sweet as sugar, although some people could detect a bitter aftertaste.

In 1944, the chemical family of cyclamates was discovered by Ludwig Frederick Audrieth and Michael Sveda. Although these compounds are only thirty to eighty times as sweet as sugar, there was no detectable aftertaste. By the mid-1960's, cyclamates had resplaced saccharin as the leading nonnutritive sweetener in the United States. Although cyclamates are still in use throughout the world, in October, 1969, FDA removed them from the list of approved food additives because of tests that indicated possible health hazards.

A POLITICAL ADDITIVE

Aspartame is the latest in artificial sweeteners that are derived from natural ingredients—in this case, two amino acids, one from milk and one from bananas. Discovered by accident in 1965 by American chemist James M. Schlatter when he licked his fingers during an experiment, aspartame is 180 times as sweet as sugar. In 1974, the FDA approved its use in dry foods such as gum and cereal and as a sugar replacement.

Shortly after its approval for this limited application, the FDA held public hearings on the safety concerns raised by John W. Olney, a professor of neuropathology at Washington University in St. Louis. There was some indication that aspartame, when combined with the common food additive monosodium glutamate, caused brain damage in children. These fears were confirmed, but the risk of brain damage was limited to a small percentage of individuals with a rare genetic disorder. At this point, the public debate took a political turn: Senator William Proxmire charged FDA Commissioner Alexander M. Schmidt with public misconduct. This controversy resulted in aspartame being taken off the market in 1975.

In 1981, the new FDA commissioner, Arthur H. Hayes, Jr., resapproved aspartame for use in the same applications: as a tabletop sweetener, as a cold-cereal additive, in chewing gum, and for other miscellaneous uses. In 1983, the FDA approved aspartame for use in carbonated beverages, its largest application to date. Later safety studies revealed that children with a rare metabolic disease, phenylketonuria, could not ingest this sweetener without severe health risks because of the presence of phenylalanine in aspartame. This condition results in a rapid buildup in phenylalanine in the blood. Laboratories simulated this condition in rats and found that high doses of aspartame inhibited the synthesis of dopamine, a neurotransmitter. Once this happens, an increase in the frequency of seizures can occur. There was no direct evidence, however, that aspartame actually caused seizures in these experiments.

Many other compounds are being tested for use as sugar replacements, the sweetest being a relative of aspartame. This compound is seventeen thousand to fifty-two thousand times sweeter than sugar.

IMPACT

The business fallout from the approval of a new low-calorie sweetener occurred over a short span of time. In 1981, sales of this artificial sweetener by G. D. Searle and Company were $74 million. In 1983, sales rose to $336 million and exceeded half a billion dollars the following year. These figures represent sales of more than 2,500 tons of this product. In 1985, 3,500 tons of aspartame were consumed. Clearly, this product's introduction was a commercial success for Searle. During this same period, the percentage of reduced-calorie carbonated beverages containing saccharin declined from 100 percent to 20 percent in an industry that had $4 billion in sales. Universally, consumers preferred products containing aspartame; the bitter aftertaste of saccharin was rejected in favor of the new, less powerful sweetener.

There is a trade-off in using these products. The FDA found evidence linking both saccharin and cyclamates to an elevated incidence of cancer. Cyclamates were banned in the United States for this reason. Public resistance to this measure caused the agency to back away from its position. The rationale was that, compared to other health risks associated with the consumption of sugar (especially for diabetics and overweight persons), the chance of getting cancer was slight and therefore a risk that many people would choose to ignore. The total domination of aspartame in the sweetener market seems to support this assumption.

See also Cyclamate; Genetically engineered insulin.

FURTHER READING

Blaylock, Russell L. *Excitotoxins: The Taste That Kills*. Santa Fe, N.Mex.: Health Press, 1998.

Hull, Janet Starr. *Sweet Poison: How the World's Most Popular Artificial Sweetener Is Killing Us—My Story*. Far Hills, N.J.: New Horizon Press, 1999.

Roberts, Hyman Jacob. *Aspartame (NutraSweet®): Is It Safe?* Philadelphia: Charles Press, 1990.

Stegink, Lewis D., and Lloyd J. Filer, *Aspartame: Physiology and Biochemistry.* New York: M. Dekker, 1984.

Stoddard, Mary Nash. *Deadly Deception: Story of Aspartame, Shocking Expose of the World's Most Controversial Sweetener.* Dallas: Odenwald Press, 1998.

ASSEMBLY LINE

THE INVENTION: A manufacturing technique pioneered in the automobile industry by Henry Ford that lowered production costs and helped bring automobile ownership within the reach of millions of Americans in the early twentieth century.

THE PEOPLE BEHIND THE INVENTION:
Henry Ford (1863-1947), an American carmaker
Eli Whitney (1765-1825), an American inventor
Elisha King Root (1808-1865), the developer of division of labor
Oliver Evans (1755-1819), the inventor of power conveyors
Frederick Winslow Taylor (1856-1915), an efficiency engineer

A PRACTICAL MAN

Henry Ford built his first "horseless carriage" by hand in his home workshop in 1896. In 1903, the Ford Motor Company was born. Ford's first product, the Model A, sold for less than one thousand dollars, while other cars at that time were priced at five to ten thousand dollars each. When Ford and his partners tried, in 1905, to sell a more expensive car, sales dropped. Then, in 1907, Ford decided that the Ford Motor Company would build "a motor car for the great multitude." It would be called the Model T.

The Model T came out in 1908 and was everything that Henry Ford said it would be. Ford's Model T was a low-priced (about $850), practical car that came in one color only: black. In the twenty years during which the Model T was built, the basic design never changed. Yet the price of the Model T, or "Tin Lizzie," as it was affectionately called, dropped over the years to less than half that of the original Model T. As the price dropped, sales increased, and the Ford Motor Company quickly became the world's largest automobile manufacturer.

The last of more than 15 million Model T's was made in 1927. Although it looked and drove almost exactly like the first Model T, these two automobiles were built in an entirely different way. The first was custom-built, while the last came off an assembly line.

At first, Ford had built his cars in the same way everyone else did: one at a time. Skilled mechanics would work on a car from start to finish, while helpers and runners brought parts to these highly paid craftsmen as they were needed. After finishing one car, the mechanics and their helpers would begin the next.

THE QUEST FOR EFFICIENCY

Custom-built products are good when there is little demand and buyers are willing to pay the high labor costs. This was not the case with the automobile. Ford realized that in order to make a large number of quality cars at a low price, he had to find a more efficient way to build cars. To do this, he looked to the past and the work of others. He found four ideas: interchangeable parts, continuous flow, division of labor, and elimination of wasted motion.

Eli Whitney, the inventor of the cotton gin, was the first person to use interchangeable parts successfully in mass production. In 1798, the United States government asked Whitney to make several thousand muskets in two years. Instead of finding and hiring gunsmiths to make the muskets by hand, Whitney used most of his time and money to design and build special machines that could make large numbers of

Model-T assembly line in the Ford Motor Company's Highland Park Factory. (Library of Congress)

identical parts—one machine for each part that was needed to build a musket. These tools, and others Whitney made for holding, measuring, and positioning the parts, made it easy for semiskilled, and even unskilled, workers to build a large number of muskets.

Production can be made more efficient by carefully arranging the different stages of production to create a "continuous flow." Ford borrowed this idea from at least two places: the meat-packing houses of Chicago and an automatic grain mill run by Oliver Evans.

Ford's idea for a moving assembly line came from Chicago's great meat-packing houses in the late 1860's. Here, the bodies of animals were moved along an overhead rail past a number of workers, each of whom made a certain cut, or handled one part of the packing job. This meant that many animals could be butchered and packaged in a single day.

Ford looked to Oliver Evans for an automatic conveyor system. In 1783, Evans had designed and operated an automatic grain mill that could be run by only two workers. As one worker poured grain into a funnel-shaped container, called a "hopper," at one end of the mill, a second worker filled sacks with flour at the other end. Everything in between was done automatically, as Evans's conveyors passed the grain through the different steps of the milling process without any help.

The idea of "division of labor" is simple: When one complicated job is divided into several easier jobs, some things can be made faster, with fewer mistakes, by workers who need fewer skills than ever before. Elisha King Root had used this principle to make the famous Colt "Six-Shooter." In 1849, Root went to work for Samuel Colt at his Connecticut factory and proved to be a manufacturing genius. By dividing the work into very simple steps, with each step performed by one worker, Root was able to make many more guns in much less time.

Before Ford applied Root's idea to the making of engines, it took one worker one day to make one engine. By breaking down the complicated job of making an automobile engine into eighty-four simpler jobs, Ford was able to make the process much more efficient. By assigning one person to each job, Ford's company was able to make 352 engines per day—an increase of more than 400 percent.

Frederick Winslow Taylor has been called the "original efficiency

HENRY FORD

Henry Ford (1863-1947) was more of a synthesizer and innovator than an inventor. Others invented the gasoline-powered automobile and the techniques of mass production, but it was Ford who brought the two together. The result was the assembly line-produced Model T that the Ford Motor Company turned out in the millions from 1908 until 1927. And it changed America profoundly.

Ford's idea was to lower production costs enough so that practically everyone could afford a car, not just the wealthy. He succeeded brilliantly. The first Model T's cost $850, rock bottom for the industry, and by 1927 the price was down to $290. Americans bought them up like no other technological marvel in the nation's history. For years, out of every one hundred cars on the road almost forty of them were Model T's. The basic version came with nothing on the dash board but an ignition switch, and the cars were quirky—so much so that an entire industry grew up to outfit them for the road and make sure they stayed running. Even then, they could only go up steep slopes backwards, and starting them was something of an art.

Americans took the Model T to heart, affectionately nicknaming it the flivver and Tin Lizzie. This "democratization of the automobile," as Ford called it, not only gave common people modern transportation and made them more mobile than every before; it started the American love affair with the car. Even after production stopped in 1927, the Model T Ford remained the archetype of American automobiles. As the great essayist E. B. White wrote in "Farewell My Lovely" (1936), his eulogy for the Model T, "...to a few million people who grew up with it, the old Ford practically *was* the American scene."

expert." His idea was that inefficiency was caused by wasted time and wasted motion. So Taylor studied ways to eliminate wasted motion. He proved that, in the long run, doing a job too quickly was as bad as doing it too slowly. "Correct speed is the speed at which men can work hour after hour, day after day, year in and year out, and remain continuously in good health," he said. Taylor also studied ways to streamline workers' movements. In this way, he was able to keep wasted motion to a minimum.

IMPACT

The changeover from custom production to mass production was an evolution rather than a revolution. Henry Ford applied the four basic ideas of mass production slowly and with care, testing each new idea before it was used. In 1913, the first moving assembly line for automobiles was being used to make Model T's. Ford was able to make his Tin Lizzies faster than ever, and his competitors soon followed his lead. He had succeeded in making it possible for millions of people to buy automobiles.

Ford's work gave a new push to the Industrial Revolution. It showed Americans that mass production could be used to improve quality, cut the cost of making an automobile, and improve profits. In fact, the Model T was so profitable that in 1914 Ford was able to double the minimum daily wage of his workers, so that they too could afford to buy Tin Lizzies.

Although Americans account for only about 6 percent of the world's population, they now own about 50 percent of its wealth. There are more than twice as many radios in the United States as there are people. The roads are crowded with more than 180 million automobiles. Homes are filled with the sounds and sights emitting from more than 150 million television sets. Never have the people of one nation owned so much. Where did all the products—radios, cars, television sets—come from? The answer is industry, which still depends on the methods developed by Henry Ford.

See also CAD/CAM; Color television; Interchangeable parts; Steelmaking process.

FURTHER READING

Abernathy, William, Kim Clark, and Alan Kantrow. *Industrial Renaissance*. New York: Basic Books, 1983.
Bruchey, Stuart. *Enterprise: The Dynamic Economy of a Free People*. Cambridge, Mass.: Harvard University Press, 1990.
Flink, James. *The Car Culture*. Cambridge, Mass.: MIT Press, 1975.
Hayes, Robert. *Restoring Our Competitive Edge*. New York: Wiley, 1984.
Olson, Sidney. *Young Henry Ford: A Picture History of the First Forty Years*. Detroit: Wayne State University Press, 1997.Wiley, 1984.

ATOMIC BOMB

THE INVENTION: A weapon of mass destruction created during World War II that utilized nuclear fission to create explosions equivalent to thousands of tons of trinitrotoluene (TNT),

THE PEOPLE BEHIND THE INVENTION:

J. Robert Oppenheimer (1904-1967), an American physicist
Leslie Richard Groves (1896-1970), an American engineer and Army general
Enrico Fermi (1900-1954), an Italian American nuclear physicist
Niels Bohr (1885-1962), a Danish physicist

ENERGY ON A LARGE SCALE

The first evidence of uranium fission (the splitting of uranium atoms) was observed by German chemists Otto Hahn and Fritz Strassmann in Berlin at the end of 1938. When these scientists discovered radioactive barium impurities in neutron-irradiated uranium, they wrote to their colleague Lise Meitner in Sweden. She and her nephew, physicist Otto Robert Frisch, calculated the large release of energy that would be generated during the nuclear fission of certain elements. This result was reported to Niels Bohr in Copenhagen.

Meanwhile, similar fission energies were measured by Frédéric Joliot and his associates in Paris, who demonstrated the release of up to three additional neutrons during nuclear fission. It was recognized immediately that if neutron-induced fission released enough additional neutrons to cause at least one more such fission, a self-sustaining chain reaction would result, yielding energy on a large scale.

While visiting the United States from January to May of 1939, Bohr derived a theory of fission with John Wheeler of Princeton University. This theory led Bohr to predict that the common isotope uranium 238 (which constitutes 99.3 percent of naturally occurring uranium) would require fast neutrons for fission, but that the rarer uranium 235 would fission with neutrons of any energy. This meant that uranium 235 would be far more suitable for use in any sort of bomb. Uranium bombardment in a cyclotron led to the discovery of plutonium in 1940 and the discovery that plutonium 239 was fissionable—and thus potentially good bomb material. Uranium 238 was then used to "breed" (create) plutonium 239, which was then separated from the uranium by chemical methods.

During 1942, the Manhattan District of the Army Corps of Engineers was formed under General Leslie Richard Groves, an engineer and Army general who contracted with E. I. Du Pont de Nemours and Company to construct three secret atomic cities at a total cost of $2 billion. At Oak Ridge, Tennessee, twenty-five thousand workers built a 1,000-kilowatt reactor as a pilot plant. A second city of sixty thousand inhabitants was built at Hanford, Washington, where three huge reactors and remotely controlled plutonium-extraction plants were completed in early 1945.

A Sustained and Awesome Roar

Studies of fast-neutron reactions for an atomic bomb were brought together in Chicago in June of 1942 under the leadership of J. Robert Oppenheimer. He soon became a personal adviser to Groves, who built for Oppenheimer a laboratory for the design and construction of the bomb at Los Alamos, New Mexico. In 1943, Oppenheimer gathered two hundred of the best scientists in what was by now being called the Manhattan Project to live and work in this third secret city.

Two bomb designs were developed. A gun-type bomb called "Little Boy" used 15 kilograms of uranium 235 in a 4,500-kilogram cylinder about 2 meters long and 0.5 meter in diameter, in which a uranium bullet could be fired into three uranium target rings to form a critical mass. An implosion-type bomb called "Fat Man" had a 5-kilogram spherical core of plutonium about the size of an orange, which could be squeezed inside a 2,300-kilogram sphere about 1.5 meters in diameter by properly shaped explosives to make the mass critical in the shorter time required for the faster plutonium fission process.

A flat scrub region 200 kilometers southeast of Alamogordo, called Trinity, was chosen for the test site, and observer bunkers were built about 10 kilometers from a 30-meter steel tower. On July 13, 1945, one of the plutonium bombs was assembled at the site; the next morning, it was raised to the top of the tower. Two days later, on July 16, after a short thunderstorm delay, the bomb was detonated at 5:30 A.M. The resulting implosion initiated a chain reaction of nearly 60 fission generations in about a microsecond. It produced an intense flash of light and a fireball that expanded to a diameter of about 600 meters in two seconds, rose to a height of more than 12 kilometers, and formed an ominous mushroom shape. Forty seconds later, an air blast hit the observer bunkers, followed by a sustained and awesome roar. Measurements confirmed that the explosion had the power of 18.6 kilotons of trinitrotoluene (TNT), nearly four times the predicted value.

IMPACT

On March 9, 1945, 325 American B-29 bombers dropped 2,000 tons of incendiary bombs on Tokyo, resulting in 100,000 deaths from the fire storms that swept the city. Nevertheless, the Japanese military refused to surrender, and American military plans called for an invasion of Japan, with estimates of up to a half million American casualties, plus as many as 2 million Japanese casualties. On August 6, 1945, after authorization by President Harry S. Truman, the B-29 *Enola Gay* dropped the uranium Little Boy bomb on Hiroshima at 8:15 A.M. On August 9, the remaining plutonium Fat Man bomb was dropped on Nagasaki. Approximately 100,000 people died at Hiroshima (out of a population of 400,000), and about 50,000 more died at Nagasaki. Japan offered to surrender on August 10, and after a brief attempt by some army officers to rebel, an official announcement by Emperor Hirohito was broadcast on August 15.

The development of the thermonuclear fusion bomb, in which hydrogen isotopes could be fused together by the force of a fission explosion to produce helium nuclei and almost unlimited energy, had been proposed early in the Manhattan Project by physicist Edward Teller. Little effort was invested in the hydrogen bomb until after the surprise explosion of a Soviet atomic bomb in September, 1949, which had been built with information stolen from the Manhattan Project. After three years of development under Teller's guidance, the first successful H-bomb was exploded on November 1, 1952, obliterating the Elugelab atoll in the Marshall Islands of the South Pacific. The arms race then accelerated until each side had stockpiles of thousands of H-bombs.

The Manhattan Project opened a Pandora's box of nuclear weapons that would plague succeeding generations, but it contributed more than merely weapons. About 19 percent of the electrical energy in the United States is generated by about 110 nuclear reactors producing more than 100,000 megawatts of power. More than 400 reactors in thirty countries provide 300,000 megawatts of the world's power. Reactors have made possible the widespread use of radioisotopes in medical diagnosis and therapy. Many of the techniques for producing and using these isotopes were developed by the hundreds of nuclear physicists who switched to the field of radiation biophysics after the war, ensuring that the benefits of their wartime efforts would reach the public.

See also Airplane; Breeder reactor; Cruise missile; Hydrogen bomb; Rocket; Stealth aircraft; V-2 rocket.

FURTHER READING

Goudsmit, Samuel Abraham, and Albert E. Moyer. *The History of Modern Physics, 1800-1950.* Los Angeles: Tomash Publishers, 1983.
Henshall, Phillip. *The Nuclear Axis: Germany, Japan, and the Atom Bomb Race, 1939-1945.* Stoud: Sutton, 2000.
Krieger, David. *Splitting the Atom: A Chronology of the Nuclear Age.* Santa Barbara, Calif.: Nuclear Age Peace Foundation, 1998.
Smith, June. *How the Atom Bombs Began, 1939-1946.* London: Brockwell, 1988.

ATOMIC CLOCK

THE INVENTION: A clock using the ammonia molecule as its oscillator that surpasses mechanical clocks in long-term stability, precision, and accuracy.

THE PERSON BEHIND THE INVENTION:
Harold Lyons (1913-1984), an American physicist

TIME MEASUREMENT

The accurate measurement of basic quantities, such as length, electrical charge, and temperature, is the foundation of science. The results of such measurements dictate whether a scientific theory is valid or must be modified or even rejected. Many experimental quantities change over time, but time cannot be measured directly. It must be measured by the occurrence of an oscillation or rotation, such as the twenty-four-hour rotation of the earth. For centuries, the rising of the Sun was sufficient as a timekeeper, but the need for more precision and accuracy increased as human knowledge grew.

Progress in science can be measured by how accurately time has been measured at any given point. In 1713, the British government, after the disastrous sinking of a British fleet in 1707 because of a miscalculation of longitude, offered a reward of 20,000 pounds for the invention of a ship's chronometer (a very accurate clock). Latitude is determined by the altitude of the Sun above the southern horizon at noon local time, but the determination of longitude requires an accurate clock set at Greenwich, England, time. The difference between the ship's clock and the local sun time gives the ship's longitude. This permits the accurate charting of new lands, such as those that were being explored in the eighteenth century. John Harrison, an English instrument maker, eventually built a chronometer that was accurate within one minute after five months at sea. He received his reward from Parliament in 1765.

ATOMIC CLOCKS PROVIDE GREATER STABILITY

A clock contains four parts: energy to keep the clock operating, an oscillator, an oscillation counter, and a display. A grandfather clock has weights that fall slowly, providing energy that powers the clock's gears. The pendulum, a weight on the end of a rod, swings back and forth (oscillates) with a regular beat. The length of the rod determines the pendulum's period of oscillation. The pendulum is attached to gears that count the oscillations and drive the display hands.

There are limits to a mechanical clock's accuracy and stability. The length of the rod changes as the temperature changes, so the period of oscillation changes. Friction in the gears changes as they wear out. Making the clock smaller increases its accuracy, precision, and stability. Accuracy is how close the clock is to telling the actual time. Stability indicates how the accuracy changes over time, while precision is the number of accurate decimal places in the display. A grandfather clock, for example, might be accurate to ten seconds per day and precise to a second, while having a stability of minutes per week.

Applying an electrical signal to a quartz crystal will make the crystal oscillate at its natural vibration frequency, which depends on its size, its shape, and the way in which it was cut from the larger crystal. Since the faster a clock's oscillator vibrates, the more precise the clock, a crystal-based clock is more precise than a large pendulum clock. By keeping the crystal under constant temperature, the clock is kept accurate, but it eventually loses its stability and slowly wears out.

In 1948, Harold Lyons and his colleagues at the National Bureau of Standards (NBS) constructed the first atomic clock, which used the ammonia molecule as its oscillator. Such a clock is called an atomic clock because, when it operates, a nitrogen atom vibrates. The pyramid-shaped ammonia molecule is composed of a triangular base; there is a hydrogen atom at each corner and a nitrogen atom at the top of the pyramid. The nitrogen atom does not remain at the top; if it absorbs radio waves of the right energy and frequency, it passes through the base to produce an upside-down pyramid and then moves back to the top. This oscillation frequency occurs at 23,870 megacycles (1 megacycle equals 1 million cycles) per second.

Lyons's clock was actually a quartz-ammonia clock, since the signal from a quartz crystal produced radio waves of the crystal's frequency that were fed into an ammonia-filled tube. If the radio waves were at 23,870 megacycles, the ammonia molecules absorbed the waves; a detector sensed this, and it sent no correction signal to the crystal. If radio waves deviated from 23,870 megacycles, the ammonia did not absorb them, the detector sensed the unabsorbed radio waves, and a correction signal was sent to the crystal. The atomic clock's accuracy and precision were comparable to those of a quartz-based clock—one part in a hundred million—but the atomic clock was more stable because molecules do not wear out.

The atomic clock's accuracy was improved by using cesium 133 atoms as the source of oscillation. These atoms oscillate at 9,192,631,770 plus or minus 20 cycles per second. They are accurate to a billionth of a second per day and precise to nine decimal places. A cesium clock is stable for years. Future developments in atomic clocks may see accuracies of one part in a million billions.

IMPACT

The development of stable, very accurate atomic clocks has far-reaching implications for many areas of science. Global positioning satellites send signals to receivers on ships and airplanes. By timing the signals, the receiver's position is calculated to within several meters of its true location.

Chemists are interested in finding the speed of chemical reactions, and atomic clocks are used for this purpose. The atomic clock led to the development of the maser (an acronym for *m*icrowave *a*mplification by *s*timulated *e*mission of *r*adiation), which is used to amplify weak radio signals, and the maser led to the development of the laser, a light-frequency maser that has more uses than can be listed here.

Atomic clocks have been used to test Einstein's theories of relativity that state that time on a moving clock, as observed by a stationary observer, slows down, and that a clock slows down near a large mass (because of the effects of gravity). Under normal conditions of low velocities and low mass, the changes in time are very small, but atomic clocks are accurate and stable enough to detect even these small changes. In such experiments, three sets of clocks were used—one group remained on Earth, one was flown west around the earth on a jet, and the last set was flown east. By comparing the times of the in-flight sets with the stationary set, the predicted slowdowns of time were observed and the theories were verified.

See also Carbon dating; Cyclotron; Electric clock; Laser; Synchrocyclotron; Tevatron accelerator.

FURTHER READING

Audoin, Claude, and Bernard Guinot. *The Measurement of Time: Time, Frequency, and the Atomic Clock.* New York: Cambridge University Press, 2001.
Barnett, Jo Ellen. *Time's Pendulum: The Quest to Capture Time—From Sundials to Atomic Clocks.* New York: Plenum Trade, 1998.
Bendick, Jeanne. *The First Book of Time.* New York: F. Watts, 1970.
"Ultra-Accurate Atomic Clock Unveiled at NIST Laboratory." *Research and Development* 42, no. 2 (February, 2000).

Atomic-powered Ship

The invention: The world's first atomic-powered merchant ship demonstrated a peaceful use of atomic power.

The people behind the invention:
Otto Hahn (1879-1968), a German chemist
Enrico Fermi (1901-1954), an Italian American physicist
Dwight D. Eisenhower (1890-1969), president of the United States, 1953-1961

Splitting the Atom

In 1938, Otto Hahn, working at the Kaiser Wilhelm Institute for Chemistry, discovered that bombarding uranium atoms with neutrons causes them to split into two smaller, lighter atoms. A large amount of energy is released during this process, which is called "fission." When one kilogram of uranium is fissioned, it releases the same amount of energy as does the burning of 3,000 metric tons of coal. The fission process also releases new neutrons.

Enrico Fermi suggested that these new neutrons could be used to split more uranium atoms and produce a chain reaction. Fermi and his assistants produced the first human-made chain reaction at the University of Chicago on December 2, 1942. Although the first use of this new energy source was the atomic bombs that were used to defeat Japan in World War II, it was later realized that a carefully controlled chain reaction could produce useful energy. The submarine *Nautilus*, launched in 1954, used the energy released from fission to make steam to drive its turbines.

U.S. President Dwight David Eisenhower proposed his "Atoms for Peace" program in December, 1953. On April 25, 1955, President Eisenhower announced that the "Atoms for Peace" program would be expanded to include the design and construction of an atomic-powered merchant ship, and he signed the legislation authorizing the construction of the ship in 1956.

SAVANNAH'S DESIGN AND CONSTRUCTION

A contract to design an atomic-powered merchant ship was awarded to George G. Sharp, Inc., on April 4, 1957. The ship was to carry approximately one hundred passengers (later reduced to sixty to reduce the ship's cost) and 10,886 metric tons of cargo while making a speed of 21 knots, about 39 kilometers per hour. The ship was to be 181 meters long and 23.7 meters wide. The reactor was to provide steam for a 20,000-horsepower turbine that would drive the ship's propeller. Most of the ship's machinery was similar to that of existing ships; the major difference was that steam came from a reactor instead of a coal- or oil-burning boiler.

New York Shipbuilding Corporation of Camden, New Jersey, won the contract to build the ship on November 16, 1957. States Marine Lines was selected in July, 1958, to operate the ship. It was christened *Savannah* and launched on July 21, 1959. The name *Savannah* was chosen to honor the first ship to use steam power while crossing an ocean. This earlier *Savannah* was launched in New York City in 1818.

Ships are normally launched long before their construction is complete, and the new *Savannah* was no exception. It was finally turned over to States Marine Lines on May 1, 1962. After extensive testing by its operators and delays caused by labor union disputes, it began its maiden voyage from Yorktown, Virginia, to Savannah, Georgia, on August 20, 1962. The original budget for design and construction was $35 million, but by this time, the actual cost was about $80 million.

Savannah's nuclear reactor was fueled with about 7,000 kilograms (15,400 pounds) of uranium. Uranium consists of two forms, or "isotopes." These are uranium 235, which can fission, and uranium 238, which cannot. Naturally occurring uranium is less than 1 percent uranium 235, but the uranium in *Savannah*'s reactor had been enriched to contain nearly 5 percent of this isotope. Thus, there was less than 362 kilograms of usable uranium in the reactor. The ship was able to travel about 800,000 kilometers on this initial fuel load. Three and a half million kilograms of water per hour flowed through the reactor under a pressure of 5,413 kilograms per square centimeter. It entered the reactor at 298.8 degrees Celsius and left at

317.7 degrees Celsius. Water leaving the reactor passed through a heat exchanger called a "steam generator." In the steam generator, reactor water flowed through many small tubes. Heat passed through the walls of these tubes and boiled water outside them. About 113,000 kilograms of steam per hour were produced in this way at a pressure of 1,434 kilograms per square centimeter and a temperature of 240.5 degrees Celsius.

Labor union disputes dogged *Savannah*'s early operations, and it did not start its first trans-Atlantic crossing until June 8, 1964. *Savannah* was never a money maker. Even in the 1960's, the trend was toward much bigger ships. It was announced that the ship would be retired in August, 1967, but that did not happen. It was finally put out of service in 1971. Later, *Savannah* was placed on permanent display at Charleston, South Carolina.

Consequences

Following the United States' lead, Germany and Japan built atomic-powered merchant ships. The Soviet Union is believed to have built several atomic-powered icebreakers. Germany's *Otto Hahn*, named for the scientist who first split the atom, began service in 1968, and Japan's *Mutsuai* was under construction as *Savannah* retired.

Numerous studies conducted in the early 1970's claimed to prove that large atomic-powered merchant ships were more profitable than oil-fired ships of the same size. Several conferences devoted to this subject were held, but no new ships were built.

Although the U.S. Navy has continued to use reactors to power submarines, aircraft carriers, and cruisers, atomic power has not been widely used for merchant-ship propulsion. Labor union problems such as those that haunted *Savannah*, high insurance costs, and high construction costs are probably the reasons. Public opinion, after the reactor accidents at Three Mile Island (in 1979) and Chernobyl (in 1986) is also a factor.

See also Gyrocompass; Hovercraft; Nuclear reactor; Supersonic passenger plane.

FURTHER READING

Epstein, Sam Epstein, Beryl (William) Epstein, and Raymond Burns. *Enrico Fermi, Father of Atomic Power.* Champaign, Ill. Garrard Publishing, 1970.

Hahn, Otto, and Wily Ley. *Otto Hahn: A Scientific Autobiography.* New York: C. Scribner's Sons, 1966.

Hoffman, Klaus. *Otto Hahn: Achievement and Responsibility.* New York: Springer, 2001.

"The Race to Power Bigger, Faster Ships." *Business Week* 2305 (November 10, 1973).

"Underway on Nuclear Power." *All Hands* 979 (November, 1998).

Autochrome Plate

The invention: The first commercially successful process in which a single exposure in a regular camera produced a color image.

The people behind the invention:
Louis Lumière (1864-1948), a French inventor and scientist
Auguste Lumière (1862-1954), an inventor, physician, physicist, chemist, and botanist
Alphonse Sèyewetz, a skilled scientist and assistant of the Lumière brothers

Adding Color

In 1882, Antoine Lumière, painter, pioneer photographer, and father of Auguste and Louis, founded a factory to manufacture photographic gelatin dry-plates. After the Lumière brothers took over the factory's management, they expanded production to include roll film and printing papers in 1887 and also carried out joint research that led to fundamental discoveries and improvements in photographic development and other aspects of photographic chemistry.

While recording and reproducing the actual colors of a subject was not possible at the time of photography's inception (about 1822), the first practical photographic process, the daguerreotype, was able to render both striking detail and good tonal quality. Thus, the desire to produce full-color images, or some approximation to realistic color, occupied the minds of many photographers and inventors, including Louis and Auguste Lumière, throughout the nineteenth century.

As researchers set out to reproduce the colors of nature, the first process that met with any practical success was based on the additive color theory expounded by the Scottish physicist James Clerk Maxwell in 1861. He believed that any color can be created by adding together red, green, and blue light in certain proportions. Maxwell, in his experiments, had taken three negatives through screens or filters of these additive primary colors. He then took slides made from these negatives and projected the slides through

ANTOINE LUMIÈRE AND SONS

Antoine Lumière was explosive in temperament, loved a good fight, and despised Americans. With these qualities—and his sons to take care of the practicalities—he turned France into a leader of the early photography and film industries.

Lumière was born into a family of wine growers in 1840 and trained to be a sign painter. Bored with his job, he learned the new art of photography, set up a studio in Lyon, and began to experiment with ways to make his own photographic plates. Failures led to frustration, and frustration ignited his temper, which often ended in his smashing the furniture and glassware nearby. His sons, Auguste, born 1862, and Louis, born 1864, came to the rescue. Louis, a science whiz as a teenager, succeeded where his father had failed. The dry plate he invented, Blue Label, was the most sensitive yet. The Lumières set up a factory to manufacture the plates and quickly found themselves wealthy, but the old man's love of extravagant spending and parties led them to the door of bankruptcy in 1882. His sons had to take control to save the family finances.

The father, an ardent French patriot, soon threw himself into a new crusade. American tariffs made it impossible for the Lumières to make a profit selling their photographic plates in the United States, which so angered the old man that he looked for revenge. He found it in the form of Thomas Edison's Kinetoscope in 1894. He got hold of samples, and soon the family factory was making motion picture film of its own and could undersell Edison in France. Louis also invented a projector, adapted from a sewing machine, that made it possible for movies to be shown to audiences.

Before Antoine Lumière died in Paris in 1911, he had the satisfaction of seeing his beloved France producing better, cheaper photographic products than those available from America, as well as becoming a pioneer in film making.

the same filters onto a screen so that their images were superimposed. As a result, he found that it was possible to reproduce the exact colors as well as the form of an object.

Unfortunately, since colors could not be printed in their tonal relationships on paper before the end of the nineteenth century,

Maxwell's experiment was unsuccessful. Although Frederick E. Ives of Philadelphia, in 1892, optically united three transparencies so that they could be viewed in proper alignment by looking through a peephole, viewing the transparencies was still not as simple as looking at a black-and-white photograph.

THE AUTOCHROME PLATE

The first practical method of making a single photograph that could be viewed without any apparatus was devised by John Joly of Dublin in 1893. Instead of taking three separate pictures through three colored filters, he took one negative through one filter minutely checkered with microscopic areas colored red, green, and blue. The filter and the plate were exactly the same size and were placed in contact with each other in the camera. After the plate was developed, a transparency was made, and the filter was permanently attached to it. The black-and-white areas of the picture allowed more or less light to shine through the filters; if viewed from a proper distance, the colored lights blended to form the various colors of nature.

In sum, the potential principles of additive color and other methods and their potential applications in photography had been discovered and even experimentally demonstrated by 1880. Yet a practical process of color photography utilizing these principles could not be produced until a truly panchromatic emulsion was available, since making a color print required being able to record the primary colors of the light cast by the subject.

Louis and Auguste Lumière, along with their research associate Alphonse Seyewetz, succeeded in creating a single-plate process based on this method in 1903. It was introduced commercially as the autochrome plate in 1907 and was soon in use throughout the world. This process is one of many that take advantage of the limited resolving power of the eye. Grains or dots too small to be recognized as separate units are accepted in their entirety and, to the sense of vision, appear as tones and continuous color.

IMPACT

While the autochrome plate remained one of the most popular color processes until the 1930's, soon this process was superseded by subtractive color processes. Leopold Mannes and Leopold Godowsky, both musicians and amateur photographic researchers who eventually joined forces with Eastman Kodak research scientists, did the most to perfect the Lumière brothers' advances in making color photography practical. Their collaboration led to the introduction in 1935 of Kodachrome, a subtractive process in which a single sheet of film is coated with three layers of emulsion, each sensitive to one primary color. A single exposure produces a color image.

Color photography is now commonplace. The amateur market is enormous, and the snapshot is almost always taken in color. Commercial and publishing markets use color extensively. Even photography as an art form, which was done in black and white through most of its history, has turned increasingly to color.

See also Color film; Instant photography; Xerography.

FURTHER READING

Collins, Douglas. *The Story of Kodak*. New York: Harry N. Abrams, 1990.

Glendinning, Peter. *Color Photography: History, Theory, and Darkroom Technique*. Englewood Cliffs, N.J.: Prentice-Hall, 1985.

Lartigue, Jacques-Henri, and Georges Herscher. *The Autochromes of J. H. Lartigue, 1912-1927*. New York: Viking Press, 1981.

Tolstoy, Ivan. *James Clerk Maxwell: A Biography*. Chicago: University of Chicago Press, 1982.

Wood, John. *The Art of the Autochrome: The Birth of Color Photography*. Iowa City: University of Iowa Press, 1993.

BASIC PROGRAMMING LANGUAGE

THE INVENTION: An interactive computer system and simple programming language that made it easier for nontechnical people to use computers.

THE PEOPLE BEHIND THE INVENTION:

John G. Kemeny (1926-1992), the chairman of Dartmouth's mathematics department

Thomas E. Kurtz (1928-   ), the director of the Kiewit Computation Center at Dartmouth

Bill Gates (1955-   ), a cofounder and later chairman of the board and chief operating officer of the Microsoft Corporation

THE EVOLUTION OF PROGRAMMING

The first digital computers were developed during World War II (1939-1945) to speed the complex calculations required for ballistics, cryptography, and other military applications. Computer technology developed rapidly, and the 1950's and 1960's saw computer systems installed throughout the world. These systems were very large and expensive, requiring many highly trained people for their operation.

The calculations performed by the first computers were determined solely by their electrical circuits. In the 1940's, The American mathematician John von Neumann and others pioneered the idea of computers storing their instructions in a program, so that changes in calculations could be made without rewiring their circuits. The programs were written in machine language, long lists of zeros and ones corresponding to on and off conditions of circuits. During the 1950's, "assemblers" were introduced that used short names for common sequences of instructions and were, in turn, transformed into the zeros and ones intelligible to the computer. The late 1950's saw the introduction of high-level languages, notably *Formula Translation* (FORTRAN), *Common Business Oriented Language* (COBOL), and *Algorithmic Language* (ALGOL), which used English words to communicate instructions to the computer. Unfortunately, these high-level languages were complicated; they required some knowledge of the computer equipment and were designed to be used by scientists, engineers, and other technical experts.

DEVELOPING BASIC

John G. Kemeny was chairman of the department of mathematics at Dartmouth College in Hanover, New Hampshire. In 1962, Thomas E. Kurtz, Dartmouth's computing director, approached Kemeny with the idea of implementing a computer system at Dartmouth College. Both men were dedicated to the idea that liberal arts students should be able to make use of computers. Although the English commands of FORTRAN and ALGOL were a tremendous improvement over the cryptic instructions of assembly language, they were both too complicated for beginners. Kemeny convinced Kurtz that they needed a completely new language, simple enough for beginners to learn quickly, yet flexible enough for many different kinds of applications.

The language they developed was known as the "*B*eginner's *A*llpurpose *S*ymbolic *I*nstruction *C*ode," or BASIC. The original language consisted of fourteen different statements. Each line of a BASIC program was preceded by a number. Line numbers were referenced by control flow statements, such as, "IF X = 9 THEN GOTO 200." Line numbers were also used as an editing reference. If line 30 of a program contained an error, the programmer could make the necessary correction merely by retyping line 30.

Programming in BASIC was first taught at Dartmouth in the fall of 1964. Students were ready to begin writing programs after two hours of classroom lectures. By June of 1968, more than 80 percent of the undergraduates at Dartmouth could write a BASIC program. Most of them were not science majors and used their programs in conjunction with other nontechnical courses.

Kemeny and Kurtz, and later others under their supervision, wrote more powerful versions of BASIC that included support for graphics on video terminals and structured programming. The creators of BASIC, however, always tried to maintain their original design goal of keeping BASIC simple enough for beginners.

CONSEQUENCES

Kemeny and Kurtz encouraged the widespread adoption of BASIC by allowing other institutions to use their computer system and by placing BASIC in the public domain. Over time, they shaped BASIC into a powerful language with numerous features added in response to the needs of its users. What Kemeny and Kurtz had not foreseen was the advent of the microprocessor chip in the early 1970's, which revolutionized computer technology. By 1975, microcomputer kits were being sold to hobbyists for well under a thousand dollars. The earliest of these was the Altair.

That same year, prelaw student William H. Gates (1955-   ) was persuaded by a friend, Paul Allen, to drop out of Harvard University and help create a version of BASIC that would run on the Altair. Gates and Allen formed a company, Microsoft Corporation, to sell their BASIC interpreter, which was designed to fit into the tiny memory of the Altair. It was about as simple as the original Dartmouth BASIC but had to depend heavily on the computer hardware. Most computers purchased for home use still include a version of Microsoft Corporation's BASIC.

See also BINAC computer; COBOL computer language; FORTRAN programming language; SAINT; Supercomputer.

FURTHER READING

Kemeney, John G., and Thomas E. Kurtz. *True BASIC: The Structured Language System for the Future. Reference Manual.* West Lebanon, N.H.: True BASIC, 1988.
Kurtz, Thomas E., and John G. Kemeney. *BASIC. 5th ed. Hanover, N.H., 1970.*
Spencer, Donald D. *Great Men and Women of Computing.* 2d ed. Ormond Beach, Fla.: Camelot Publishing, 1999.

Bathyscaphe

THE INVENTION: A submersible vessel capable of exploring the deepest trenches of the world's oceans.

THE PEOPLE BEHIND THE INVENTION:

William Beebe (1877-1962), an American biologist and explorer
Auguste Piccard (1884-1962), a Swiss-born Belgian physicist
Jacques Piccard (1922-   ), a Swiss ocean engineer

EARLY EXPLORATION OF THE DEEP SEA

The first human penetration of the deep ocean was made by William Beebe in 1934, when he descended 923 meters into the Atlantic Ocean near Bermuda. His diving chamber was a 1.5-meter steel ball that he named *Bathysphere*, from the Greek word *bathys* (deep) and the word *sphere*, for its shape. He found that a sphere resists pressure in all directions equally and is not easily crushed if it is constructed of thick steel. The bathysphere weighed 2.5 metric tons. It had no buoyancy and was lowered from a surface ship on a single 2.2-centimeter cable; a broken cable would have meant certain death for the bathysphere's passengers.

Numerous deep dives by Beebe and his engineer colleague, Otis Barton, were the first uses of submersibles for science. Through two small viewing ports, they were able to observe and photograph many deep-sea creatures in their natural habitats for the first time. They also made valuable observations on the behavior of light as the submersible descended, noting that the green surface water became pale blue at 100 meters, dark blue at 200 meters, and nearly black at 300 meters. A technique called "contour diving" was particularly dangerous. In this practice, the bathysphere was slowly towed close to the seafloor. On one such dive, the bathysphere narrowly missed crashing into a coral crag, but the explorers learned a great deal about the submarine geology of Bermuda and the biology of a coral-reef community. Beebe wrote several popular and scientific books about his adventures that did much to arouse interest in the ocean.

Testing the Bathyscaphe

The next important phase in the exploration of the deep ocean was led by the Swiss physicist Auguste Piccard. In 1948, he launched a new type of deep-sea research craft that did not require a cable and that could return to the surface by means of its own buoyancy. He called the craft a bathyscaphe, which is Greek for "deep boat." Piccard began work on the bathyscaphe in 1937, supported by a grant from the Belgian National Scientific Research Fund. The German occupation of Belgium early in World War II cut the project short, but Piccard continued his work after the war. The finished bathyscaphe was named *FNRS 2*, for the initials of the Belgian fund that had sponsored the project. The vessel was ready for testing in the fall of 1948.

The first bathyscaphe, as well as later versions, consisted of two basic components: first, a heavy steel cabin to accommodate observers, which looked somewhat like an enlarged version of Beebe's bathysphere; and second, a light container called a float, filled with gasoline, that provided lifting power because it was lighter than water. Enough iron shot was stored in silos to cause the vessel to descend. When this ballast was released, the gasoline in the float gave the bathyscaphe sufficient buoyancy to return to the surface.

Piccard's bathyscaphe had a number of ingenious devices. Jacques-Yves Cousteau, inventor of the Aqualung six years earlier, contributed a mechanical claw that was used to take samples of rocks, sediment, and bottom creatures. A seven-barreled harpoon gun, operated by water pressure, was attached to the sphere to capture specimens of giant squids or other large marine animals for study. The harpoons had electrical-shock heads to stun the "sea monsters," and if that did not work, the harpoon could give a lethal injection of strychnine poison. Inside the sphere were various instruments for measuring the deep-sea environment, including a Geiger counter for monitoring cosmic rays. The air-purification system could support two people for up to twenty-four hours. The bathyscaphe had a radar mast to broadcast its location as soon as it surfaced. This was essential because there was no way for the crew to open the sphere from the inside.

AUGUSTE PICCARD

Auguste Piccard used balloons to set records in altitude both above sea level and below sea level. However, setting records was not his purpose: He went where no one had gone before for the sake of science.

Born in Basel, Switzerland, in 1884, Auguste and his twin brother, Jean-Félix Piccard, studied in Zurich. After university in 1913, Auguste, a physicist, and Jean-Félix, a chemist, took up hot-air ballooning, and they joined the balloon section of the Swiss Army in 1915.

(Library of Congress)

Auguste moved to Brussels, Belgium, in 1922 to take a professorship of applied physics, and there he continued his ballooning. His subject of interest was cosmic rays, and in order to study them he had to get above the thick lower layer of atmosphere. Accordingly, he designed hydrogen-filled balloons that could reach high altitude. A ball-shaped, pressurized gondola carried him, his instruments, and one colleague to 51,775 feet altitude in 1931 and to 53,152 feet in 1932. Both were records.

Auguste, working with his son Jacques, then turned his attention to the sea. In order to explore the largely unknown world underwater, he built the bathyscaphe. It was really just another type of balloon, one which was made of steel and carried him inside. His dives with his son in various models of bathyscaphe set record after record. Their 1953 dive down 10,300 feet into the Mediterranean Sea was the deepest until Jacques, accompanied by a U.S. Navy officer, descended to the deepest spot on Earth seven years later.

The *FNRS* 2 was first tested off the Cape Verde Islands with the assistance of the French navy. Although Piccard descended to only 25 meters, the dive demonstrated the potential of the bathyscaphe. On the second dive, the vessel was severely damaged by waves, and further tests were suspended. A redesigned and rebuilt bathyscaphe, renamed *FNRS* 3 and operated by the French navy, descended to a depth of 4,049 meters off Dakar, Senegal, on the west coast of Africa in early 1954.

In August, 1953, Auguste Piccard, with his son Jacques, launched a greatly improved bathyscaphe, the *Trieste*, which they named for the Italian city in which it was built. In September of the same year, the *Trieste* successfully dived to 3,150 meters in the Mediterranean Sea. The Piccards glimpsed, for the first time, animals living on the seafloor at that depth. In 1958, the U.S. Navy purchased the *Trieste* and transported it to California, where it was equipped with a new cabin designed to enable the vessel to reach the seabed of the great oceanic trenches. Several successful descents were made in the Pacific by Jacques Piccard, and on January 23, 1960, Piccard, accompanied by Lieutenant Donald Walsh of the U.S. Navy, dived a record 10,916 meters to the bottom of the Mariana Trench near the island of Guam.

IMPACT

The oceans have always raised formidable barriers to humanity's curiosity and understanding. In 1960, two events demonstrated the ability of humans to travel underwater for prolonged periods and to observe the extreme depths of the ocean. The nuclear submarine *Triton* circumnavigated the world while submerged, and Jacques Piccard and Lieutenant Donald Walsh descended nearly 11 kilometers to the bottom of the ocean's greatest depression aboard the *Trieste*. After sinking for four hours and forty-eight minutes, the *Trieste* landed in the Challenger Deep of the Mariana Trench, the deepest known spot on the ocean floor. The explorers remained on the bottom for only twenty minutes, but they answered one of the biggest questions about the sea: Can animals live in the immense cold and pressure of the deep trenches? Observations of red shrimp and flatfishes proved that the answer was yes.

The *Trieste* played another important role in undersea exploration when, in 1963, it located and photographed the wreckage of the nuclear submarine *Thresher*. The *Thresher* had mysteriously disappeared on a test dive off the New England coast, and the Navy had been unable to find a trace of the lost submarine using surface vessels equipped with sonar and remote-control cameras on cables. Only the *Trieste* could actually search the bottom. On its third dive, the bathyscaphe found a piece of the wreckage, and it eventually photographed a 3,000-meter trail of debris that led to *Thresher's* hull, at a depth of 2.5 kilometers.

These exploits showed clearly that scientific submersibles could be used anywhere in the ocean. Piccard's work thus opened the last geographic frontier on Earth.

See also Aqualung; Bathysphere; Sonar; Ultrasound.

FURTHER READING

Ballard, Robert D., and Will Hively. *The Eternal Darkness: A Personal History of Deep-Sea Exploration*. Princeton, N.J.: Princeton University Press, 2000.
Piccard, Jacques, and Robert S. Dietz. *Seven Miles Down: The Story of the Bathyscaphe Trieste*. New York: Longmans, 1962.
Welker, Robert Henry. *Natural Man: The Life of William Beebe*. Bloomington: Indiana University Press, 1975.

BATHYSPHERE

THE INVENTION: The first successful chamber for manned deep-sea
diving missions.

THE PEOPLE BEHIND THE INVENTION:
William Beebe (1877-1962), an American naturalist and curator
of ornithology
Otis Barton (1899-   ), an American engineer
John Tee-Van (1897-1967), an American general associate with
the New York Zoological Society
Gloria Hollister Anable (1903?-1988), an American research
associate with the New York Zoological Society

INNER SPACE

Until the 1930's, the vast depths of the oceans had remained
largely unexplored, although people did know something of the
ocean's depths. Soundings and nettings of the ocean bottom had
been made many times by a number of expeditions since the 1870's.
Diving helmets had allowed humans to descend more than 91 meters below the surface, and the submarine allowed them to reach a
depth of nearly 120 meters. There was no firsthand knowledge,
however, of what it was like in the deepest reaches of the ocean: inner space.

The person who gave the world the first account of life at great
depths was William Beebe. When he announced in 1926 that he was
attempting to build a craft to explore the ocean, he was already a
well-known naturalist. Although his only degrees had been honorary doctorates, he was graduated as a special student in the Department of Zoology of Columbia University in 1898. He began his life-
long association with the New York Zoological Society in 1899.

It was during a trip to the Galápagos Islands off the west coast of
South America that Beebe turned his attention to oceanography. He
became the first scientist to use a diving helmet in fieldwork, swimming in the shallow waters. He continued this shallow-water work
at the new station he established in 1928, with the permission of En-

glish authorities, on the tiny island of Nonesuch in the Bermudas. Beebe realized, however, that he had reached the limits of the current technology and that to study the animal life of the ocean depths would require a new approach.

A NEW APPROACH

While he was considering various cylindrical designs for a new deep-sea exploratory craft, Beebe was introduced to Otis Barton. Barton, a young New Englander who had been trained as an engineer at Harvard University, had turned to the problems of ocean diving while doing postgraduate work at Columbia University. In December, 1928, Barton brought his blueprints to Beebe. Beebe immediately saw that Barton's design was what he was looking for, and the two went ahead with the construction of Barton's craft.

The "bathysphere," as Beebe named the device, weighed 2,268 kilograms and had a diameter of 1.45 meters and steel walls 3.8 centimeters thick. The door, weighing 180 kilograms, would be fastened over a manhole with ten bolts. Four windows, made of fused quartz, were ordered from the General Electric Company at a cost of $500 each. A 250-watt water spotlight lent by the Westinghouse Company provided the exterior illumination, and a telephone lent by the Bell Telephone Laboratory provided a means of communicating with the surface. The breathing apparatus consisted of two oxygen tanks that allowed 2 liters of oxygen per minute to escape into the sphere. During the dive, the carbon dioxide and moisture were removed, respectively, by trays containing soda lime and calcium chloride. A winch would lower the bathysphere on a steel cable.

In early July, 1930, after several test dives, the first manned dive commenced. Beebe and Barton descended to a depth of 244 meters. A short circuit in one of the switches showered them with sparks momentarily, but the descent was largely a success. Beebe and Barton had descended farther than any human.

Two more days of diving yielded a final dive record of 435 meters below sea level. Beebe and the other members of his staff (ichthyologist John Tee-Van and zoologist Gloria Hollister Anable) saw many species of fish and other marine life that previously had been seen only after being caught in nets. These first dives proved that an undersea exploratory craft had potential value, at least for deep water. After 1932, the bathysphere went on display at the Century of Progress Exhibition in Chicago.

In late 1933, the National Geographic Society offered to sponsor another series of dives. Although a new record was not a stipulation, Beebe was determined to supply one. The bathysphere was completely refitted before the new dives.

An unmanned test dive to 920 meters was made on August 7, 1934, once again off Nonesuch Island. Minor adjustments were made, and on the morning of August 11, the first dive commenced, attaining a depth of 765 meters and recording a number of new scientific observations. Several days later, on August 15, the weather was again right for the dive.

This dive also paid rich dividends in the number of species of deep-sea life observed. Finally, with only a few turns of cable left on the winch spool, the bathysphere reached a record depth of 923 meters—almost a kilometer below the ocean's surface.

IMPACT

Barton continued to work on the bathysphere design for some years. It was not until 1948, however, that his new design, the benthoscope, was finally constructed. It was similar in basic design to the bathysphere, though the walls were increased to withstand greater pressures. Other improvements were made, but the essential strengths and weaknesses remained. On August 16, 1949, Barton, diving alone, broke the record he and Beebe had set earlier, reaching a depth of 1,372 meters off the coast of Southern California.

The bathysphere effectively marked the end of the tethered exploration of the deep, but it pointed the way to other possibilities. The first advance in this area came in 1943, when undersea explorer Jacques-Yves Cousteau and engineer Émile Gagnan developed the Aqualung underwater breathing apparatus, which made possible unfettered and largely unencumbered exploration down to about 60 meters. This was by no means deep diving, but it was clearly a step along the lines that Beebe had envisioned for underwater research.

A further step came in the development of the bathyscaphe by

Auguste Piccard, the renowned Swiss physicist, who, in the 1930's, had conquered the stratosphere in high-altitude balloons. The bathyscaphe was a balloon that operated in reverse. A spherical steel passenger cabin was attached beneath a large float filled with gasoline for buoyancy. Several tons of iron pellets held by electromagnets acted as ballast. The bathyscaphe would sink slowly to the bottom of the ocean, and when its passengers wished to return, the ballast would be dumped. The craft would then slowly rise to the surface. On September 30, 1953, Piccard touched bottom off the coast of Italy, some 3,000 meters below sea level.

See also Aqualung; Bathyscaphe; Sonar; Ultrasound.

FURTHER READING

Ballard, Robert D., and Will Hively. *The Eternal Darkness: A Personal History of Deep-Sea Exploration.* Princeton, N.J.: Princeton University Press, 2000.
Forman, Will. *The History of American Deep Submersible Operations, 1775-1995.* Flagstaff, Ariz.: Best, 1999.
Welker, Robert Henry. *Natural Man: The Life of William Beebe.* Bloomington: Indiana University Press, 1975.

BINAC COMPUTER

THE INVENTION: The world's first electronic general-purpose digital computer.

THE PEOPLE BEHIND THE INVENTION:
John Presper Eckert (1919-1995), an American electrical engineer
John W. Mauchly (1907-1980), an American physicist
John von Neumann (1903-1957), a Hungarian American
mathematician
Alan Mathison Turing (1912-1954), an English mathematician

COMPUTER EVOLUTION

In the 1820's, there was a need for error-free mathematical and astronomical tables for use in navigation, unreliable versions of which were being produced by human "computers." The problem moved English mathematician and inventor Charles Babbage to design and partially construct some of the earliest prototypes of modern computers, with substantial but inadequate funding from the British government. In the 1880's, the search by the U.S. Bureau of the Census for a more efficient method of compiling the 1890 census led American inventor Herman Hollerith to devise a punched-card calculator, a machine that reduced by several years the time required to process the data.

The emergence of modern electronic computers began during World War II (1939-1945), when there was an urgent need in the American military for reliable and quickly produced mathematical tables that could be used to aim various types of artillery. The calculation of very complex tables had progressed somewhat since Babbage's day, and the human computers were being assisted by mechanical calculators. Still, the growing demand for increased accuracy and efficiency was pushing the limits of these machines. Finally, in 1946, following three years of intense work at the University of Pennsylvania's Moore School of Engineering, John Presper Eckert and John W. Mauchly presented their solution to the problems in the form of the Electronic Numerical Integrator and Calculator (ENIAC) the world's first electronic general-purpose digital computer.

The ENIAC, built under a contract with the Army's Ballistic Research Laboratory, became a great success for Eckert and Mauchly, but even before it was completed, they were setting their sights on loftier targets. The primary drawback of the ENIAC was the great difficulty involved in programming it. Whenever the operators needed to instruct the machine to shift from one type of calculation to another, they had to reset a vast array of dials and switches, unplug and replug numerous cables, and make various other adjustments to the multiple pieces of hardware involved. Such a mode of operation was deemed acceptable for the ENIAC because, in computing firing tables, it would need reprogramming only occasionally. Yet if instructions could be stored in a machine's memory, along with the data, such a machine would be able to handle a wide range of calculations with ease and efficiency.

THE TURING CONCEPT

The idea of a stored-program computer had first appeared in a paper published by English mathematician Alan Mathison Turing in 1937. In this paper, Turing described a hypothetical machine of quite simple design that could be used to solve a wide range of logical and mathematical problems. One significant aspect of this imaginary Turing machine was that the tape that would run through it would contain both information to be processed and instructions on how to process it. The tape would thus be a type of memory device, storing both the data and the program as sets of symbols that the machine could "read" and understand. Turing never attempted to construct this machine, and it was not until 1946 that he developed a design for an electronic stored-program computer, a prototype of which was built in 1950.

In the meantime, John von Neumann, a Hungarian American mathematician acquainted with Turing's ideas, joined Eckert and Mauchly in 1944 and contributed to the design of ENIAC's successor, the Electronic Discrete Variable Automatic Computer (EDVAC), another project financed by the Army. The EDVAC was the first computer designed to incorporate the concept of the stored program.

In March of 1946, Eckert and Mauchly, frustrated by a controversy over patent rights for the ENIAC, resigned from the Moore School. Several months later, they formed the Philadelphia-based Electronic Control Company on the strength of a contract from the National Bureau of Standards and the Census Bureau to build a much grander computer, the Universal Automatic Computer (UNIVAC). They thus abandoned the EDVAC project, which was finally completed by the Moore School in 1952, but they incorporated the main features of the EDVAC into the design of the UNIVAC.

Building the UNIVAC, however, proved to be much more involved and expensive than anticipated, and the funds provided by the original contract were inadequate. Eckert and Mauchly, therefore, took on several other smaller projects in an effort to raise funds. On October 9, 1947, they signed a contract with the Northrop Corporation of Hawthorne, California, to produce a relatively small computer to be used in the guidance system of a top-secret missile called the Snark, which Northrop was building for the Air Force. This computer, the Binary Automatic Computer (BINAC), turned out to be Eckert and Mauchly's first commercial sale and the first stored-program computer completed in the United States.

The BINAC was designed to be at least a preliminary version of a compact, airborne computer. It had two main processing units. These contained a total of fourteen hundred vacuum tubes, a drastic reduction from the eighteen thousand used in the ENIAC. There were also two memory units, as well as two power supplies, an input converter unit, and an input console, which used either a typewriter keyboard or an encoded magnetic tape (the first time such tape was used for computer input). Because of its dual processing, memory, and power units, the BINAC was actually two computers, each of which would continually check its results against those of the other in an effort to identify errors.

The BINAC became operational in August, 1949. Public demonstrations of the computer were held in Philadelphia from August 18 through August 20.

IMPACT

The design embodied in the BINAC is the real source of its significance. It demonstrated successfully the benefits of the dual processor design for minimizing errors, a feature adopted in many subsequent computers. It showed the suitability of magnetic tape as an input-output medium. Its most important new feature was its ability to store programs in its relatively spacious memory, the principle that Eckert, Mauchly, and von Neumann had originally designed into the EDVAC. In this respect, the BINAC was a direct descendant of the EDVAC.

In addition, the stored-program principle gave electronic computers new powers, quickness, and automatic control that, as they have continued to grow, have contributed immensely to the aura of intelligence often associated with their operation.

The BINAC successfully demonstrated some of these impressive new powers in August of 1949 to eager observers from a number of major American corporations. It helped to convince many influential leaders of the commercial segment of society of the promise of electronic computers. In doing so, the BINAC helped to ensure the further evolution of computers.

See also Apple II computer; BINAC computer; Colossus computer; ENIAC computer; IBM Model 1401 computer; Personal computer; Supercomputer; UNIVAC computer.

FURTHER READING

Macrae, Norman. *John von Neumann: The Scientific Genius Who Pioneered the Modern Computer, Game Theory, Nuclear Deterrence, and Much More.* New York: Pantheon Books, 1992.

Spencer, Donald D. *Great Men and Women of Computing.* 2d ed. Ormond Beach, Fla.: Camelot Publishing, 1999.

Zientara, Marguerite. *The History of Computing: A Biographical Portrait of the Visionaries Who Shaped the Destiny of the Computer Industry.* Framingham, Mass.: CW Communications, 1981.

Birth control pill

THE INVENTION: An orally administered drug that inhibits ovulation in women, thereby greatly reducing the chance of pregnancy.

THE PEOPLE BEHIND THE INVENTION:
Gregory Pincus (1903-1967), an American biologist
Min-Chueh Chang (1908-1991), a Chinese-born reproductive biologist
John Rock (1890-1984), an American gynecologist
Celso-Ramon Garcia (1921-   ), a physician
Edris Rice-Wray (1904-   ), a physician
Katherine Dexter McCormick (1875-1967), an American millionaire
Margaret Sanger (1879-1966), an American activist

AN ARDENT CRUSADER

Margaret Sanger was an ardent crusader for birth control and family planning. Having decided that a foolproof contraceptive was necessary, Sanger met with her friend, the wealthy socialite Katherine Dexter McCormick. A 1904 graduate in biology from the Massachusetts Institute of Technology, McCormick had the knowledge and the vision to invest in biological research. Sanger arranged a meeting between McCormick and Gregory Pincus, head of the Worcester Institutes of Experimental Biology. After listening to Sanger's pleas for an effective contraceptive and McCormick's offer of financial backing, Pincus agreed to focus his energies on finding a pill that would prevent pregnancy.

Pincus organized a team to conduct research on both laboratory animals and humans. The laboratory studies were conducted under the direction of Min-Chueh Chang, a Chinese-born scientist who had been studying sperm biology, artificial insemination, and in vitro fertilization. The goal of his research was to see whether pregnancy might be prevented by manipulation of the hormones usually found in a woman.

It was already known that there was one time when a woman could not become pregnant—when she was already pregnant. In 1921, Ludwig Haberlandt, an Austrian physiologist, had transplanted the ovaries from a pregnant rabbit into a nonpregnant one. The latter failed to produce ripe eggs, showing that some substance from the ovaries of a pregnant female prevents ovulation. This substance was later identified as the hormone progesterone by George W. Corner, Jr., and Willard M. Allen in 1928.

If progesterone could inhibit ovulation during pregnancy, maybe progesterone treatment could prevent ovulation in nonpregnant females as well. In 1937, this was shown to be the case by scientists from the University of Pennsylvania, who prevented ovulation in rabbits with injections of progesterone. It was not until 1951, however, when Carl Djerassi and other chemists devised inexpensive ways of producing progesterone in the laboratory, that serious consideration was given to the medical use of progesterone. The synthetic version of progesterone was called "progestin."

TESTING THE PILL

In the laboratory, Chang tried more than two hundred different progesterone and progestin compounds, searching for one that would inhibit ovulation in rabbits and rats. Finally, two compounds were chosen: progestins derived from the root of a wild Mexican yam. Pincus arranged for clinical tests to be carried out by Celso-Ramon Garcia, a physician, and John Rock, a gynecologist.

Rock had already been conducting experiments with progesterone as a treatment for infertility. The treatment was effective in some women but required that large doses of expensive progesterone be injected daily. Rock was hopeful that the synthetic progestin that Chang had found effective in animals would be helpful in infertile women as well. With Garcia and Pincus, Rock treated another group of fifty infertile women with the synthetic progestin. After treatment ended, seven of these previously infertile women became pregnant within half a year. Garcia, Pincus, and Rock also took several physiological measurements of the women while they were taking the progestin and were able to conclude that ovulation did not occur while the women were taking the progestin pill.

Margaret Sanger

Margaret Louise Higgins saw her mother die at the age of only fifty. The cause was tuberculosis, but Margaret, the sixth of eleven children, was convinced her mother's string of pregnancies was what killed her. Her crusade to liberate women from the burden of unwanted, dangerous pregnancies lasted the rest of her life.

Born in Corning, New York, in 1879, she went to Claverack College and Hudson River Institute and joined a nursing program at White Plains Hospital, graduating in 1900.

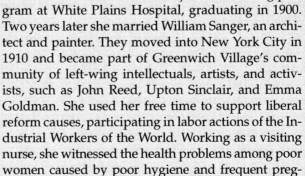

Two years later she married William Sanger, an architect and painter. They moved into New York City in 1910 and became part of Greenwich Village's community of left-wing intellectuals, artists, and activists, such as John Reed, Upton Sinclair, and Emma Goldman. She used her free time to support liberal reform causes, participating in labor actions of the Industrial Workers of the World. Working as a visiting nurse, she witnessed the health problems among poor women caused by poor hygiene and frequent pregnancies. In 1912 she test this began a newspaper column, "What Every Girl Should Know," about reproductive health and education. The authorities tried to suppress some of the columns as obscene—for instance, one explaining venereal disease—but Sanger was undaunted. In 1914, she launched *The Woman Rebel*, a magazine promoting women's liberation and birth control. From then on, although threatened with legal action and jail, she vigorously fought the political battles for birth control She published books, lectured, took part in demonstrations, opened a birth control clinic in Brooklyn (the nation's first), started the Birth Control Federation of American (later renamed Planned Parenthood Federation of America), and traveled overseas to promote birth control in order to improve the standard of living in Third World countries and to curb population growth.

Sanger was not an inventor, but she contributed ideas to the invention of various birth control devices and in the 1950's found the money needed for the research and development of oral contraceptives at the Worcester Foundation for Experimental Biology, which produced the first birth control pill. She died in Tucson, Arizona, in 1966.

(Library of Congess)

Having shown that the hormone could effectively prevent ovulation in both animals and humans, the investigators turned their attention back to birth control. They were faced with several problems: whether side effects might occur in women using progestins for a long time, and whether women would remember to take the pill day after day, for months or even years. To solve these problems, the birth control pill was tested on a large scale. Because of legal problems in the United States, Pincus decided to conduct the test in Puerto Rico.

The test started in April of 1956. Edris Rice-Wray, a physician, was responsible for the day-to-day management of the project. As director of the Puerto Rico Family Planning Association, she had seen firsthand the need for a cheap, reliable contraceptive. The women she recruited for the study were married women from a low-income population living in a housing development in Río Piedras, a suburb of San Juan. Word spread quickly, and soon women were volunteering to take the pill that would prevent pregnancy. In the first study, 221 women took a pill containing 10 milligrams of progestin and 0.15 milligrams of estrogen. (The estrogen was added to help control breakthrough bleeding.)

Results of the test were reported in 1957. Overall, the pill proved highly effective in preventing conception. None of the women who took the pill according to directions became pregnant, and most women who wanted to get pregnant after stopping the pill had no difficulty. Nevertheless, 17 percent of the women had some unpleasant reactions, such as nausea or dizziness. The scientists believed that these mild side effects, as well as one death from congestive heart failure, were unrelated to the use of the pill.

Even before the final results were announced, additional field tests were begun. In 1960, the U.S. Food and Drug Administration (FDA) approved the use of the pill developed by Pincus and his collaborators as an oral contraceptive.

CONSEQUENCES

Within two years of approval by the FDA, more than a million women in the United States were using the birth control pill. New contraceptives were developed in the 1960's and 1970's, but the birth control pill remains the most widely used method of preventing pregnancy. More than 60 million women use the pill worldwide.

The greatest impact of the pill has been in the social and political world. Before Sanger began the push for the pill, birth control was regarded often as socially immoral and often illegal as well. Women in those post-World War II years were expected to have a lifelong career as a mother to their many children. With the advent of the pill, a radical change occurred in society's attitude toward women's work. Women had in-

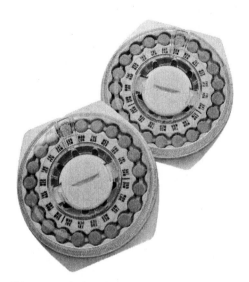

Dispensers designed to help users keep track of the days on which they take their pills. (Image Club Graphics)

creased freedom to work and enter careers previously closed to them because of fears that they might get pregnant. Women could control more precisely when they would get pregnant and how many children they would have. The women's movement of the 1960's—with its change to more liberal social and sexual values—gained much of its strength from the success of the birth control pill.

See also Abortion pill; Amniocentesis; Artificial hormone; Genetically engineered insulin; Mammography; Syphilis test; Ultrasound.

FURTHER READING

DeJauregui, Ruth. *One Hundred Medical Milestones That Shaped World History.* San Mateo, Calif.: Bluewood Books, 1998.
Tone, Andrea. *Devices and Desires: A History of Contraceptives in America.* New York: Hill and Wang, 2001.
Watkins, Elizabeth Siegel. *On the Pill: A Social History of Oral Contraceptives, 1950-1970.* Baltimore: Johns Hopkins University Press, 1998.

Blood transfusion

The invention: A technique that greatly enhanced surgery patients' chances of survival by replenishing the blood they lose in surgery with a fresh supply.

The people behind the invention:
Charles Drew (1904-1950), American pioneer in blood transfusion techniques
George Washington Crile (1864-1943), an American surgeon, author, and brigadier general in the U.S. Army Medical Officers' Reserve Corps
Alexis Carrel (1873-1944), a French surgeon
Samuel Jason Mixter (1855-1923), an American surgeon

Nourishing Blood Transfusions

It is impossible to say when and where the idea of blood transfusion first originated, although descriptions of this procedure are found in ancient Egyptian and Greek writings. The earliest documented case of a blood transfusion is that of Pope Innocent VII. In April, 1492, the pope, who was gravely ill, was transfused with the blood of three young boys. As a result, all three boys died without bringing any relief to the pope.

In the centuries that followed, there were occasional descriptions of blood transfusions, but it was not until the middle of the seventeenth century that the technique gained popularity following the English physician and anatomist William Harvey's discovery of the circulation of the blood in 1628. In the medical thought of those times, blood transfusion was considered to have a nourishing effect on the recipient. In many of those experiments, the human recipient received animal blood, usually from a lamb or a calf. Blood transfusion was tried as a cure for many different diseases, mainly those that caused hemorrhages, as well as for other medical problems and even for marital problems.

Blood transfusions were a dangerous procedure, causing many deaths of both donor and recipient as a result of excessive blood loss, infection, passage of blood clots into the circulatory systems of the recipients, passage of air into the blood vessels (air embolism), and transfusion reaction as a result of incompatible blood types. In the mid-nineteenth century, blood transfusions from animals to humans stopped after it was discovered that the serum of one species agglutinates and dissolves the blood cells of other species. A sharp drop in the use of blood transfusion came with the introduction of physiologic salt solution in 1875. Infusion of salt solution was simple and was safer than blood transfusion.

DIRECT-CONNECTION BLOOD TRANSFUSIONS

In 1898, when George Washington Crile began his work on blood transfusions, the major obstacle he faced was solving the problem of blood clotting during transfusions. He realized that salt solutions were not helpful in severe cases of blood loss, when there is a need to restore the patient to consciousness, steady the heart action, and raise the blood pressure. At that time, he was experimenting with indirect blood transfusions by drawing the blood of the donor into a vessel, then transferring it into the recipient's vein by tube, funnel, and cannula, the same technique used in the infusion of saline solution.

The solution to the problem of blood clotting came in 1902 when Alexis Carrel developed the technique of surgically joining blood vessels without exposing the blood to air or germs, either of which can lead to clotting. Crile learned this technique from Carrel and used it to join the peripheral artery in the donor to a peripheral vein of the recipient. Since the transfused blood remained sealed in the inner lining of the vessels, blood clotting did not occur.

The first human blood transfusion of this type was performed by Crile in December, 1905. The patient, a thirty-five-year-old woman, was transfused by her husband but died a few hours after the procedure.

The second, but first successful, transfusion was performed on August 8, 1906. The patient, a twenty-three-year-old male, suffered from severe hemorrhaging following surgery to remove kidney stones. After all attempts to stop the bleeding were exhausted with no results, and the patient was dangerously weak, transfusion was considered as a last resort. One of the patient's brothers was the do-

CHARLES DREW

While he was still in medical school, Charles Richard Drew saw a man's life saved with a blood transfusion. He also saw patients die because suitable donors could not be found. Impressed by both the life-saving power of transfusions and the dire need for more of them, Drew devoted his career to improving the nation's blood supply. His inventions saved untold thousands of lives, especially during World War II, before artificial blood was developed.

Born in 1904 in Washington, D.C., Drew was a star athlete in high school, in Amherst College—from which he graduated in 1926—and even in medical school at McGill University in Montreal from 1928 to 1933. He returned to the U.S. capital to become a resident in Freedmen's Hospital of Howard University. While there he invented a method for separating plasma from whole blood and discovered that it was not necessary to recombine the plasma and red blood cells for transfusion. Plasma alone was sufficient, and by drying or and freezing it, the plasma remained fresh enough over long periods to act as an emergency reserve. In 1938 Drew took a fellowship in blood research at Columbia Presbyterian Hospital in New York City. Employing his plasma preservation methods, he opened the first blood bank and wrote a dissertation on his techniques. He became the first African American to earn a Doctor of Science degree from Columbia University in 1940.

He organized another blood bank, this one in Great Britain, and in 1941 was appointed director of the American Red Cross blood donor project. However, Drew learned to his disgust that the Red Cross and U.S. government would not allow blood from African Americans and Caucasians to be mixed in the blood bank. There was no scientific reason for such segregation. Bias prevailed. Drew angrily denounced the policy at a press conference and resigned from the Red Cross.

He went back to Howard University as head of surgery and, later, director of Freedmen's Hospital. Drew died in 1950 following an automobile accident.

(Associated Publishers)

nor. Following the transfusion, the patient showed remarkable recovery and was strong enough to withstand surgery to remove the kidney and stop the bleeding. When his condition deteriorated a few days later, another transfusion was done. This time, too, he showed remarkable improvement, which continued until his complete recovery.

For his first transfusions, Crile used the Carrel suture method, which required using very fine needles and thread. It was a very delicate and time-consuming procedure. At the suggestion of Samuel Jason Mixter, Crile developed a new method using a short tubal device with an attached handle to connect the blood vessels. By this method, 3 or 4 centimeters of the vessels to be connected were surgically exposed, clamped, and cut, just as under the previous method. Yet, instead of suturing of the blood vessels, the recipient's vein was passed through the tube and then cuffed back over the tube and tied to it. Then the donor's artery was slipped over the cuff. The clamps were opened, and blood was allowed to flow from the donor to the recipient. In order to accommodate different-sized blood vessels, tubes of four different sizes were made, ranging in diameter from 1.5 to 3 millimeters.

IMPACT

Crile's method was the preferred method of blood transfusion for a number of years. Following the publication of his book on transfusion, a number of modifications to the original method were published in medical journals. In 1913, Edward Lindeman developed a method of transfusing blood simply by inserting a needle through the patient's skin and into a surface vein, making it for the first time a nonsurgical method. This method allowed one to measure the exact quantity of blood transfused. It also allowed the donor to serve in multiple transfusions. This development opened the field of transfusions to all physicians. Lindeman's needle and syringe method also eliminated another major drawback of direct blood transfusion: the need to have both donor and recipient right next to each other.

See also Coronary artery bypass surgery; Electrocardiogram; Electroencephalogram; Heart-lung machine.

FURTHER READING

English, Peter C. *Shock, Physiological Surgery, and George Washington Crile: Medical Innovation in the Progressive Era*. Westport, Conn.: Greenwood Press, 1980.

Le Vay, David, and Roy Porter. *Alexis Carrel: The Perfectibility of Man*. Rockville, Md.: Kabel Publishers, 1996.

Malinin, Theodore I. *Surgery and Life: The Extraordinary Career of Alexis Carrel*. New York: Harcourt Brace Jovanovich, 1979.

May, Angelo M., and Alice G. May. *The Two Lions of Lyons: The Tale of Two Surgeons, Alexis Carrel and René Leriche*. Rockville, Md.: Kabel Publishers, 1992.

Breeder reactor

THE INVENTION: A plant that generates electricity from nuclear fission while creating new fuel.

THE PERSON BEHIND THE INVENTION:
Walter Henry Zinn (1906-2000), the first director of the Argonne National Laboratory

PRODUCING ELECTRICITY WITH MORE FUEL

The discovery of nuclear fission involved both the discovery that the nucleus of a uranium atom would split into two lighter elements when struck by a neutron and the observation that additional neutrons, along with a significant amount of energy, were released at the same time. These neutrons might strike other atoms and cause them to fission (split) also. That, in turn, would release more energy and more neutrons, triggering a chain reaction as the process continued to repeat itself, yielding a continuing supply of heat.

Besides the possibility that an explosive weapon could be constructed, early speculation about nuclear fission included its use in the generation of electricity. The occurrence of World War II (1939-1945) meant that the explosive weapon would be developed first.

Both the weapons technology and the basic physics for the electrical reactor had their beginnings in Chicago with the world's first nuclear chain reaction. The first self-sustaining nuclear chain reaction occurred in a laboratory at the University of Chicago on December 2, 1942.

It also became apparent at that time that there was more than one way to build a bomb. At this point, two paths were taken: One was to build an atomic bomb with enough fissionable uranium in it to explode when detonated, and another was to generate fissionable plutonium and build a bomb. Energy was released in both methods, but the second method also produced another fissionable substance.

The observation that plutonium and energy could be produced together meant that it would be possible to design electric power systems that would produce fissionable plutonium in quantities as large as, or larger than, the amount of fissionable material consumed. This is the breeder concept, the idea that while using up fissionable uranium 235, another fissionable element can be made. The full development of this concept for electric power was delayed until the end of World War II.

ELECTRICITY FROM ATOMIC ENERGY

On August 1, 1946, the Atomic Energy Commission (AEC) was established to control the development and explore the peaceful uses of nuclear energy. The Argonne National Laboratory was assigned the major responsibilities for pioneering breeder reactor technologies. Walter Henry Zinn was the laboratory's first director. He led a team that planned a modest facility (Experimental Breeder Reactor I, or EBR-I) for testing the validity of the breeding principle.

Planning for this had begun in late 1944 and grew as a natural extension of the physics that developed the plutonium atomic bomb. The conceptual design details for a breeder-electric reactor were reasonably complete by late 1945. On March 1, 1949, the AEC announced the selection of a site in Idaho for the National Reactor Station (later to be named the Idaho National Engineering Laboratory, or INEL). Construction at the INEL site in Arco, Idaho, began in October, 1949. Critical mass was reached in August, 1951. ("Critical mass" is the amount and concentration of fissionable material required to produce a self-sustaining chain reaction.)

The system was brought to full operating power, 1.1 megawatts of thermal power, on December 19, 1951. The next day, December 20, at 11:00 A.M., steam was directed to a turbine generator. At 1:23 P.M., the generator was connected to the electrical grid at the site, and "electricity flowed from atomic energy," in the words of Zinn's console log of that day. Approximately 200 kilowatts of electric power were generated most of the time that the reactor was run. This was enough to satisfy the needs of the EBR-I facilities. The reactor was shut down in 1964 after five years of use primarily as a test facility. It had also produced the first pure plutonium.

With the first fuel loading, a conversion ratio of 1.01 was achieved, meaning that more new fuel was generated than was consumed by about 1 percent. When later fuel loadings were made with plutonium, the conversion ratios were more favorable, reaching as high as 1.27. EBR-I was the first reactor to generate its own fuel and the first power reactor to use plutonium for fuel.

The use of EBR-I also included pioneering work on fuel recovery and reprocessing. During its five-year lifetime, EBR-I operated with four different fuel loadings, each designed to establish specific benchmarks of breeder technology. This reactor was seen as the first in a series of increasingly large reactors in a program designed to develop breeder technology. The reactor was replaced by EBR-II, which had been proposed in 1953 and was constructed from 1955 to 1964. EBR-II was capable of producing 20 megawatts of electrical power. It was approximately fifty times more powerful than EBR-I but still small compared to light-water commercial reactors of 600 to 1,100 megawatts in use toward the end of the twentieth century.

Consequences

The potential for peaceful uses of nuclear fission were dramatized with the start-up of EBR-I in 1951: It was the first in the world to produce electricity, while also being the pioneer in a breeder reactor program. The breeder program was not the only reactor program being developed, however, and it eventually gave way to the light-water reactor design for use in the United States. Still, if energy resources fall into short supply, it is likely that the technologies first developed with EBR-I will find new importance. In France and Japan, commercial reactors make use of breeder reactor technology; these reactors require extensive fuel reprocessing.

Following the completion of tests with plutonium loading in 1964, EBR-I was shut down and placed in standby status. In 1966, it was declared a national historical landmark under the stewardship of the U.S. Department of the Interior. The facility was opened to the public in June, 1975.

See also Atomic bomb; Geothermal power; Nuclear power plant; Nuclear reactor; Solar thermal engine; Tidal power plant.

FURTHER READING

"Breeder Trouble." *Technology Review* 91, no. 5 (July, 1988).
Hippel, Frank von, and Suzanne Jones. "Birth of the Breeder." *Bulletin of the Atomic Scientists* 53, no. 5 (September/October, 1997).
Krieger, David. *Splitting the Atom: A Chronology of the Nuclear Age.* Santa Barbara, Calif.: Nuclear Age Peace foundation, 1998.

Broadcaster guitar

The invention: The first commercially manufactured solid-body electric guitar, the Broadcaster revolutionized the guitar industry and changed the face of popular music

The people behind the invention:
Leo Fender (1909-1991), designer of affordable and easily mass-produced solid-body electric guitars
Les Paul (Lester William Polfuss, 1915-   ), a legendary guitarist and designer of solid-body electric guitars
Charlie Christian (1919-1942), an influential electric jazz guitarist of the 1930's

Early Electric Guitars

It has been estimated that between 1931 and 1937, approximately twenty-seven hundred electric guitars and amplifiers were sold in the United States. The Electro String Instrument Company, run by Adolph Rickenbacker and his designer partners, George Beauchamp and Paul Barth, produced two of the first commercially manufactured electric guitars—the Rickenbacker A-22 and A-25—in 1931. The Rickenbacker models were what are known as "lap steel" or Hawaiian guitars. A Hawaiian guitar is played with the instrument lying flat across a guitarist's knees. By the mid-1930's, the Gibson company had introduced an electric Spanish guitar, the ES-150. Legendary jazz guitarist Charlie Christian made this model famous while playing for Benny Goodman's orchestra. Christian was the first electric guitarist to be heard by a large American audience. He became an inspiration for future electric guitarists, because he proved that the electric guitar could have its own unique solo sound. Along with Christian, the other electric guitar figures who put the instrument on the musical map were blues guitarist T-Bone Walker, guitarist and inventor Les Paul, and engineer and inventor Leo Fender.

Early electric guitars were really no more than acoustic guitars, with the addition of one or more pickups, which convert string vibrations to electrical signals that can be played through a speaker. Amplification of a guitar made it a more assertive musical instrument. The electrification of the guitar ultimately would make it more flexible, giving it a more prominent role in popular music. Les Paul, always a compulsive inventor, began experimenting with ways of producing an electric solid-body guitar in the late 1930's. In 1929, at the age of thirteen, he had amplified his first acoustic guitar. Another influential inventor of the 1940's was Paul Bigsby. He built a prototype solid-body guitar for country music star Merle Travis in 1947. It was Leo Fender who revolutionized the electric guitar industry by producing the first commercially viable solid-body electric guitar, the Broadcaster, in 1948.

LEO FENDER

Leo Fender was born in the Anaheim, California, area in 1909. As a teenager, he began to build and repair guitars. By the 1930's, Fender was building and renting out public address systems for group gatherings. In 1937, after short tenures of employment with the Division of Highways and the U.S. Tire Company, he opened a radio repair company in Fullerton, California. Always looking to expand and invent new and exciting electrical gadgets, Fender and Clayton Orr "Doc" Kauffman started the K & F Company in 1944. Kauffman was a musician and a former employee of the Electro String Instrument Company. The K & F Company lasted until 1946 and produced steel guitars and amplifiers. After that partnership ended, Fender founded the Fender Electric Instruments Company.

With the help of George Fullerton, who joined the company in 1948, Fender developed the Fender Broadcaster. The body of the Broadcaster was made of a solid plank of ash wood. The corners of the ash body were rounded. There was a cutaway located under the joint with the solid maple neck, making it easier for the guitarist to access the higher frets. The maple neck was bolted to the body of the guitar, which was unusual, since most guitar necks prior to the Broadcaster had been glued to the body. Frets were positioned directly into designed cuts made in the maple of the neck. The guitar had two pickups.

The Fender Electric Instruments Company made fewer than one

thousand Broadcasters. In 1950, the name of the guitar was changed from the Broadcaster to the Telecaster, as the Gretsch company had already registered the name Broadcaster for some of its drums and banjos. Fender decided not to fight in court over use of the name.

Leo Fender has been called the Henry Ford of the solid-body electric guitar, and the Telecaster became known as the Model T of the industry. The early Telecasters sold for $189.50. Besides being inexpensive, the Telecaster was a very durable instrument. Basically, the Telecaster was a continuation of the Broadcaster. Fender did not file for a patent on its unique bridge pickup until January 13, 1950, and he did not file for a patent on the Telecaster's unique body shape until April 3, 1951.

In the music industry during the late 1940's, it was important for a company to unveil new instruments at trade shows. At this time, there was only one important trade show, sponsored by the National Association of Music Merchants. The Broadcaster was first sprung on the industry at the 1948 trade show in Chicago. The industry had seen nothing like this guitar ever before. This new guitar existed only to be amplified; it was not merely an acoustic guitar that had been converted.

IMPACT

The Telecaster, as it would be called after 1950, remained in continuous production for more years than any other guitar of its type and was one of the industry's best sellers. From the beginning, it looked and sounded unique. The electrified acoustic guitars had a mellow woody tone, whereas the Telecaster had a clean twangy tone. This tone made it popular with country and blues guitarists. The Telecaster could also be played at higher volume than previous electric guitars.

Because Leo Fender attempted something revolutionary by introducing an electric solid-body guitar, there was no guarantee that his business venture would succeed. Fender Electric Instruments Company had fifteen employees in 1947. At times, during the early years of the company, it looked as though Fender's dreams would not come to fruition, but the company persevered and grew. Between 1948 and 1955 with an increase of employees, the company was able to produce ten thousand Broadcaster/Telecaster guitars. Fender had taken a big risk, but it paid off enormously. Between 1958 and the mid-1970's, Fender produced more than 250,000 Telecasters. Other guitar manufacturers were placed in a position of having to catch up. Fender had succeeded in developing a process by which electric solid-body guitars could be manufactured profitably on a large scale.

EARLY GUITAR PICKUPS

The first pickups used on a guitar can be traced back to the 1920's and the efforts of Lloyd Loar, but there was not strong interest on the part of the American public for the guitar to be amplified. The public did not become intrigued until the 1930's. Charlie Christian's electric guitar performances with Benny Goodman woke up the public to the potential of this new and exciting sound. It was not until the 1950's, though, that the electric guitar became firmly established. Leo Fender was the right man in the right place. He could not have known that his Fender guitars would help to usher in a whole new musical landscape. Since the electric guitar was the newest member of the family of guitars, it took some time for musical audiences to fully appreciate what it could do. The electric solid-body guitar has been called a dangerous, uncivilized instrument. The youth culture of the 1950's found in this new guitar a voice for their rebellion. Fender unleashed a revolution not only in the construction of a guitar but also in the way popular music would be approached henceforth.

Because of the ever-increasing demand for the Fender product, Fender Sales was established as a separate distribution company in 1953 by Don Randall. Fender Electric Instruments Company had fifteen employees in 1947, but by 1955, the company employed fifty people. By 1960, the number of employees had risen to more than one hundred. Before Leo Fender sold the company to CBS on January 4, 1965, for $13 million, the company occupied twenty-seven buildings and employed more than five hundred workers.

Always interested in finding new ways of designing a more nearly perfect guitar, Leo Fender again came up with a remarkable guitar in 1954, with the Stratocaster. There was talk in the guitar industry that

CHARLIE CHRISTIAN

Charlie Christian (1919-1942) did not invent the electric guitar, but he did pioneer its use. He was born to music, and for jazz aficionados he quickly developed into a legend, not only establishing a new solo instrument but also helping to invent a whole new type of jazz.

Christian grew up in Texas, surrounded by a family of professional musicians. His parents and two brothers played trumpet, guitar, and piano, and sang, and Charlie was quick to imitate them. As a boy he made his own guitars out of cigar boxes and, according to a childhood friend, novelist Ralph Ellison, wowed his friends at school with his riffs. When he first heard an electric guitar in the mid-1930's, he made that his own, too.

The acoustic guitar had been only a backup instrument in jazz because it was too quiet to soar in solos. In 1935, Eddie Durham found that electric guitars could swing side by side with louder instruments. Charlie, already an experienced performer with acoustic guitar and bass, immediately recognized the power and range of subtle expression possible with the electrified instrument. He bought a Gibson ES-150 and began to make musical history with his improvisations.

He impressed producer John Hammond, who introduced him to big-band leader Benny Goodman in 1939. Notoriously hard to please, Goodman rejected Christian after an audition. However, Hammond later sneaked him on stage while the Goodman band was performing. Outraged, Goodman segued into a tune he was sure Christian did not know, "Rose Room." Christian was undaunted. He delivered an astonishingly inventive solo, and Goodman was won over despite himself. Christian's ensuing tenure with Goodman's band brought electric guitar solos into the limelight.

However, it was during after-hours jam sessions at the Hotel Cecil in New York that Christian left his stylistic imprint on jazz. Including such jazz greats as Joe Guy, Thelonious Monk, and Kenny Clarke, the groups played around with new sounds. Out of these sessions bebop was born, and Christian was a central figure. Sick with tuberculosis, he had to quit playing in 1941 and died the following spring, only twenty-five years old.

Fender had gone too far with the introduction of the Stratocaster, but it became a huge success because of its versatility. It was the first commercial solid-body electric guitar to have three pickups and a vibrato bar. It was also easier to play than the Telecaster because of its double cutaway, contoured body, and scooped back. The Stratocaster sold for $249.50. Since its introduction, the Stratocaster has undergone some minor changes, but Fender and his staff basically got it right the first time.

The Gibson company entered the solid-body market in 1952 with the unveiling of the "Les Paul" model. After the Telecaster, the Les Paul guitar was the next significant solid-body to be introduced. Les Paul was a legendary guitarist who also had been experimenting with electric guitar designs for many years. The Gibson designers came up with a striking model that produced a thick rounded tone. Over the years, the Les Paul model has won a loyal following.

THE PRECISION BASS

In 1951, Leo Fender introduced another revolutionary guitar, the Precision bass. At a cost of $195.50, the first electric bass would go on to dominate the market. The Fender company has manufactured numerous guitar models over the years, but the three that stand above all others in the field are the Telecaster, the Precision bass, and the Stratocaster. The Telecaster is considered to be more of a workhorse, whereas the Stratocaster is thought of as the thoroughbred of electric guitars. The Precision bass was in its own right a revolutionary guitar. With a styling that had been copied from the Telecaster, the Precision freed musicians from bulky oversized acoustic basses, which were prone to feedback. The name Precision had meaning. Fender's electric bass made it possible, with its frets, for the precise playing of notes; many acoustic basses were fretless. The original Precision bass model was manufactured from 1951 to 1954. The next version lasted from 1954 until June of 1957. The Precision bass that went into production in June, 1957, with its split humbucking pickup, continued to be the standard electric bass on the market into the 1990's.

By 1964, the Fender Electric Instruments Company had grown enormously. In addition to Leo Fender, a number of crucial people worked for the organization, including George Fullerton and Don

Randall. Fred Tavares joined the company's research and development team in 1953. In May, 1954, Forrest White became Fender's plant manager. All these individuals played vital roles in the success of Fender, but the driving force behind the scene was always Leo Fender. As Fender's health deteriorated, Randall commenced negotiations with CBS to sell the Fender company. In January, 1965, CBS bought Fender for $13 million. Eventually, Leo Fender regained his health, and he was hired as a technical adviser by CBS/Fender. He continued in this capacity until 1970. He remained determined to create more guitar designs of note. Although he never again produced anything that could equal his previous success, he never stopped trying to attain a new perfection of guitar design.

Fender died on March 21, 1991, in Fullerton, California. He had suffered for years from Parkinson's disease, and he died of complications from the disease. He is remembered for his Broadcaster/Telecaster, Precision bass, and Stratocaster, which revolutionized popular music. Because the Fender company was able to mass produce these and other solid-body electric guitars, new styles of music that relied on the sound made by an electric guitar exploded onto the scene. The electric guitar manufacturing business grew rapidly after Fender introduced mass production. Besides American companies, there are guitar companies that have flourished in Europe and Japan.

The marriage between rock music and solid-body electric guitars was initiated by the Fender guitars. The Telecaster, Precision bass, and Stratocaster become synonymous with the explosive character of rock and roll music. The multi-billion-dollar music business can point to Fender as the pragmatic visionary who put the solid-body electric guitar into the forefront of the musical scene. His innovative guitars have been used by some of the most important guitarists of the rock era, including Jimi Hendrix, Eric Clapton, and Jeff Beck. More important, Fender guitars have remained bestsellers with the public worldwide. Amateur musicians purchased them by the thousands for their own entertainment. Owning and playing a Fender guitar, or one of the other electric guitars that followed, allowed these amateurs to feel closer to their musician idols. A large market for sheet music from popular artists also developed.

In 1992, Fender was inducted into the Rock and Roll Hall of

Fame. He is one of the few non-musicians ever to be inducted. The sound of an electric guitar is the sound of exuberance, and since the Broadcaster was first unveiled in 1948, that sound has grown to be pervasive and enormously profitable.

See also Cassette recording; Dolby noise reduction; Electronic synthesizer.

FURTHER READING

Bacon, Tony, and Paul Day. *The Fender Book*. San Francisco: GPI Books, 1992.

Brosnac, Donald, ed. *Guitars Made by the Fender Company*. Westport, Conn.: Bold Strummer, 1986.

Freeth, Nick. *The Electric Guitar*. Philadelphia: Courage Books, 1999.

Trynka, Paul. *The Electric Guitar: An Illustrated History*. San Francisco: Chronicle Books, 1995.

Wheeler, Tom. *American Guitars: An Illustrated History*. New York: Harper & Row, 1982.

_____. "Electric Guitars." In *The Guitar Book: A Handbook for Electric and Acoustic Guitarists*. New York: Harper & Row, 1974.

Brownie camera

The invention: The first inexpensive and easy-to-use camera available to the general public, the Brownie revolutionized photography by making it possible for every person to become a photographer.

The people behind the invention:
George Eastman (1854-1932), founder of the Eastman Kodak Company
Frank A. Brownell, a camera maker for the Kodak Company who designed the Brownie
Henry M. Reichenbach, a chemist who worked with Eastman to develop flexible film
William H. Walker, a Rochester camera manufacturer who collaborated with Eastman

A New Way to Take Pictures

In early February of 1900, the first shipments of a new small box camera called the Brownie reached Kodak dealers in the United States and England. George Eastman, eager to put photography within the reach of everyone, had directed Frank Brownell to design a small camera that could be manufactured inexpensively but that would still take good photographs.

Advertisements for the Brownie proclaimed that everyone—even children—could take good pictures with the camera. The Brownie was aimed directly at the children's market, a fact indicated by its box, which was decorated with drawings of imaginary elves called "Brownies" created by the Canadian illustrator Palmer Cox. Moreover, the camera cost only one dollar.

The Brownie was made of jute board and wood, with a hinged back fastened by a sliding catch. It had an inexpensive two-piece glass lens and a simple rotary shutter that allowed both timed and instantaneous exposures to be made. With a lens aperture of approximately f14 and a shutter speed of approximately 1/50 of a second, the Brownie was certainly capable of taking acceptable snapshots. It had no viewfinder; however, an optional clip-on reflecting viewfinder was available. The camera came loaded with a six-exposure roll of Kodak film that produced square negatives 2.5 inches on a side. This film could be developed, printed, and mounted for forty cents, and a new roll could be purchased for fifteen cents.

George Eastman's first career choice had been banking, but when he failed to receive a promotion he thought he deserved, he decided to devote himself to his hobby, photography. Having worked with a rigorous wet-plate process, he knew why there were few amateur photographers at the time—the whole process, from plate preparation to printing, was too expensive and too much trouble. Even so, he had already begun to think about the commercial possibilities of photography; after reading of British experiments with dry-plate technology, he set up a small chemical laboratory and came up with a process of his own. The Eastman Dry Plate Company became one of the most successful producers of gelatin dry plates.

Dry-plate photography had attracted more amateurs, but it was still a complicated and expensive hobby. Eastman realized that the number of photographers would have to increase considerably if the market for cameras and supplies were to have any potential. In the early 1880's, Eastman first formulated the policies that would make the Eastman Kodak Company so successful in years to come: mass production, low prices, foreign and domestic distribution, and selling through extensive advertising and by demonstration.

In his efforts to expand the amateur market, Eastman first tackled the problem of the glass-plate negative, which was heavy, fragile, and expensive to make. By 1884, his experiments with paper negatives had been successful enough that he changed the name of his company to The Eastman Dry Plate and Film Company. Since flexible roll film needed some sort of device to hold it steady in the camera's focal plane, Eastman collaborated with William Walker to develop the Eastman-Walker roll-holder. Eastman's pioneering manufacture and use of roll films led to the appearance on the market in the 1880's of a wide array of hand cameras from a number of different companies. Such cameras were called "detective cameras" because they were small and could be used surreptitiously. The most famous of these, introduced by Eastman in 1888, was named the "Kodak"—a word he coined to be terse, distinctive, and easily pronounced in any language. This camera's simplicity of operation was appealing to the general public and stimulated the growth of amateur photography.

THE CAMERA

The Kodak was a box about seven inches long and four inches wide, with a one-speed shutter and a fixed-focus lens that produced reasonably sharp pictures. It came loaded with enough roll film to make one hundred exposures. The camera's initial price of twenty-five dollars included the cost of processing the first roll of film; the camera also came with a leather case and strap. After the film was exposed, the camera was mailed, unopened, to the company's plant in Rochester, New York, where the developing and printing were done. For an additional ten dollars, the camera was reloaded and sent back to the customer.

The Kodak was advertised in mass-market publications, rather than in specialized photographic journals, with the slogan: "You press the button, we do the rest." With his introduction of a camera that was easy to use and a service that eliminated the need to know anything about processing negatives, Eastman revolutionized the photographic market. Thousands of people no longer depended upon professional photographers for their portraits but instead learned to make their own. In 1892, the Eastman Dry Plate and Film Company became the Eastman Kodak Company, and by the mid-1890's, one hundred thousand Kodak cameras had been manufactured and sold, half of them in Europe by Kodak Limited.

Having popularized photography with the first Kodak, in 1900 Eastman turned his attention to the children's market with the introduction of the Brownie. The first five thousand cameras sent to dealers were sold immediately; by the end of the following year, almost a quarter of a million had been sold. The Kodak Company organized Brownie camera clubs and held competitions specifically for young photographers. The Brownie came with an instruction booklet that gave children simple directions for taking successful pictures, and "The Brownie Boy," an appealing youngster who loved photography, became a standard feature of Kodak's advertisements.

IMPACT

Eastman followed the success of the first Brownie by introducing several additional models between 1901 and 1917. Each was a more elaborate version of the original. These Brownie box cameras were on the market until the early 1930's, and their success inspired other companies to manufacture box cameras of their own. In 1906, the Ansco company produced the Buster Brown camera in three sizes that corresponded to Kodak's Brownie camera range; in 1910 and 1914, Ansco made three more versions. The Seneca company's Scout box camera, in three sizes, appeared in 1913, and Sears Roebuck's Kewpie cameras, in five sizes, were sold beginning in 1916. In England, the Houghtons company introduced its first Scout camera in 1901, followed by another series of four box cameras in 1910 sold under the Ensign trademark. Other English manufacturers of box cameras included the James Sinclair company, with its Traveller Una of 1909, and the Thornton-Pickard company, with a Filma camera marketed in four sizes in 1912.

After World War I ended, several series of box cameras were manufactured in Germany by companies that had formerly concentrated on more advanced and expensive cameras. The success of box cameras in other countries, led by Kodak's Brownie, undoubtedly prompted this trend in the German photographic industry. The Ernemann Film K series of cameras in three sizes, introduced in 1919, and the all-metal Trapp Little Wonder of 1922 are examples of popular German box cameras.

In the early 1920's, camera manufacturers began making box-camera bodies from metal rather than from wood and cardboard. Machine-formed metal was less expensive than the traditional hand-worked materials. In 1924, Kodak's two most popular Brownie sizes appeared with aluminum bodies.

In 1928, Kodak Limited of England added two important new features to the Brownie—a built-in portrait lens, which could be brought in front of the taking lens by pressing a lever, and camera bodies in a range of seven different fashion colors. The Beau Brownie cameras, made in 1930, were the most popular of all the colored box cameras. The work of Walter Dorwin Teague, a leading American designer, these cameras had an Art Deco geometric pattern on the front panel, which was enameled in a color matching the leatherette covering of the camera body. Several other companies, including Ansco, again followed Kodak's lead and introduced their own lines of colored cameras.

In the 1930's, several new box cameras with interesting features appeared, many manufactured by leading film companies. In France, the Lumiere Company advertised a series of box cameras—the Luxbox, Scoutbox, and Lumibox—that ranged from a basic camera to one with an adjustable lens and shutter. In 1933, the German Agfa company restyled its entire range of box cameras, and in 1939, the Italian Ferrania company entered the market with box cameras in two sizes. In 1932, Kodak redesigned its Brownie series to take the new 620 roll film, which it had just introduced. This film and the new Six-20 Brownies inspired other companies to experiment with variations of their own; some box cameras, such as the Certo Double-box, the Coronet Every Distance, and the Ensign E-20 cameras, offered a choice of two picture formats.

Another new trend was a move toward smaller-format cameras using standard 127 roll film. In 1934, Kodak marketed the small Baby Brownie. Designed by Teague and made from molded black plastic, this little camera with a folding viewfinder sold for only one dollar—the price of the original Brownie in 1900.

The Baby Brownie, the first Kodak camera made of molded plastic, heralded the move to the use of plastic in camera manufacture. Soon many others, such as the Altissa series of box cameras and the Voigtlander Brilliant V/6 camera, were being made from this new material.

Later Trends

By the late 1930's, flashbulbs had replaced flash powder for taking pictures in low light; again, the Eastman Kodak Company led the way in introducing this new technology as a feature on the inexpensive box camera. The Falcon Press-Flash, marketed in 1939, was the first mass-produced camera to have flash synchronization and was followed the next year by the Six-20 Flash Brownie, which had a detachable flash gun. In the early 1940's, other companies, such as Agfa-Ansco, introduced this feature on their own box cameras.

GEORGE EASTMAN

Frugal, bold, practical, generous to those who were loyal, impatient with dissent, and possessing a steely determination, George Eastman (1854-1932) rose to become one of the richest people of his generation. He abhorred poverty and did his best to raise others from it as well.

At age fourteen, when his father died, Eastman had to drop out of school to support his mother and sister. The missed opportunity for an education and the struggle to earn a living shaped his outlook. He worked at an insurance agency and then at a bank, keeping careful record of the money he earned. By the time he was twenty-five he had saved three thousand dollars and found his job as a banker to be unrewarding.

(Smithsonian Institution)

As a teenager, he had taught himself photography. However, that was only a start. He taught himself the physics and chemistry of photography too—and enough French and German to read the latest foreign scientific journals. His purpose was practical, to make cameras cheap and easy to use so that average people could own them. This launched him on the career of invention and business that took him away from banking and made his fortune. At the same time he remembered his origins and family. Out of his first earnings, he bought photographs for his mother and a favorite teacher. He never stopped giving. At the company he founded, he gave substantial raises to employees, reduced their hours, and installed safety equipment, a medical department, and a lunch room. He gave millions to the Hampton Institute, Tuskegee Institute, Massachusetts Institute of Technology, and University of Rochester, while also establishing dental clinics for the poor.

In his old age he found he could no longer keep up with his younger scientific and business colleagues. In 1932, leaving behind a note that asked, simply, "My work is done, why wait?" he committed suicide. Even then he continued to give. His will left most of his vast fortune to charities.

In the years after World War II, the box camera evolved into an eye-level camera, making it more convenient to carry and use. Many amateur photographers, however, still had trouble handling paper-backed roll film and were taking their cameras back to dealers to be unloaded and reloaded. Kodak therefore developed a new system of film loading, using the Kodapak cartridge, which could be mass-produced with a high degree of accuracy by precision plastic-molding techniques. To load the camera, the user simply opened the camera back and inserted the cartridge. This new film was introduced in 1963, along with a series of Instamatic cameras designed for its use. Both were immediately successful.

The popularity of the film cartridge ended the long history of the simple and inexpensive roll film camera. The last English Brownie was made in 1967, and the series of Brownies made in the United States was discontinued in 1970. Eastman's original marketing strategy of simplifying photography in order to increase the demand for cameras and film continued, however, with the public's acceptance of cartridge-loading cameras such as the Instamatic.

From the beginning, Eastman had recognized that there were two kinds of photographers other than professionals. The first, he declared, were the true amateurs who devoted time enough to acquire skill in the complex processing procedures of the day. The second were those who merely wanted personal pictures or memorabilia of their everyday lives, families, and travels. The second class, he observed, outnumbered the first by almost ten to one. Thus, it was to this second kind of amateur photographer that Eastman had appealed, both with his first cameras and with his advertising slogan, "You press the button, we do the rest." Eastman had done much more than simply invent cameras and films; he had invented a system and then developed the means for supporting that system. This is essentially what the Eastman Kodak Company continued to accomplish with the series of Instamatics and other descendants of the original Brownie. In the decade between 1963 and 1973, for example, approximately sixty million Instamatics were sold throughout the world.

The research, manufacturing, and marketing activities of the Eastman Kodak Company have been so complex and varied that no one would suggest that the company's prosperity rests solely on the success of its line of inexpensive cameras and cartridge films, although these have continued to be important to the company. Like Kodak, however, most large companies in the photographic industry have expanded their research to satisfy the ever-growing demand from amateurs. The amateurism that George Eastman recognized and encouraged at the beginning of the twentieth century thus still flourished at its end.

See also Autochrome plate; Color film; Instant photography.

FURTHER READING

Brooke-Ball, Peter. *George Eastman and Kodak*. Watford: Exley, 1994.
Collins, Douglas. *The Story of Kodak*. New York: Harry N. Abrams, 1990.
Freund, Gisele. *Photography and Society*. Boston: David R. Godine, 1980.
Wade, John. *A Short History of the Camera*. Watford, England: Fountain Press, 1979.
West, Nancy Martha. *Kodak and the Lens of Nostalgia*. Charlottesville: University Press of Virginia, 2000.

Bubble memory

The invention: An early nonvolatile medium for storing information on computers.

The person behind the invention:
Andrew H. Bobeck (1926-   ), a Bell Telephone Laboratories scientist

Magnetic Technology

The fanfare over the commercial prospects of magnetic bubbles was begun on August 8, 1969, by a report appearing in both *The New York Times* and *The Wall Street Journal*. The early 1970's would see the anticipation mount (at least in the computer world) with each prediction of the benefits of this revolution in information storage technology.

Although it was not disclosed to the public until August of 1969, magnetic bubble technology had held the interest of a small group of researchers around the world for many years. The organization that probably can claim the greatest research advances with respect to computer applications of magnetic bubbles is Bell Telephone Laboratories (later part of American Telephone and Telegraph). Basic research into the properties of certain ferrimagnetic materials started at Bell Laboratories shortly after the end of World War II (1939-1945).

Ferrimagnetic substances are typically magnetic iron oxides. Research into the properties of these and related compounds accelerated after the discovery of ferrimagnetic garnets in 1956 (these are a class of ferrimagnetic oxide materials that have the crystal structure of garnet). Ferrimagnetism is similar to ferromagnetism, the phenomenon that accounts for the strong attraction of one magnetized body for another. The ferromagnetic materials most suited for bubble memories contain, in addition to iron, the element yttrium or a metal from the rare earth series.

It was a fruitful collaboration between scientist and engineer, between pure and applied science, that produced this promising breakthrough in data storage technology. In 1966, Bell Laboratories scientist Andrew H. Bobeck and his coworkers were the first to realize the data storage potential offered by the strange behavior of thin slices of magnetic iron oxides under an applied magnetic field. The first U.S. patent for a memory device using magnetic bubbles was filed by Bobeck in the fall of 1966 and issued on August 5, 1969.

BUBBLES FULL OF MEMORIES

The three basic functional elements of a computer are the central processing unit, the input/output unit, and memory. Most implementations of semiconductor memory require a constant power source to retain the stored data. If the power is turned off, all stored data are lost. Memory with this characteristic is called "volatile." Disks and tapes, which are typically used for secondary memory, are "nonvolatile." Nonvolatile memory relies on the orientation of magnetic domains, rather than on electrical currents, to sustain its existence.

One can visualize by analogy how this will work by taking a group of permanent bar magnets that are labeled with *N* for north at one end and *S* for south at the other. If an arrow is painted starting from the north end with the tip at the south end on each magnet, an orientation can then be assigned to a magnetic domain (here one whole bar magnet). Data are "stored" with these bar magnets by arranging them in rows, some pointing up, some pointing down. Different arrangements translate to different data. In the binary world of the computer, all information is represented by two states. A stored data item (known as a "bit," or binary digit) is either on or off, up or down, true or false, depending on the physical representation. The "on" state is commonly labeled with the number 1 and the "off" state with the number 0. This is the principle behind magnetic disk and tape data storage.

Now imagine a thin slice of a certain type of magnetic material in the shape of a 3-by-5-inch index card. Under a microscope, using a special source of light, one can see through this thin slice in many regions of the surface. Darker, snakelike regions can also be seen, representing domains of an opposite orientation (polarity) to the transparent regions. If a weak external magnetic field is then applied by placing a permanent magnet of the same shape as the card on the underside of the slice, a strange thing happens to the dark serpentine pattern—the long domains shrink and eventually contract into "bubbles," tiny magnetized spots. Viewed from the side of the slice, the bubbles are cylindrically shaped domains having a polarity opposite to that of the material on which they rest. The presence or absence of a bubble indicates either a 0 or a 1 bit. Data bits are stored by moving the bubbles in the thin film. As long as the field is applied by the permanent magnet substrate, the data will be retained. The bubble is thus a nonvolatile medium for data storage.

CONSEQUENCES

Magnetic bubble memory created quite a stir in 1969 with its splashy public introduction. Most of the manufacturers of computer chips immediately instituted bubble memory development projects. Texas Instruments, Philips, Hitachi, Motorola, Fujitsu, and International Business Machines (IBM) joined the race with Bell Laboratories to mass-produce bubble memory chips. Texas Instruments became the first major chip manufacturer to mass-produce bubble memories in the mid-to-late 1970's. By 1990, however, almost all the research into magnetic bubble technology had shifted to Japan. Hitachi and Fujitsu began to invest heavily in this area.

Mass production proved to be the most difficult task. Although the materials it uses are different, the process of producing magnetic bubble memory chips is similar to the process applied in producing semiconductor-based chips such as those used for random access memory (RAM). It is for this reason that major semiconductor manufacturers and computer companies initially invested in this technology. Lower fabrication yields and reliability issues plagued early production runs, however, and, although these problems have mostly been solved, gains in the performance characteristics of competing conventional memories have limited the impact that magnetic bubble technology has had on the marketplace. The materials used for magnetic bubble memories are costlier and possess more complicated structures than those used for semiconductor or disk memory.

Speed and cost of materials are not the only bases for comparison. It is possible to perform some elementary logic with magnetic bubbles. Conventional semiconductor-based memory offers storage only. The capability of performing logic with magnetic bubbles puts bubble technology far ahead of other magnetic technologies with respect to functional versatility.

A small niche market for bubble memory developed in the 1980's. Magnetic bubble memory can be found in intelligent terminals, desktop computers, embedded systems, test equipment, and similar microcomputer-based systems.

See also Computer chips; Floppy disk; Hard disk; Optical disk; Personal computer.

FURTHER READING

"Bubble Memory's Ruggedness Revives Interest for Military Use." *Aviation Week and Space Technology* 130, no. 3 (January 16, 1989).
Graff, Gordon. "Better Bubbles." *Popular Science* 232, no. 2 (February, 1988).
McLeod, Jonah. "Will Bubble Memories Make a Comeback?" *Electronics* 61, no. 14 (August, 1988).
Nields, Megan. "Bubble Memory Bursts into Niche Markets." *Mini-Micro Systems* 20, no. 5 (May, 1987).

Bullet train

The invention: An ultrafast passenger railroad system capable of moving passengers at speeds double or triple those of ordinary trains.

The people behind the invention:

Ikeda Hayato (1899-1965), Japanese prime minister from 1960 to 1964, who pushed for the expansion of public expenditures

Shinji Sogo (1901-1971), the president of the Japanese National Railways, the "father of the bullet train"

Building a Faster Train

By 1900, Japan had a world-class railway system, a logical result of the country's dense population and the needs of its modernizing economy. After 1907, the government controlled the system through the Japanese National Railways (JNR). In 1938, JNR engineers first suggested the idea of a train that would travel 125 miles per hour from Tokyo to the southern city of Shimonoseki. Construction of a rapid train began in 1940 but was soon stopped because of World War II.

The 311-mile railway between Tokyo and Osaka, the Tokaido Line, has always been the major line in Japan. By 1957, a business express along the line operated at an average speed of 57 miles per hour, but the double-track line was rapidly reaching its transport capacity. The JNR established two investigative committees to explore alternative solutions. In 1958, the second committee recommended the construction of a high-speed railroad on a separate double track, to be completed in time for the Tokyo Olympics of 1964. The Railway Technical Institute of the JNR concluded that it was feasible to design a line that would operate at an average speed of about 130 miles per hour, cutting time for travel between Tokyo and Osaka from six hours to three hours.

By 1962, about 17 miles of the proposed line were completed for test purposes. During the next two years, prototype trains were tested to correct flaws and make improvements in the design. The entire project was completed on schedule in July, 1964, with total construction costs of more than $1 billion, double the original estimates.

THE SPEEDING BULLET

Service on the Shinkansen, or New Trunk Line, began on October 1, 1964, ten days before the opening of the Olympic Games. Commonly called the "bullet train" because of its shape and speed, the Shinkansen was an instant success with the public, both in Japan and abroad. As promised, the time required to travel between Tokyo and Osaka was cut in half. Initially, the system provided daily services of sixty trains consisting of twelve cars each, but the number of scheduled trains was almost doubled by the end of the year.

The Shinkansen was able to operate at its unprecedented speed because it was designed and operated as an integrated system, making use of countless technological and scientific developments. Tracks followed the standard gauge of 56.5 inches, rather than the more narrow gauge common in Japan. For extra strength, heavy

Japanese bullet trains. (PhotoDisc)

welded rails were attached directly onto reinforced concrete slabs. The minimum radius of a curve was 8,200 feet, except where sharper curves were mandated by topography. In many ways similar to modern airplanes, the railway cars were made airtight in order to prevent ear discomfort caused by changes in pressure when trains enter tunnels.

The Shinkansen trains were powered by electric traction motors, with four 185-kilowatt motors on each car—one motor attached to each axle. This design had several advantages: It provided an even distribution of axle load for reducing strain on the tracks; it allowed the application of dynamic brakes (where the motor was used for braking) on all axles; and it prevented the failure of one or two units from interrupting operation of the entire train. The 25,000-volt electrical current was carried by trolley wire to the cars, where it was rectified into a pulsating current to drive the motors.

The Shinkansen system established a casualty-free record because of its maintenance policies combined with its computerized Centralized Traffic Control system. The control room at Tokyo Station was designed to maintain timely information about the location of all trains and the condition of all routes. Although train operators had some discretion in determining speed, automatic brakes also operated to ensure a safe distance between trains. At least once each month, cars were thoroughly inspected; every ten days, an inspection train examined the conditions of tracks, communication equipment, and electrical systems.

IMPACT

Public usage of the Tokyo-Osaka bullet train increased steadily because of the system's high speed, comfort, punctuality, and superb safety record. Businesspeople were especially happy that the rapid service allowed them to make the round-trip without the necessity of an overnight stay, and continuing modernization soon allowed nonstop trains to make a one-way trip in two and one-half hours, requiring speeds of 160 miles per hour in some stretches. By the early 1970's, the line was transporting a daily average of 339,000 passengers in 240 trains, meaning that a train departed from Tokyo about every ten minutes.

The popularity of the Shinkansen system quickly resulted in demands for its extension into other densely populated regions. In 1972, a 100-mile stretch between Osaka and Okayama was opened for service. By 1975, the line was further extended to Hakata on the island of Kyushu, passing through the Kammon undersea tunnel. The cost of this 244-mile stretch was almost $2.5 billion. In 1982, lines were completed from Tokyo to Niigata and from Tokyo to Morioka. By 1993, the system had grown to 1,134 miles of track. Since high usage made the system extremely profitable, the sale of the JNR to private companies in 1987 did not appear to produce adverse consequences.

The economic success of the Shinkansen had a revolutionary effect on thinking about the possibilities of modern rail transportation, leading one authority to conclude that the line acted as "a savior of the declining railroad industry." Several other industrial countries were stimulated to undertake large-scale railway projects; France, especially, followed Japan's example by constructing high-speed electric railroads from Paris to Nice and to Lyon. By the mid-1980's, there were experiments with high-speed trains based on magnetic levitation and other radical innovations, but it was not clear whether such designs would be able to compete with the Shinkansen model.

See also Airplane; Atomic-powered ship; Diesel locomotive; Supersonic passenger plane.

FURTHER READING

French, Howard W. "Japan's New Bullet Train Draws Fire." *New York Times* (September 24, 2000).

Frew, Tim. *Locomotives: From the Steam Locomotive to the Bullet Train.* New York: Mallard Press, 1990.

Holley, David. "Faster Than a Speeding Bullet: High-Speed Trains Are Japan's Pride, Subject of Debate." *Los Angeles Times* (April 10, 1994).

O'Neill, Bill. "Beating the Bullet Train." *New Scientist* 140, no. 1893 (October 2, 1993).

Raoul, Jean-Claude. "How High-Speed Trains Make Tracks." *Scientific American* 277 (October, 1997).

Buna Rubber

THE INVENTION: The first practical synthetic rubber product developed, Buna inspired the creation of other other synthetic substances that eventually replaced natural rubber in industrial applications.

THE PEOPLE BEHIND THE INVENTION:
Charles de la Condamine (1701-1774), a French naturalist
Charles Goodyear (1800-1860), an American inventor
Joseph Priestley (1733-1804), an English chemist
Charles Greville Williams (1829-1910), an English chemist

A New Synthetic Rubber

The discovery of natural rubber is often credited to the French scientist Charles de la Condamine, who, in 1736, sent the French Academy of Science samples of an elastic material used by Peruvian Indians to make balls that bounced. The material was primarily a curiosity until 1770, when Joseph Priestley, an English chemist, discovered that it rubbed out pencil marks, after which he called it "rubber." Natural rubber, made from the sap of the rubber tree (*Hevea brasiliensis*), became important after Charles Goodyear discovered in 1830 that heating rubber with sulfur (a process called "vulcanization") made it more elastic and easier to use. Vulcanized natural rubber came to be used to make raincoats, rubber bands, and motor vehicle tires.

Natural rubber is difficult to obtain (making one tire requires the amount of rubber produced by one tree in two years), and wars have often cut off supplies of this material to various countries. Therefore, efforts to manufacture synthetic rubber began in the late eighteenth century. Those efforts followed the discovery by English chemist Charles Greville Williams and others in the 1860's that natural rubber was composed of thousands of molecules of a chemical called isoprene that had been joined to form giant, necklace-like molecules. The first successful synthetic rubber, Buna, was patented by Germany's I. G. Farben Industrie in 1926. The success of this rubber led to the development of many other synthetic rubbers, which are now used in place of natural rubber in many applications.

CHARLES GOODYEAR

It was an accident that finally showed Charles Goodyear (1800-1860) how to make rubber into a durable, practical material. For years he had been experimenting at home looking for ways to improve natural rubber—and producing stenches that drove his family and neighbors to distraction—when in 1839 he dropped a piece of rubber mixed with sulfur onto a hot stove. When he examined the charred specimen, he discovered it was not sticky, as hot natural rubber always is, and when he took it outside into the cold, it did not become brittle.

The son of an inventor, Goodyear invented much more than his vulcanizing process for rubber. He also patented a spring-lever faucet, pontoon boat, hay fork, and air pump, but he was never successful in making money from his inventions. Owner of a hardware store, he went broke during a financial panic in 1830 and had to spend time in debtor's prison. He was never financially stable afterwards, often having to borrow money and sell his family's belongings to support his experiments. And he had a large family—twelve children, of whom only half lived beyond childhood.

(Smithsonian Institution)

Even vulcanized rubber did not make Goodyear's fortune. He delayed patenting it until Thomas Hancock, an Englishman, replicated Goodyear's method of vulcanizing and began producing rubber in England. Goodyear sued and lost. Others stole his method, and although he won one large case, legal expenses took away most of the settlement. He borrowed more and more money to advertise his product, with some success. For example, Emperor Napoleon III awarded Goodyear the Cross of the Legion of Honor for his display at the 1851 Crystal Palace Exhibition in London. Nevertheless, Goodyear died deeply in debt.

Despite all the imitators, vulcanized rubber remained associated with Goodyear. Thirty-eight years after he died, the world's larger rubber manufacturer took his name for the company's title.

FROM ERASERS TO GAS PUMPS

Natural rubber belongs to the group of chemicals called "polymers." A polymer is a giant molecule that is made up of many simpler chemical units ("monomers") that are attached chemically to form long strings. In natural rubber, the monomer is isoprene (dimethylbutadiene). The first efforts to make a synthetic rubber used the discovery that isoprene could be made and converted into an elastic polymer. The synthetic rubber that was created from isoprene was, however, inferior to natural rubber. The first Buna rubber, which was patented by I. G. Farben in 1926, was better, but it was still less than ideal. Buna rubber was made by polymerizing the monomer butadiene in the presence of sodium. The name Buna comes from the first two letters of the words "butadiene" and "natrium" (German for sodium). Natural and Buna rubbers are called homopolymers because they contain only one kind of monomer.

The ability of chemists to make Buna rubber, along with its successful use, led to experimentation with the addition of other monomers to isoprene-like chemicals used to make synthetic rubber. Among the first great successes were materials that contained two alternating monomers; such materials are called "copolymers." If the two monomers are designated A and B, part of a polymer molecule can be represented as (ABABABABABABABABAB). Numerous synthetic copolymers, which are often called "elastomers," now replace natural rubber in applications where they have superior properties. All elastomers are rubbers, since objects made from them both stretch greatly when pulled and return quickly to their original shape when the tension is released.

Two other well-known rubbers developed by I. G. Farben are the copolymers called Buna-N and Buna-S. These materials combine butadiene and the monomers acrylonitrile and styrene, respectively. Many modern motor vehicle tires are made of synthetic rubber that differs little from Buna-S rubber. This rubber was developed after the United States was cut off in the 1940's, during World War II, from its Asian source of natural rubber. The solution to this problem was the development of a synthetic rubber industry based on GR-S rubber (government rubber plus styrene), which was essentially Buna-S rubber. This rubber is still widely used.

Buna-S rubber is often made by mixing butadiene and styrene in huge tanks of soapy water, stirring vigorously, and heating the mixture. The polymer contains equal amounts of butadiene and styrene (BSBSBSBSBSBSBSBS). When the molecules of the Buna-S polymer reach the desired size, the polymerization is stopped and the rubber is coagulated (solidified) chemically. Then, water and all the unused starting materials are removed, after which the rubber is dried and shipped to various plants for use in tires and other products. The major difference between Buna-S and GR-S rubber is that the method of making GR-S rubber involves the use of low temperatures.

Buna-N rubber is made in a fashion similar to that used for Buna-S, using butadiene and acrylonitrile. Both Buna-N and the related neoprene rubber, invented by Du Pont, are very resistant to gasoline and other liquid vehicle fuels. For this reason, they can be used in gas-pump hoses. All synthetic rubbers are vulcanized before they are used in industry.

IMPACT

Buna rubber became the basis for the development of the other modern synthetic rubbers. These rubbers have special properties that make them suitable for specific applications. One developmental approach involved the use of chemically modified butadiene in homopolymers such as neoprene. Made of chloroprene (chlorobutadiene), neoprene is extremely resistant to sun, air, and chemicals. It is so widely used in machine parts, shoe soles, and hoses that more than 400 million pounds are produced annually.

Another developmental approach involved copolymers that alternated butadiene with other monomers. For example, the successful Buna-N rubber (butadiene and acrylonitrile) has properties similar to those of neoprene. It differs sufficiently from neoprene, however, to be used to make items such as printing press rollers. About 200 million pounds of Buna-N are produced annually. Some 4 billion pounds of the even more widely used polymer Buna-S/GR-S are produced annually, most of which is used to make tires.

Several other synthetic rubbers have significant industrial applications, and efforts to make copolymers for still other purposes continue.

See also Neoprene; Nylon; Orlon; Plastic; Polyester; Polyethylene; Polystyrene; Silicones; Teflon; Velcro.

FURTHER READING

Herbert, Vernon. *Synthetic Rubber: A Project That Had to Succeed.* Westport, Conn.: Greenwood Press, 1985.
Mossman, S. T. I., and Peter John Turnbull Morris. *The Development of Plastics.* Cambridge: Royal Society of Chemistry, 1994.
Von Hagen, Victor Wolfgang. *South America Called Them: Explorations of the Great Naturalists, La Condamine, Humboldt, Darwin, Spruce.* New York: A. A. Knopf, 1945.

CAD/CAM

THE INVENTION: Computer-Aided Design (CAD) and Computer-Aided Manufacturing (CAM) enhanced flexibility in engineering design, leading to higher quality and reduced time for manufacturing

THE PEOPLE BEHIND THE INVENTION:

Patrick Hanratty, a General Motors Research Laboratory worker who developed graphics programs

Jack St. Clair Kilby (1923-   ), a Texas Instruments employee who first conceived of the idea of the integrated circuit

Robert Noyce (1927-1990), an Intel Corporation employee who developed an improved process of manufacturing integrated circuits on microchips

Don Halliday, an early user of CAD/CAM who created the Made-in-America car in only four months by using CAD and project management software

Fred Borsini, an early user of CAD/CAM who demonstrated its power

SUMMARY OF EVENT

Computer-Aided Design (CAD) is a technique whereby geometrical descriptions of two-dimensional (2-D) or three-dimensional (3-D) objects can be created and stored, in the form of mathematical models, in a computer system. Points, lines, and curves are represented as graphical coordinates. When a drawing is requested from the computer, transformations are performed on the stored data, and the geometry of a part or a full view from either a two- or a three-dimensional perspective is shown. CAD systems replace the tedious process of manual drafting, and computer-aided drawing and redrawing that can be retrieved when needed has improved drafting efficiency. A CAD system is a combination of computer hardware and software that facilitates the construction of geometric models and, in many cases, their analysis. It allows a wide variety of visual representations of those models to be displayed.

Computer-Aided Manufacturing (CAM) refers to the use of computers to control, wholly or partly, manufacturing processes. In practice, the term is most often applied to computer-based developments of numerical control technology; robots and flexible manufacturing systems (FMS) are included in the broader use of CAM systems. A CAD/CAM interface is envisioned as a computerized database that can be accessed and enriched by either design or manufacturing professionals during various stages of the product development and production cycle.

In CAD systems of the early 1990's, the ability to model solid objects became widely available. The use of graphic elements such as lines and arcs and the ability to create a model by adding and subtracting solids such as cubes and cylinders are the basic principles of CAD and of simulating objects within a computer. CAD systems enable computers to simulate both taking things apart (sectioning) and putting things together for assembly. In addition to being able to construct prototypes and store images of different models, CAD systems can be used for simulating the behavior of machines, parts, and components. These abilities enable CAD to construct models that can be subjected to nondestructive testing; that is, even before engineers build a physical prototype, the CAD model can be subjected to testing and the results can be analyzed. As another example, designers of printed circuit boards have the ability to test their circuits on a CAD system by simulating the electrical properties of components.

During the 1950's, the U.S. Air Force recognized the need for reducing the development time for special aircraft equipment. As a result, the Air Force commissioned the Massachusetts Institute of Technology to develop numerically controlled (NC) machines that were programmable. A workable demonstration of NC machines was made in 1952; this began a new era for manufacturing. As the speed of an aircraft increased, the cost of manufacturing also increased because of stricter technical requirements. This higher cost provided a stimulus for the further development of NC technology, which promised to reduce errors in design before the prototype stage.

The early 1960's saw the development of mainframe computers. Many industries valued computing technology for its speed and for its accuracy in lengthy and tedious numerical operations in design, manufacturing, and other business functional areas. Patrick Hanratty, working for General Motors Research Laboratory, saw other potential applications and developed graphics programs for use on mainframe computers. The use of graphics in software aided the development of CAD/CAM, allowing visual representations of models to be presented on computer screens and printers.

The 1970's saw an important development in computer hardware, namely the development and growth of personal computers (PCs). Personal computers became smaller as a result of the development of integrated circuits. Jack St. Clair Kilby, working for Texas Instruments, first conceived of the integrated circuit; later, Robert Noyce, working for Intel Corporation, developed an improved process of manufacturing integrated circuits on microchips. Personal computers using these microchips offered both speed and accuracy at costs much lower than those of mainframe computers.

Five companies offered integrated commercial computer-aided design and computer-aided manufacturing systems by the first half of 1973. Integration meant that both design and manufacturing were contained in one system. Of these five companies—Applicon, Computervision, Gerber Scientific, Manufacturing and Consulting Services (MCS), and United Computing—four offered turnkey systems exclusively. Turnkey systems provide design, development, training, and implementation for each customer (company) based on the contractual agreement; they are meant to be used as delivered, with no need for the purchaser to make significant adjustments or perform programming.

The 1980's saw a proliferation of mini- and microcomputers with a variety of platforms (processors) with increased speed and better graphical resolution. This made the widespread development of computer-aided design and computer-aided manufacturing possible and practical. Major corporations spent large research and development budgets developing CAD/CAM systems that would automate manual drafting and machine tool movements. Don Halliday, working for Truesports Inc., provided an early example of the benefits of CAD/CAM. He created the Made-in-America car in only four months by using CAD and project management software. In the late 1980's, Fred Borsini, the president of Leap Technologies in

Michigan, brought various products to market in record time through the use of CAD/CAM.

In the early 1980's, much of the CAD/CAM industry consisted of software companies. The cost for a relatively slow interactive system in 1980 was close to $100,000. The late 1980's saw the demise of minicomputer-based systems in favor of Unix work stations and PCs based on 386 and 486 microchips produced by Intel. By the time of the International Manufacturing Technology show in September, 1992, the industry could show numerous CAD/CAM innovations including tools, CAD/CAM models to evaluate manufacturability in early design phases, and systems that allowed use of the same data for a full range of manufacturing functions.

IMPACT

In 1990, CAD/CAM hardware sales by U.S. vendors reached $2.68 billion. In software alone, $1.42 billion worth of CAD/CAM products and systems were sold worldwide by U.S. vendors, according to International Data Corporation figures for 1990. CAD/CAM systems were in widespread use throughout the industrial world. Development lagged in advanced software applications, particularly in image processing, and in the communications software and hardware that ties processes together.

A reevaluation of CAD/CAM systems was being driven by the industry trend toward increased functionality of computer-driven numerically controlled machines. Numerical control (NC) software enables users to graphically define the geometry of the parts in a product, develop paths that machine tools will follow, and exchange data among machines on the shop floor. In 1991, NC configuration software represented 86 percent of total CAM sales. In 1992, the market shares of the five largest companies in the CAD/CAM market were 29 percent for International Business Machines, 17 percent for Intergraph, 11 percent for Computervision, 9 percent for Hewlett-Packard, and 6 percent for Mentor Graphics.

General Motors formed a joint venture with Ford and Chrysler to develop a common computer language in order to make the next generation of CAD/CAM systems easier to use. The venture was aimed particularly at problems that posed barriers to speeding up the design of new automobiles. The three car companies all had sophisticated computer systems that allowed engineers to design parts on computers and then electronically transmit specifications to tools that make parts or dies.

CAD/CAM technology was expected to advance on many fronts. As of the early 1990's, different CAD/CAM vendors had developed systems that were often incompatible with one another, making it difficult to transfer data from one system to another. Large corporations, such as the major automakers, developed their own interfaces and network capabilities to allow different systems to communicate. Major users of CAD/CAM saw consolidation in the industry through the establishment of standards as being in their interests.

Resellers of CAD/CAM products also attempted to redefine their markets. These vendors provide technical support and service to users. The sale of CAD/CAM products and systems offered substantial opportunities, since demand remained strong. Resellers worked most effectively with small and medium-sized companies, which often were neglected by the primary sellers of CAD/CAM equipment because they did not generate a large volume of business. Some projections held that by 1995 half of all CAD/CAM systems would be sold through resellers, at a cost of $10,000 or less for each system. The CAD/CAM market thus was in the process of dividing into two markets: large customers (such as aerospace firms and automobile manufacturers) that would be served by primary vendors, and small and medium-sized customers that would be serviced by resellers.

CAD will find future applications in marketing, the construction industry, production planning, and large-scale projects such as shipbuilding and aerospace. Other likely CAD markets include hospitals, the apparel industry, colleges and universities, food product manufacturers, and equipment manufacturers. As the linkage between CAD and CAM is enhanced, systems will become more productive. The geometrical data from CAD will be put to greater use by CAM systems.

CAD/CAM already had proved that it could make a big difference in productivity and quality. Customer orders could be changed much faster and more accurately than in the past, when a change could require a manual redrafting of a design. Computers could do automatically in minutes what once took hours manually. CAD/ CAM saved time by reducing, and in some cases eliminating, human error. Many flexible manufacturing systems (FMS) had machining centers equipped with sensing probes to check the accuracy of the machining process. These self-checks can be made part of numerical control (NC) programs. With the technology of the early 1990's, some experts estimated that CAD/CAM systems were in many cases twice as productive as the systems they replaced; in the long run, productivity is likely to improve even more, perhaps up to three times that of older systems or even higher. As costs for CAD/ CAM systems concurrently fall, the investment in a system will be recovered more quickly. Some analysts estimated that by the mid-1990's, the recovery time for an average system would be about three years.

Another frontier in the development of CAD/CAM systems is expert (or knowledge-based) systems, which combine data with a human expert's knowledge, expressed in the form of rules that the computer follows. Such a system will analyze data in a manner mimicking intelligence. For example, a 3-D model might be created from standard 2-D drawings. Expert systems will likely play a pivotal role in CAM applications. For example, an expert system could determine the best sequence of machining operations to produce a component.

Continuing improvements in hardware, especially increased speed, will benefit CAD/CAM systems. Software developments, however, may produce greater benefits. Wider use of CAD/CAM systems will depend on the cost savings from improvements in hardware and software as well as on the productivity of the systems and the quality of their product. The construction, apparel, automobile, and aerospace industries have already experienced increases in productivity, quality, and profitability through the use of CAD/CAM. A case in point is Boeing, which used CAD from start to finish in the design of the 757.

See also Differential analyzer; Mark I calculator; Personal computer; SAINT; Virtual machine; Virtual reality.

FURTHER READING

Groover, Mikell P., and Emory W. Zimmers, Jr. *CAD/CAM: Computer-Aided Design and Manufacturing.* Englewood Cliffs, N.J.: Prentice-Hall, 1984.

Jurgen, Ronald K. *Computers and Manufacturing Productivity.* New York: Institute of Electrical and Electronics Engineers, 1987.

McMahon, Chris, and Jimmie Browne. *CAD/CAM: From Principles to Practice.* Reading, Mass.: Addison-Wesley, 1993.

_____. *CAD/CAM: Principles, Practice, and Manufacturing Management.* 2d ed. Harlow, England: Addison-Wesley, 1998.

Medland, A. J., and Piers Burnett. *CAD/CAM in Practice.* New York: John Wiley & Sons, 1986.

CARBON DATING

THE INVENTION: A technique that measures the radioactive decay of carbon 14 in organic substances to determine the ages of artifacts as old as ten thousand years.

THE PEOPLE BEHIND THE INVENTION:

Willard Frank Libby (1908-1980), an American chemist who won the 1960 Nobel Prize in Chemistry

Charles Wesley Ferguson (1922-1986), a scientist who demonstrated that carbon 14 dates before 1500 B.C. needed to be corrected

ONE IN A TRILLION

Carbon dioxide in the earth's atmosphere contains a mixture of three carbon isotopes (isotopes are atoms of the same element that contain different numbers of neutrons), which occur in the following percentages: about 99 percent carbon 12, about 1 percent carbon 13, and approximately one atom in a trillion of radioactive carbon 14. Plants absorb carbon dioxide from the atmosphere during photosynthesis, and then animals eat the plants, so all living plants and animals contain a small amount of radioactive carbon.

When a plant or animal dies, its radioactivity slowly decreases as the radioactive carbon 14 decays. The time it takes for half of any radioactive substance to decay is known as its "half-life." The half-life for carbon 14 is known to be about fifty-seven hundred years. The carbon 14 activity will drop to one-half after one half-life, one-fourth after two half-lives, one-eighth after three half-lives, and so forth. After ten or twenty half-lives, the activity becomes too low to be measurable. Coal and oil, which were formed from organic matter millions of years ago, have long since lost any carbon 14 activity. Wood samples from an Egyptian tomb or charcoal from a prehistoric fireplace a few thousand years ago, however, can be dated with good reliability from the leftover radioactivity.

In the 1940's, the properties of radioactive elements were still being discovered and were just beginning to be used to solve problems. Scientists still did not know the half-life of carbon 14, and archaeologists still depended mainly on historical evidence to determine the ages of ancient objects.

In early 1947, Willard Frank Libby started a crucial experiment in testing for radioactive carbon. He decided to test samples of methane gas from two different sources. One group of samples came from the sewage disposal plant at Baltimore, Maryland, which was rich in fresh organic matter. The other sample of methane came from an oil refinery, which should have contained only ancient carbon from fossils whose radioactivity should have completely decayed. The experimental results confirmed Libby's suspicions: The methane from fresh sewage was radioactive, but the methane from oil was not. Evidently, radioactive carbon was present in fresh organic material, but it decays away eventually.

TREE-RING DATING

In order to establish the validity of radiocarbon dating, Libby analyzed known samples of varying ages. These included tree-ring samples from the years 575 and 1075 and one redwood from 979 B.C.E., as well as artifacts from Egyptian tombs going back to about 3000 B.C.E. In 1949, he published an article in the journal *Science* that contained a graph comparing the historical ages and the measured radiocarbon ages of eleven objects. The results were accurate within 10 percent, which meant that the general method was sound.

The first archaeological object analyzed by carbon dating, obtained from the Metropolitan Museum of Art in New York, was a piece of cypress wood from the tomb of King Djoser of Egypt. Based on historical evidence, the age of this piece of wood was about forty-six hundred years. A small sample of carbon obtained from this wood was deposited on the inside of Libby's radiation counter, giving a count rate that was about 40 percent lower than that of modern organic carbon. The resulting age of the wood calculated from its residual radioactivity was about thirty-eight hundred years, a difference of eight hundred years. Considering that this was the first object to be analyzed, even such a rough agreement with the historic age was considered to be encouraging.

The validity of radiocarbon dating depends on an important assumption—namely, that the abundance of carbon 14 in nature has

WILLARD FRANK LIBBY

Born in 1908, Willard Frank Libby came from a family of farmers in Grand View, Colorado. They moved to Sebastopol, California, where Libby went through public school. He entered the University of California, Berkeley, in 1927, earning a bachelor of science degree in 1931 and a doctorate in 1933. He stayed on at Berkeley as an instructor of chemistry until he won the first of his three Guggenheim Fellowships in 1941. He moved to Princeton University to study, but World War II cut short his fellowship. Instead, he joined the Manhattan Project, helping design the atomic bomb at Columbia University's Division of War Research.

After the war Libby became a professor of chemistry at the University of Chicago, where he conducted his research on carbon-14 dating. A leading expert in radiochemistry, he also investigated isotope tracers and the effects of fallout. However, his career saw as much public service as research. In 1954 President Dwight Eisenhower appointed him to the Atomic Energy Commission as its first chemist, and Libby directed the administration's international Atoms for Peace program. He resigned in 1959 to take an appointment at the University of California, Los Angeles, as professor of chemistry and then in 1962 as director of the Institute of Geophysics and Planetary Physics, a position he held until he died in 1980.

Libby received the Nobel Prize in Chemistry in 1960 for developing carbon-14 dating. Among his many other honors were the American Chemical Society's Willard Gibbs Award in 1958, the Albert Einstein Medal in 1959, and the Day Medal of the Geological Society of America in 1961. He was a member of the Advisory Board of the Guggenheim Memorial Foundation, the Office of Civil and Defense Mobilization, the National Science Foundation's General Commission on Science, and the Academic Institution and also a director of Douglas Aircraft Company.

been constant for many thousands of years. If carbon 14 was less abundant at some point in history, organic samples from that era would have started with less radioactivity. When analyzed today, their reduced activity would make them appear to be older than they really are.

Charles Wesley Ferguson from the Tree-Ring Research Laboratory at the University of Arizona tackled this problem. He measured the age of bristlecone pine trees both by counting the rings and by using carbon 14 methods. He found that carbon 14 dates before 1500 B.C.E. needed to be corrected. The results show that radiocarbon dates are older than tree-ring counting dates by as much as several hundred years for the oldest samples. He knew that the number of tree rings had given him the correct age of the pines, because trees accumulate one ring of growth for every year of life. Apparently, the carbon 14 content in the atmosphere has not been constant. Fortunately, tree-ring counting gives reliable dates that can be used to correct radiocarbon measurements back to about 6000 B.C.E.

IMPACT

Some interesting samples were dated by Libby's group. The Dead Sea Scrolls had been found in a cave by an Arab shepherd in 1947, but some Bible scholars at first questioned whether they were genuine. The linen wrapping from the Book of Isaiah was tested for carbon 14, giving a date of 100 B.C.E., which helped to establish its authenticity. Human hair from an Egyptian tomb was determined to be nearly five thousand years old. Well-preserved sandals from a cave in eastern Oregon were determined to be ninety-three hundred years old. A charcoal sample from a prehistoric site in western South Dakota was found to be about seven thousand years old.

The Shroud of Turin, located in Turin, Italy, has been a controversial object for many years. It is a linen cloth, more than four meters long, which shows the image of a man's body, both front and back. Some people think it may have been the burial shroud of Jesus Christ after his crucifixion. A team of scientists in 1978 was permitted to study the shroud, using infrared photography, analysis of possible blood stains, microscopic examination of the linen fibers, and other methods. The results were ambiguous. A carbon 14 test was not permitted because it would have required cutting a piece about the size of a handkerchief from the shroud.

A new method of measuring carbon 14 was developed in the late 1980's. It is called "accelerator mass spectrometry," or AMS. Unlike Libby's method, it does not count the radioactivity of carbon. Instead, a mass spectrometer directly measures the ratio of carbon 14 to ordinary carbon. The main advantage of this method is that the sample size needed for analysis is about a thousand times smaller than before. The archbishop of Turin permitted three laboratories with the appropriate AMS apparatus to test the shroud material. The results agreed that the material was from the fourteenth century, not from the time of Christ. The figure on the shroud may be a watercolor painting on linen.

Since Libby's pioneering experiments in the late 1940's, carbon 14 dating has established itself as a reliable dating technique for archaeologists and cultural historians. Further improvements are expected to increase precision, to make it possible to use smaller samples, and to extend the effective time range of the method back to fifty thousand years or earlier.

See also Atomic clock; Geiger counter; Richter scale.

FURTHER READING

Goldberg, Paul, Vance T. Holliday, and C. Reid Ferring. *Earth Sciences and Archaeology*. New York: Kluwer Academic Plenum, 2001.

Libby, Willard Frank. "Radiocarbon Dating" [Nobel lecture]. In *Chemistry, 1942-1962*. River Edge, N.J.: World Scientific, 1999.

Lowe, John J. *Radiocarbon Dating: Recent Applications and Future Potential*. New York: John Wiley and Sons, 1996.

Cassette recording

THE INVENTION: Self-contained system making it possible to record and repeatedly play back sound without having to thread tape through a machine.

THE PERSON BEHIND THE INVENTION:

Fritz Pfleumer, a German engineer whose work on audiotapes paved the way for audiocassette production

SMALLER IS BETTER

The introduction of magnetic audio recording tape in 1929 was met with great enthusiasm, particularly in the entertainment industry, and specifically among radio broadcasters. Although somewhat practical methods for recording and storing sound for later playback had been around for some time, audiotape was much easier to use, store, and edit, and much less expensive to produce.

It was Fritz Pfleumer, a German engineer, who in 1929 filed the first audiotape patent. His detailed specifications indicated that tape could be made by bonding a thin coating of oxide to strips of either paper or film. Pfleumer also suggested that audiotape could be attached to filmstrips to provide higher-quality sound than was available with the film sound technologies in use at that time. In 1935, the German electronics firm AEG produced a reliable prototype of a record-playback machine based on Pfleumer's idea. By 1947, the American company 3M had refined the concept to the point where it was able to produce a high-quality tape using a plastic-based backing and red oxide. The tape recorded and reproduced sound with a high degree of clarity and dynamic range and would soon become the standard in the industry.

Still, the tape was sold and used in a somewhat inconvenient open-reel format. The user had to thread it through a machine and onto a take-up reel. This process was somewhat cumbersome and complicated for the layperson. For many years, sound-recording technology remained a tool mostly for professionals.

In 1963, the first audiocassette was introduced by the Netherlands-based Philips NV company. This device could be inserted into a machine without threading. Rewind and fast-forward were faster, and it made no difference where the tape was stopped prior to the ejection of the cassette. By contrast, open-reel audiotape required that the tape be wound fully onto one or the other of the two reels before it could be taken off the machine.

Technical advances allowed the cassette tape to be much narrower than the tape used in open reels and also allowed the tape speed to be reduced without sacrificing sound quality. Thus, the cassette was easier to carry around, and more sound could be recorded on a cassette tape. In addition, the enclosed cassette decreased wear and tear on the tape and protected it from contamination.

CREATING A MARKET

One of the most popular uses for audiocassettes was to record music from radios and other audio sources for later playback. During the 1970's, many radio stations developed "all music" formats in which entire albums were often played without interruption. That gave listeners an opportunity to record the music for later playback. At first, the music recording industry complained about this practice, charging that unauthorized recording of music from the radio was a violation of copyright laws. Eventually, the issue died down as the same companies began to recognize this new, untapped market for recorded music on cassette.

Audiocassettes, all based on the original Philips design, were being manufactured by more than sixty companies within only a few years of their introduction. In addition, spin-offs of that design were being used in many specialized applications, including dictation, storage of computer information, and surveillance. The emergence of videotape resulted in a number of formats for recording and playing back video based on the same principle. Although each is characterized by different widths of tape, each uses the same technique for tape storage and transport.

The cassette has remained a popular means of storing and retrieving information on magnetic tape for more than a quarter of a century. During the early 1990's, digital technologies such as audio CDs (compact discs) and the more advanced CD-ROM (compact discs that reproduce sound, text, and images via computer) were beginning to store information in revolutionary new ways. With the development of this increasingly sophisticated technology, need for the audiocassette, once the most versatile, reliable, portable, and economical means of recording, storing, and playing-back sound, became more limited.

CONSEQUENCES

The cassette represented a new level of convenience for the audiophile, resulting in a significant increase in the use of recording technology in all walks of life. Even small children could operate cassette recorders and players, which led to their use in schools for a variety of instructional tasks and in the home for entertainment. The recording industry realized that audiotape cassettes would allow consumers to listen to recorded music in places where record players were impractical: in automobiles, at the beach, even while camping.

The industry also saw the need for widespread availability of music and information on cassette tape. It soon began distributing albums on audiocassette in addition to the long-play vinyl discs, and recording sales increased substantially. This new technology put recorded music into automobiles for the first time, again resulting in a surge in sales for recorded music. Eventually, information, including language instruction and books-on-tape, became popular commuter fare.

With the invention of the microchip, audiotape players became available in smaller and smaller sizes, making them truly portable. Audiocassettes underwent another explosion in popularity during the early 1980's, when the Sony Corporation introduced the Walkman, an extremely compact, almost weightless cassette player that could be attached to clothing and used with lightweight earphones virtually anywhere. At the same time, cassettes were suddenly being used with microcomputers for backing up magnetic data files.

Home video soon exploded onto the scene, bringing with it new applications for cassettes. As had happened with audiotape, video camera-recorder units, called "camcorders," were miniaturized to the point where 8-millimeter videocassettes capable of recording up to 90 minutes of live action and sound were widely available. These cassettes closely resembled the audiocassette first introduced in 1963.

See also Compact disc; Dolby noise reduction; Electronic synthesizer; FM radio; Transistor radio; Walkman cassette player.

FURTHER READING

Miller, Christopher. "The One Hundred Greatest Inventions: Audio and Video." *Popular Science* 254, no. 4 (April, 1999).

Praag, Phil van. *Evolution of the Audio Recorder.* Waukesha, Wis.: EC Designs, 1997.

Stark, Craig. "Thirty Five Years of Tape Recording." *Stereo Review* 58 (September, 1993).

CAT SCANNER

THE INVENTION: A technique that collects X-ray data from solid, opaque masses such as human bodies and uses a computer to construct a three-dimensional image.

THE PEOPLE BEHIND THE INVENTION:

Godfrey Newbold Hounsfield (1919-   ), an English electronics engineer who shared the 1979 Nobel Prize in Physiology or Medicine

Allan M. Cormack (1924-1998), a South African-born American physicist who shared the 1979 Nobel Prize in Physiology or Medicine

James Ambrose, an English radiologist

A SIGNIFICANT MERGER

Computerized axial tomography (CAT) is a technique that collects X-ray data from an opaque, solid mass such as a human body and uses a sophisticated computer to assemble those data into a three-dimensional image. This sophisticated merger of separate technologies led to another name for CAT, computer-assisted tomography (it came to be called computed tomography, or CT). CAT is a technique of medical radiology, an area of medicine that began after the German physicist Wilhelm Conrad Röntgen's 1895 discovery of the high-energy electromagnetic radiations he named "X rays." Röntgen and others soon produced X-ray images of parts of the human body, and physicians were quick to learn that these images were valuable diagnostic aids.

In the late 1950's and early 1960's, Allan M. Cormack, a physicist at Tufts University in Massachusetts, pioneered a mathematical method for obtaining detailed X-ray absorption patterns in opaque samples meant to model biological samples. His studies used narrow X-ray beams that were passed through samples at many different angles. Because the technique probed test samples from many different points of reference, it became possible—by using the proper mathematics—to reconstruct the interior structure of a thin slice of the object being studied.

Cormack published his data but received almost no recognition because computers that could analyze the data in an effective fashion had not yet been developed. Nevertheless, X-ray tomography—the process of using X-rays to produce detailed images of thin sections of solid objects—had been born. It remained for Godfrey Newbold Hounsfield of England's Electrical and Musical Instruments (EMI) Limited (independently, and reportedly with no knowledge of Cormack's work) to design the first practical CAT scanner.

A Series of Thin Slices

Hounsfield, like Cormack, realized that X-ray tomography was the most practical approach to developing a medical body imager. It could be used to divide any three-dimensional object into a series of thin slices that could be reconstructed into images by using appropriate computers. Hounsfield developed another mathematical approach to the method. He estimated that the technique would make possible the very accurate reconstruction of images of thin body sections with a sensitivity well above that of the X-ray methodology then in use. Moreover, he proposed that his method would enable

Medical technicians studying CAT scan results. (PhotoDisc)

GODFREY NEWBOLD HOUNSFIELD

On his family farm outside Newark, Nottinghamshire, England, Godfrey Newbold Hounsfield (born 1919), the youngest of five children, was usually left to his own devices. The farm, he later wrote, offered an infinite variety of diversions, and his favorites were the many mechanical and electrical gadgets. By his teen years, he was making his own gadgets, such as an electrical recording machine, and experimenting with homemade gliders and water-propelled rockets. All these childhood projects taught him the fundamentals of practical reasoning.

During World War II he joined the Royal Air Force, where his talent with gadgets got him a position as an instructor at the school for radio mechanics. There, on his own, he built his an oscilloscope and demonstration equipment. This initiative caught the eye of a high-ranking officer, who after the war arranged a scholarship so that Hounsfield could attend the Faraday Electrical Engineering College in London. Upon graduating in 1951, he took a research position with Electrical and Musical Instruments, Limited (EMI). His first assignments involved radar and guided weapons, but he also developed an interest in computers and in 1958 led the design team that put together England's first all-transistor computer, the EMIDEC 1100. This experience, in turn, prepared him to follow through on his idea for computed tomography, which came to him in 1967.

EMI released its first CT scanner in 1971, and it so impressed the medical world that in 1979 Hounsfield and Allan M. Cormack shared the Nobel Prize in Physiology or Medicine for the invention. Hounsfield, who continued to work on improved computed tomography and other diagnostic imagining techniques, was knighted in 1981.

researchers and physicians to distinguish between normal and diseased tissue. Hounsfield was correct about that.

The prototype instrument that Hounsfield developed was quite slow, requiring nine days to scan an object. Soon, he modified the scanner so that its use took only nine hours, and he obtained successful tomograms of preserved human brains and the fresh brains of cattle. The further development of the CAT scanner then proceeded quickly, yielding an instrument that required four and one-half minutes to gather tomographic data and twenty minutes to produce the tomographic image.

In late 1971, the first clinical CAT scanner was installed at Atkinson Morley's Hospital in Wimbledon, England. By early 1972, the first patient, a woman with a suspected brain tumor, had been examined, and the resultant tomogram identified a dark, circular cyst in her brain. Additional data collection from other patients soon validated the technique. Hounsfield and EMI patented the CAT scanner in 1972, and the findings were reported at that year's annual meeting of the British Institute of Radiology.

Hounsfield published a detailed description of the instrument in 1973. Hounsfield's clinical collaborator, James Ambrose, published on the clinical aspects of the technique. Neurologists all around the world were ecstatic about the new tool that allowed them to locate tissue abnormalities with great precision.

The CAT scanner consisted of an X-ray generator, a scanner unit composed of an X-ray tube and a detector in a circular chamber about which they could be rotated, a computer that could process all the data obtained, and a cathode-ray tube on which tomograms were viewed. To produce tomograms, the patient was placed on a couch, head inside the scanner chamber, and the emitter-detector was rotated 1 degree at a time. At each position, 160 readings were taken, converted to electrical signals, and fed into the computer. In the 180 degrees traversed, 28,800 readings were taken and processed. The computer then converted the data into a tomogram (a cross-sectional representation of the brain that shows the differences in tissue density). A Polaroid picture of the tomogram was then taken and interpreted by the physician in charge.

CONSEQUENCES

Many neurologists agree that CAT is the most important method developed in the twentieth century to facilitate diagnosis of disorders of the brain. Even the first scanners could distinguish between brain tumors and blood clots and help physicians to diagnose a variety of brain-related birth defects. In addition, the scanners are believed to have saved many lives by allowing physicians to avoid the dangerous exploratory brain surgery once required in many cases and by replacing more dangerous techniques, such as pneumoencephalography, which required a physician to puncture the head for diagnostic purposes.

By 1975, improvements, including quicker reaction time and more complex emitter-detector systems, made it possible for EMI to introduce full-body CAT scanners to the world market. Then it became possible to examine other parts of the body—including the lungs, the heart, and the abdominal organs—for cardiovascular problems, tumors, and other structural health disorders. The technique became so ubiquitous that many departments of radiology changed their names to departments of medical imaging.

The use of CAT scanners has not been problem-free. Part of the reason for this is the high cost of the devices—ranging from about $300,000 for early models to $1 million for modern instruments—and resultant claims by consumer advocacy groups that the scanners are unnecessarily expensive toys for physicians. Still, CAT scanners have become important everyday diagnostic tools in many areas of medicine. Furthermore, continuation of the efforts of Hounsfield and others has led to more improvements of CAT scanners and to the use of nonradiologic nuclear magnetic resonance imaging in such diagnoses.

See also Amniocentesis; Electrocardiogram; Electroencephalogram; Mammography; Nuclear magnetic resonance; Pap test; Ultrasound; X-ray image intensifier.

FURTHER READING

Gambarelli, J. *Computerized Axial Tomography: An Anatomic Atlas of Serial Sections of the Human Body: Anatomy—Radiology—Scanner.* New York: Springer Verlag, 1977.
Raju, Tones N. K. "The Nobel Chronicles." *Lancet* 354, no. 9190 (November 6, 1999).
Thomas, Robert McG., Jr. "Allan Cormack, Seventy Four, Nobelist Who Helped Invent CAT Scan." *New York Times* (May 9, 1998).

172

Cell phone

THE INVENTION: Mobile telephone system controlled by computers to use a region's radio frequencies, or channels, repeatedly, thereby accommodating large numbers of users.

THE PEOPLE BEHIND THE INVENTION:

William Oliver Baker (1915-   ), the president of Bell Laboratories

Richard H. Fefrenkiel, the head of the mobile systems engineering department at Bell

The First Radio Telephones

The first recorded attempt to use radio technology to provide direct access to a telephone system took place in 1920. It was not until 1946, however, that Bell Telephone established the first such commercial system in St. Louis. The system had a number of disadvantages; users had to contact an operator who did the dialing and the connecting, and the use of a single radio frequency prevented simultaneous talking and listening. In 1949, a system was developed that used two radio frequencies (a "duplex pair"), permitting both the mobile unit and the base station to transmit and receive simultaneously and making a more normal sort of telephone conversation possible. This type of service, known as Mobile Telephone Service (MTS), was the norm in the field for many years.

The history of MTS is one of continuously increasing business usage. The development of the transistor made possible the design and manufacture of reasonably light, compact, and reliable equipment, but the expansion of MTS was slowed by the limited number of radio frequencies; there is nowhere near enough space on the radio spectrum for each user to have a separate frequency. In New York City, for example, New York Telephone Company was limited to just twelve channels for its more than seven hundred mobile subscribers, meaning that only twelve conversations could be carried on at once. In addition, because of possible interference, none of those channels could be reused in nearby cities; only fifty-four channels were available nationwide. By the late 1970's, most of the systems in major cities were considered full, and new subscribers were placed on a waiting list; some people had been waiting for as long as ten years to become subscribers. Mobile phone users commonly experienced long delays in getting poor-quality channels.

The Cellular Breakthrough

In 1968, the Federal Communications Commission (FCC) requested proposals for the creation of high-capacity, spectrum-efficient mobile systems.

A dominant trend in cell phone design is smaller and lighter units. (PhotoDisc)

Bell Telephone had already been lobbying for the creation of such a system for some years. In the early 1970's, both Motorola and Bell Telephone proposed the use of cellular technology to solve the problems posed by mobile telephone service. Cellular systems involve the use of a computer to make it possible to use an area's frequencies, or channels, repeatedly, allowing such systems to accommodate many more users.

A two-thousand-customer, 2100-square-mile cellular telephone system called the Advanced Mobile Phone Service, built by the AMPS Corporation, an AT&T subsidiary, became operational in Chicago in 1978. The Illinois Bell Telephone Company was allowed to make a limited commercial offering and obtained about fourteen hundred subscribers. American Radio Telephone Service was allowed to conduct a similar test in the Baltimore/Washington area. These first systems showed the technological feasibility and affordability of cellular service.

In 1979, Bell Labs of Murray Hill, New Jersey, received a patent

WILLIAM OLIVER BAKER

For great discoveries and inventions to be possible in the world of high technology, inventors need great facilities—laboratories and workshops—with brilliant colleagues. These must be managed by imaginative administrators.

One of the best was William Oliver Baker (b. 1915), who rose to become president of the legendary Bell Labs. Baker started out as one of the most promising scientists of his generation. After earning a Ph.D. in chemistry at Princeton University, he joined the research section at Bell Telephone Laboratories in 1939. He studied the physics and chemistry of polymers, especially for use in electronics and telecommunications. During his research career he helped develop synthetic rubber and radar, found uses for polymers in communications and power cables, and participated in the discovery of microgels. In 1954 he ranked among the top-ten scientists in American industry and asked to chair a National Research Council committee studying heat shields for missiles and satellites.

Administration suited him. The following year he took over as leader of research at Bell Labs and served as president from 1973 until 1979. Under his direction, basic discoveries and inventions poured out of the lab that later transformed the way people live and work: satellite communications, principles for programming high-speed computers, the technology for modern electronic communications, the superconducting solenoid, the maser, and the laser. His scientists won Nobel Prizes and legions of other honors, as did Baker himself, who received dozens of medals, awards, and honorary degrees. Moreover, he was an original member of the President's Science Advisory Board, became the first chair of the National Science Information Council, and served on the National Science Board. His influence on American science and technology was deep and lasting.

for such a system. The inventor was Richard H. Fefrenkiel, head of the mobile systems engineering department under the leadership of Labs president William Baker. The patented method divides a city into small coverage areas called "cells," each served by low-power transmitter-receivers. When a vehicle leaves the coverage of one cell, calls are switched to the antenna and channels of an adjacent cell; a conversation underway is automatically transferred and continues without interruption. A channel used in one cell can be reused a few cells away for a different conversation. In this way, a few hundred channels can serve hundreds of thousands of users. Computers control the call-transfer process, effectively reducing the amount of radio spectrum required. Cellular systems thus actually use radio frequencies to transmit conversations, but because the equipment is so telephone-like, "cellular telephone" (or "cell phone") became the accepted term for the new technology.

Each AMPS cell station is connected by wire to a central switching office, which determines when a mobile phone should be transferred to another cell as the transmitter moves out of range during a conversation. It does this by monitoring the strength of signals received from the mobile unit by adjacent cells, "handing off" the call when a new cell receives a stronger signal; this change is imperceptible to the user.

IMPACT

In 1982, the FCC began accepting applications for cellular system licenses in the thirty largest U.S. cities. By the end of 1984, there were about forty thousand cellular customers in nearly two dozen cities. Cellular telephone ownership boomed to 9 million by 1992.

As cellular telephones became more common, they also became cheaper and more convenient to buy and to use. New systems developed in the 1990's continued to make smaller, lighter, and cheaper cellular phones even more accessible. Since the cellular telephone was made possible by the marriage of communications and computers, advances in both these fields have continued to change the industry at a rapid rate.

Cellular phones have proven ideal for many people who need or want to keep in touch with others at all times. They also provide convenient emergency communication devices for travelers and field-workers. On the other hand, ownership of a cellular phone can also have its drawbacks; many users have found that they can never be out of touch—even when they would rather be.

See also Internet; Long-distance telephone; Rotary dial telephone; Telephone switching; Touch-tone telephone.

FURTHER READING

Carlo, George Louis, and Martin Schram. *Cell Phones: Invisible Hazards in the Wireless Age.* New York: Carroll and Graf, 2001.

"The Cellular Phone." *Newsweek* 130, 24A (Winter 1997/1998).

Oliphant, Malcolm W. "How Mobile Telephony Got Going." *IEEE Spectrum* 36, no. 8 (August, 1999).

Young, Peter. *Person to Person: The International Impact of the Telephone.* Cambridge: Granta Editions, 1991.

CLONING

THE INVENTION: Experimental technique for creating exact duplicates of living organisms by recreating their DNA.

THE PEOPLE BEHIND THE INVENTION:
Ian Wilmut, an embryologist with the Roslin Institute
Keith H. S. Campbell, an experiment supervisor with the Roslin
Institute
J. McWhir, a researcher with the Roslin Institute
W. A. Ritchie, a researcher with the Roslin Institute

MAKING COPIES

On February 22, 1997, officials of the Roslin Institute, a biological research institution near Edinburgh, Scotland, held a press conference to announce startling news: They had succeeded in creating a clone—a biologically identical copy—from cells taken from
an adult sheep. Although cloning had been performed previously
with simpler organisms, the Roslin Institute experiment marked
the first time that a large, complex mammal had been successfully
cloned.

Cloning, or the production of genetically identical individuals,
has long been a staple of science fiction and other popular literature.
Clones do exist naturally, as in the example of identical twins. Scientists have long understood the process by which identical twins
are created, and agricultural researchers have often dreamed of a
method by which cheap identical copies of superior livestock could
be created.

The discovery of the double helix structure of deoxyribonucleic
acid (DNA), or the genetic code, by James Watson and Francis Crick
in the 1950's led to extensive research into cloning and genetic engineering. Using the discoveries of Watson and Crick, scientists were
soon able to develop techniques to clone laboratory mice; however,
the cloning of complex, valuable animals such as livestock proved
to be hard going.

Early versions of livestock cloning were technical attempts at dupli-

Ian Wilmut

Ian Wilmut was born in Hampton Lucey, not far from Warwick in central England, in 1944. He found his life's calling in embryology—and especially animal genetic engineering— while he was studying at the University of Nottingham, where his mentor was G. Eric Lamming, a leading expert on reproduction. After receiving his undergraduate degree, he attended Darwin College, Cambridge University. He completed his doctorate in 1973 upon submitting a thesis about freezing boar sperm. This came after he produced a viable calf, named Frosty, from the frozen semen, the first time anyone had done so.

Soon afterward he joined the Animal Breeding Research Station, which later became the Roslin Institute in Roslin, Scotland. He immersed himself in research, seldom working fewer than nine hours a day. During the 1980's he experimented with the insertion of genes into sheep embryos but concluded that cloning would be less time-consuming and less prone to failure. Joined by Keith Campbell in 1990, he cloned two Welsh mountain sheep from differentiated embryo cells, a feat similar to those of other reproductive experimenters. However, Dolly, who was cloned from adult cells, shook the world when her birth was announced in 1997. That same year Wilmut and Campbell produced another cloned sheep, Polly. Cloned from fetal skin cells, she was genetically altered to carry a human gene.

Wilmut's technique for cloning from adult cells, which the laboratory patented, was a fundamentally new method of reproduction, but he had a loftier purpose in mind than simply establishing a first. He believed that animals genetically engineered to include human genes can produce proteins needed by people who because of genetic diseases cannot make the proteins themselves. The production of new treatments for old diseases, he told an astonished public after the revelation of Dolly, was his goal.

cating the natural process of fertilized egg splitting that leads to the birth of identical twins. Artificially inseminated eggs were removed, split, and then reinserted into surrogate mothers. This method proved to be overly costly for commercial purposes, a situation aggravated by a low success rate.

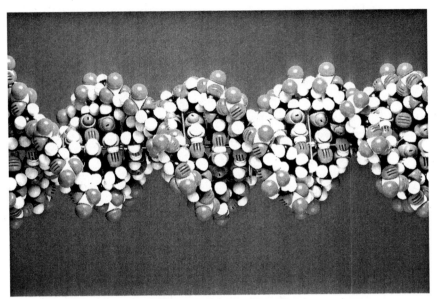

Model of a double helix. (PhotoDisc)

NUCLEAR TRANSFER

Researchers at the Roslin Institute found these earlier attempts to be fundamentally flawed. Even if the success rate could be improved, the number of clones created (of sheep, in this case) would still be limited. The Scots, led by embryologist Ian Wilmut and experiment supervisor Keith Campbell, decided to take an entirely different approach. The result was the first live birth of a mammal produced through a process known as "nuclear transfer."

Nuclear transfer involves the replacement of the nucleus of an immature egg with a nucleus taken from another cell. Previous attempts at nuclear transfer had cells from a single embryo divided up and implanted into an egg. Because a sheep embryo has only about forty usable cells, this method also proved limiting.

The Roslin team therefore decided to grow their own cells in a laboratory culture. They took more mature embryonic cells than those previously used, and they experimented with the use of a nutrient mixture. One of their breakthroughs occurred when they discovered that these "cell lines" grew much more quickly when certain nutrients were absent.

Using this technique, the Scots were able to produce a theoretically unlimited number of genetically identical cell lines. The next step was to transfer the cell lines of the sheep into the nucleus of unfertilized sheep eggs.

First, 277 nuclei with a full set of chromosomes were transferred to the unfertilized eggs. An electric shock was then used to cause the eggs to begin development, the shock performing the duty of fertilization. Of these eggs, twenty-nine developed enough to be inserted into surrogate mothers.

All the embryos died before birth except one: a ewe the scientists named "Dolly." Her birth on July 5, 1996, was witnessed by only a veterinarian and a few researchers. Not until the clone had survived the critical earliest stages of life was the success of the experiment disclosed; Dolly was more than seven months old by the time her birth was announced to a startled world.

IMPACT

The news that the cloning of sophisticated organisms had left the realm of science fiction and become a matter of accomplished scientific fact set off an immediate uproar. Ethicists and media commentators quickly began to debate the moral consequences of the use—and potential misuse—of the technology. Politicians in numerous countries responded to the news by calling for legal restrictions on cloning research. Scientists, meanwhile, speculated about the possible benefits and practical limitations of the process.

The issue that stirred the imagination of the broader public and sparked the most spirited debate was the possibility that similar experiments might soon be performed using human embryos. Although most commentators seemed to agree that such efforts would be profoundly immoral, many experts observed that they would be virtually impossible to prevent. "Could someone do this tomorrow morning on a human embryo?" Arthur L. Caplan, the director of the University of Pennsylvania's bioethics center, asked reporters. "Yes. It would not even take too much science. The embryos are out there."

Such observations conjured visions of a future that seemed marvelous to some, nightmarish to others. Optimists suggested that the best and brightest of humanity could be forever perpetuated, creating an endless supply of Albert Einsteins and Wolfgang Amadeus Mozarts. Pessimists warned of a world overrun by clones of self-serving narcissists and petty despots, or of the creation of a secondary class of humans to serve as organ donors for their progenitors.

The Roslin Institute's researchers steadfastly proclaimed their own opposition to human experimentation. Moreover, most scientists were quick to point out that such scenarios were far from realization, noting the extremely high failure rate involved in the creation of even a single sheep. In addition, most experts emphasized more practical possible uses of the technology: improving agricultural stock by cloning productive and disease-resistant animals, for example, or regenerating endangered or even extinct species. Even such apparently benign schemes had their detractors, however, as other observers remarked on the potential dangers of thus narrowing a species' genetic pool.

Even prior to the Roslin Institute's announcement, most European nations had adopted a bioethics code that flatly prohibited genetic experiments on human subjects. Ten days after the announcement, U.S. president Bill Clinton issued an executive order that banned the use of federal money for human cloning research, and he called on researchers in the private sector to refrain from such experiments voluntarily. Nevertheless, few observers doubted that Dolly's birth marked only the beginning of an intriguing—and possibly frightening—new chapter in the history of science.

See also Amniocentesis; Artificial chromosome; Artificial insemination; Genetic "fingerprinting"; In vitro plant culture; Rice and wheat strains.

FURTHER READING

Facklam, Margery, Howard Facklam, and Paul Facklam. *From Cell to Clone: The Story of Genetic Engineering*. New York: Harcourt Brace Jovanovich, 1979.
Gillis, Justin. "Cloned Cows Are Fetching Big Bucks: Dozens of Genetic Duplicates Ready to Take Up Residence on U.S. Farms." *Washington Post* (March 25, 2001).

Kolata, Gina Bari. *Clone: The Road to Dolly, and the Path Ahead.* New York: William Morrow, 1998.

Regalado, Antonio. "Clues Are Sought for Cloning's Fail Rate: Researchers Want to Know Exactly How an Egg Reprograms Adult DNA." *Wall Street Journal* (November 24, 2000).

Winslow, Ron. "Scientists Clone Pigs, Lifting Prospects of Replacement Organs for Humans." *Wall Street Journal* (August 17, 2000).

Cloud seeding

THE INVENTION: Technique for inducing rainfall by distributing dry ice or silver nitrate into reluctant rainclouds.

THE PEOPLE BEHIND THE INVENTION:

Vincent Joseph Schaefer (1906-1993), an American chemist and meteorologist

Irving Langmuir (1881-1957), an American physicist and chemist who won the 1932 Nobel Prize in Chemistry

Bernard Vonnegut (1914-1997), an American physical chemist and meteorologist

Praying for Rain

Beginning in 1943, an intense interest in the study of clouds developed into the practice of weather "modification." Working for the General Electric Research Laboratory, Nobel laureate Irving Langmuir and his assistant researcher and technician, Vincent Joseph Schaefer, began an intensive study of precipitation and its causes.

Past research and study had indicated two possible ways that clouds produce rain. The first possibility is called "coalescing," a process by which tiny droplets of water vapor in a cloud merge after bumping into one another and become heavier and fatter until they drop to earth. The second possibility is the "Bergeron process" of droplet growth, named after the Swedish meteorologist Tor Bergeron. Bergeron's process relates to supercooled clouds, or clouds that are at or below freezing temperatures and yet still contain both ice crystals and liquid water droplets. The size of the water droplets allows the droplets to remain liquid despite freezing temperatures; while small droplets can remain liquid only down to 4 degrees Celsius, larger droplets may not freeze until reaching –15 degrees Celsius. Precipitation occurs when the ice crystals become heavy enough to fall. If the temperature at some point below the cloud is warm enough, it will melt the ice crystals before they reach the earth, producing rain. If the temperature remains at the freezing point, the ice crystals retain their form and fall as snow.

Schaefer used a deep-freezing unit in order to observe water droplets in pure cloud form. In order to observe the droplets better, Schaefer lined the chest with black velvet and concentrated a beam of light inside. The first agent he introduced inside the supercooled freezer was his own breath. When that failed to form the desired ice crystals, he proceeded to try other agents. His hope was to form ice crystals that would then cause the moisture in the surrounding air to condense into more ice crystals, which would produce a miniature snowfall.

He eventually achieved success when he tossed a handful of dry ice inside and was rewarded with the long-awaited snow. The freezer was set at the freezing point of water, 0 degrees Celsius, but not all the particles were ice crystals, so when the dry ice was introduced all the stray water droplets froze instantly, producing ice crystals, or snowflakes.

PLANTING THE FIRST SEEDS

On November 13, 1946, Schaefer took to the air over Mount Greylock with several pounds of dry ice in order to repeat the experiment in nature. After he had finished sprinkling, or seeding, a supercooled cloud, he instructed the pilot to fly underneath the cloud he had just seeded. Schaefer was greeted by the sight of snow. By the time it reached the ground, it had melted into the first-ever human-made rainfall.

Independently of Schaefer and Langmuir, another General Electric scientist, Bernard Vonnegut, was also seeking a way to cause rain. He found that silver iodide crystals, which have the same size and shape as ice crystals, could "fool" water droplets into condensing on them. When a certain chemical mixture containing silver iodide is heated on a special burner called a "generator," silver iodide crystals appear in the smoke of the mixture. Vonnegut's discovery allowed seeding to occur in a way very different from seeding with dry ice, but with the same result. Using Vonnegut's process, the seeding is done from the ground. The generators are placed outside and the chemicals are mixed. As the smoke wafts upward, it carries the newly formed silver iodide crystals with it into the clouds.

The results of the scientific experiments by Langmuir, Vonnegut, and Schaefer were alternately hailed and rejected as legitimate. Critics argue that the process of seeding is too complex and would have to require more than just the addition of dry ice or silver nitrate in order to produce rain. One of the major problems surrounding the question of weather modification by cloud seeding is the scarcity of knowledge about the earth's atmosphere. A journey begun about fifty years ago is still a long way from being completed.

IMPACT

Although the actual statistical and other proofs needed to support cloud seeding are lacking, the discovery in 1946 by the General Electric employees set off a wave of interest and demand for information that far surpassed the interest generated by the discovery of nuclear fission shortly before. The possibility of ending drought and, in the process, hunger excited many people. The discovery also prompted both legitimate and false "rainmakers" who used the information gathered by Schaefer, Langmuir, and Vonnegut to set up cloud-seeding businesses. Weather modification, in its current stage of development, cannot be used to end worldwide drought. It does, however, have beneficial results in some cases on the crops of smaller farms that have been affected by drought.

In order to understand the advances made in weather modification, new instruments are needed to record accurately the results of further experimentation. The storm of interest—both favorable and nonfavorable—generated by the discoveries of Schaefer, Langmuir, and Vonnegut has had and will continue to have far-reaching effects on many aspects of society.

See also Airplane; Artificial insemination; In vitro plant culture; Weather satellite.

FURTHER READING

Cole, Stephen. "Mexico Results Spur New Looking at Rainmaking." *Washington Post* (January 22, 2001).

Havens, Barrington S., James E. Jiusto, and Bernard Vonnegut. *Early History of Cloud Seeding*. Socorro, N.Mex.: Langmuir Laboratory, New Mexico Institute of Mining and Technology, 1978.

"Science and Technology: Cloudbusting." *The Economist* (August 21, 1999).

Villiers, Marq de. *Water: The Fate of Our Most Precious Resource*. Boston: Houghton Mifflin, 2000.

COBOL COMPUTER LANGUAGE

THE INVENTION: The first user-friendly computer programming language, COBOL was originally designed to solve ballistics problems.

THE PEOPLE BEHIND THE INVENTION:
Grace Murray Hopper (1906-1992), an American
    mathematician
Howard Hathaway Aiken (1900-1973), an American
    mathematician

PLAIN SPEAKING

Grace Murray Hopper, a mathematician, was a faculty member at Vassar College when World War II (1939-1945) began. She enlisted in the Navy and in 1943 was assigned to the Bureau of Ordnance Computation Project, where she worked on ballistics problems. In 1944, the Navy began using one of the first electronic computers, the Automatic Sequence Controlled Calculator (ASCC), designed by an International Business Machines (IBM) Corporation team of engineers headed by Howard Hathaway Aiken, to solve ballistics problems. Hopper became the third programmer of the ASCC.

Hopper's interest in computer programming continued after the war ended. By the early 1950's, Hopper's work with programming languages had led to her development of FLOW-MATIC, the first English-language data processing compiler. Hopper's work on FLOW-MATIC paved the way for her later work with COBOL (Common Business Oriented Language).

Until Hopper developed FLOW-MATIC, digital computer programming was all machine-specific and was written in machine code. A program designed for one computer could not be used on another. Every program was both machine-specific and problem-specific in that the programmer would be told what problem the machine was going to be asked and then would write a completely new program for that specific problem in the machine code.

GRACE MURRAY HOPPER

Grace Brewster Murray was born in New York City in 1906. As a child she revered her great-grandfather, a U.S. Navy admiral, and her grandfather, an engineer. Her career melded their professions.

She studied mathematics and physics at Vassar College, earning a bachelor's degree in 1928 and a master's degree in 1930, when she married Vincent Foster Hopper. She accepted a teaching post at Vassar but continued her studies, completing a doctorate at Yale University in 1934. In 1943 she left academia for the Navy and was assigned to the Bureau of Ordnance Computation Project at Harvard University. She worked on the nation's first modern computer, the Mark I, and contributed to the development of major new models afterward, including Sperry Corporation's ENIAC and UNIVAC. While still with the Navy project at Harvard, Hopper participated in a minor incident that forever marked computer slang. One day a moth became caught in a switch, causing the computer to malfunction. She and other technicians found it and ever after referred to correcting mechanical glitches as "debugging."

Hopper joined Sperry Corporation after the war and carried out her seminal work with the FLOW-MATIC and COBOL computer languages. Meanwhile, she retained her commission in the Naval Reserves, helping the service incorporate computers and COBOL into its armaments and administration systems. She retired from the Navy in 1966 and from Sperry in 1971, but the Navy soon had her out of retirement on temporary active duty to help with its computer systems. After her second retirement, the Navy, grateful for her tireless service, promoted her to rear admiral in 1985, the nation's first woman admiral. She was also awarded the Distinguished Service Cross by the Department of Defense, the National Medal of Technology, and the Legion of Merit. She became an inductee into the Engineering and Science Hall of Fame in 1991. Hopper, nicknamed Amazing Grace, died a year later.

Machine code was based on the programmer's knowledge of the physical characteristics of the computer as well as the requirements of the problem to be solved; that is, the programmer had to know what was happening within the machine as it worked through a series of calculations, which relays tripped when and in what order, and what mathematical operations were necessary to solve the problem. Programming was therefore a highly specialized skill requiring a unique combination of linguistic, reasoning, engineering, and mathematical abilities that not even all the mathematicians and electrical engineers who designed and built the early computers possessed.

While every computer still operates in response to the programming, or instructions, built into it, which are formatted in machine code, modern computers can accept programs written in nonmachine code—that is, in various automatic programming languages. They are able to accept nonmachine code programs because specialized programs now exist to translate those programs into the appropriate machine code. These translating programs are known as "compilers," or "assemblers," and FLOW-MATIC was the first such program.

Hopper developed FLOW-MATIC after realizing that it would be necessary to eliminate unnecessary steps in programming to make computers more efficient. FLOW-MATIC was based, in part, on Hopper's recognition that certain elements, or commands, were common to many different programming applications. Hopper theorized that it would not be necessary to write a lengthy series of instructions in machine code to instruct a computer to begin a series of operations; instead, she believed that it would be possible to develop commands in an assembly language in such a way that a programmer could write one command, such as the word *add*, that would translate into a sequence of several commands in machine code. Hopper's successful development of a compiler to translate programming languages into machine code thus meant that programming became faster and easier. From assembly languages such as FLOW-MATIC, it was a logical progression to the development of high-level computer languages, such as FORTRAN (*Formula Translation*) and COBOL.

THE LANGUAGE OF BUSINESS

Between 1955 (when FLOW-MATIC was introduced) and 1959, a number of attempts at developing a specific business-oriented language were made. IBM and Remington Rand believed that the only way to market computers to the business community was through the development of a language that business people would be comfortable using. Remington Rand officials were especially committed to providing a language that resembled English. None of the attempts to develop a business-oriented language succeeded, however, and by 1959 Hopper and other members of the U.S. Department of Defense had persuaded representatives of various companies of the need to cooperate.

On May 28 and 29, 1959, a conference sponsored by the Department of Defense was held at the Pentagon to discuss the problem of establishing a common language for the adaptation of electronic computers for data processing. As a result, the first distribution of COBOL was accomplished on December 17, 1959. Although many people were involved in the development of COBOL, Hopper played a particularly important role. She not only found solutions to technical problems but also succeeded in selling the concept of a common language from an administrative and managerial point of view. Hopper recognized that while the companies involved in the commercial development of computers were in competition with one another, the use of a common, business-oriented language would contribute to the growth of the computer industry as a whole, as well as simplify the training of computer programmers and operators.

CONSEQUENCES

COBOL was the first compiler developed for business data processing operations. Its development simplified the training required for computer users in business applications and demonstrated that computers could be practical tools in government and industry as well as in science. Prior to the development of COBOL, electronic computers had been characterized as expensive, oversized adding machines that were adequate for performing time-consuming mathematics but lacked the flexibility that business people required.

In addition, the development of COBOL freed programmers not only from the need to know machine code but also from the need to understand the physical functioning of the computers they were using. Programming languages could be written that were both machine-independent and almost universally convertible from one computer to another.

Finally, because Hopper and the other committee members worked under the auspices of the Department of Defense, the software was not copyrighted, and in a short period of time COBOL became widely available to anyone who wanted to use it. It diffused rapidly throughout the industry and contributed to the widespread adaptation of computers for use in countless settings.

See also BASIC programming language; Colossus computer; ENIAC computer; FORTRAN programming language; SAINT.

Further Reading

Cohen, Bernard I., Gregory W. Welch, and Robert V. D. Campbell. *Makin' Numbers: Howard Aiken: and the Computer.* Cambridge, Mass.: MIT Press, 1999.

Cohen, Bernard I. *Howard Aiken: Portrait of a Computer Pioneer.* Cambridge, Mass.: MIT Press, 1999.

Ferguson, David E. "The Roots of COBOL." *Systems 3X World and As World* 17, no. 7 (July, 1989).

Yount, Lisa. *A to Z of Women in Science and Math.* New York: Facts on File, 1999.

COLOR FILM

THE INVENTION: A photographic medium used to take full-color pictures.

THE PEOPLE BEHIND THE INVENTION:

Rudolf Fischer (1881-1957), a German chemist

H. Siegrist (1885-1959), a German chemist and Fischer's collaborator

Benno Homolka (1877-1949), a German chemist

THE PROCESS BEGINS

Around the turn of the twentieth century, Arthur-Louis Ducos du Hauron, a French chemist and physicist, proposed a tripack (three-layer) process of film development in which three color negatives would be taken by means of superimposed films. This was a subtractive process. (In the "additive method" of making color pictures, the three colors are added in projection—that is, the colors are formed by the mixture of colored light of the three primary hues. In the "subtractive method," the colors are produced by the superposition of prints.) In Ducos du Hauron's process, the blue-light negative would be taken on the top film of the pack; a yellow filter below it would transmit the yellow light, which would reach a green-sensitive film and then fall upon the bottom of the pack, which would be sensitive to red light. Tripacks of this type were unsatisfactory, however, because the light became diffused in passing through the emulsion layers, so the green and red negatives were not sharp.

To obtain the real advantage of a tripack, the three layers must be coated one over the other so that the distance between the blue-sensitive and red-sensitive layers is a small fraction of a thousandth of an inch. Tripacks of this type were suggested by the early pioneers of color photography, who had the idea that the packs would be separated into three layers for development and printing. The manipulation of such systems proved to be very difficult in practice. It was also suggested, however, that it might be possible to develop such tripacks as a unit and then, by chemical treatment, convert the silver images into dye images.

FISCHER'S THEORY

One of the earliest subtractive tripack methods that seemed to hold great promise was that suggested by Rudolf Fischer in 1912. He proposed a tripack that would be made by coating three emulsions on top of one another; the lowest one would be red-sensitive, the middle one would be green-sensitive, and the top one would be blue-sensitive. Chemical substances called "couplers," which would produce dyes in the development process, would be incorporated into the layers. In this method, the molecules of the developing agent, after becoming oxidized by developing the silver image, would react with the unoxidized form (the coupler) to produce the dye image.

The two types of developing agents described by Fischer are paraminophenol and paraphenylenediamine (or their derivatives). The five types of dye that Fischer discovered are formed when silver images are developed by these two developing agents in the presence of suitable couplers. The five classes of dye he used (indophenols, indoanilines, indamines, indothiophenols, and azomethines) were already known when Fischer did his work, but it was he who discovered that the photographic latent image could be used to promote their formulation from "coupler" and "developing agent." The indoaniline and azomethine types have been found to possess the necessary properties, but the other three suffer from serious defects. Because only p-phenylenediamine and its derivatives can be used to form the indoaniline and azomethine dyes, it has become the most widely used color developing agent.

IMPACT

In the early 1920's, Leopold Mannes and Leopold Godowsky made a great advance beyond the Fischer process. Working on a new process of color photography, they adopted coupler development, but instead of putting couplers into the emulsion as Fischer had, they introduced them during processing. Finally, in 1935, the film was placed on the market under the name "Kodachrome," a name that had been used for an early two-color process.

The first use of the new Kodachrome process in 1935 was for 16-millimeter film. Color motion pictures could be made by the Kodachrome process as easily as black-and-white pictures, because the complex work involved (the color development of the film) was done under precise technical control. The definition (quality of the image) given by the process was soon sufficient to make it practical for 8-millimeter pictures, and in 1936, Kodachrome film was introduced in a 35-millimeter size for use in popular miniature cameras.

Soon thereafter, color processes were developed on a larger scale and new color materials were rapidly introduced. In 1940, the Kodak Research Laboratories worked out a modification of the Fischer process in which the couplers were put into the emulsion layers. These couplers are not dissolved in the gelatin layer itself, as the Fischer couplers are, but are carried in small particles of an oily material that dissolves the couplers, protects them from the gelatin, and protects the silver bromide from any interaction with the couplers. When development takes place, the oxidation product of the developing agent penetrates into the organic particles and reacts with the couplers so that the dyes are formed in small particles that are dispersed throughout the layers. In one form of this material, Ektachrome (originally intended for use in aerial photography), the film is reversed to produce a color positive. It is first developed with a black-and-white developer, then reexposed and developed with a color developer that recombines with the couplers in each layer to produce the appropriate dyes, all three of which are produced simultaneously in one development.

In summary, although Fischer did not succeed in putting his theory into practice, his work still forms the basis of most modern color photographic systems. Not only did he demonstrate the general principle of dye-coupling development, but the art is still mainly confined to one of the two types of developing agent, and two of the five types of dye, described by him.

See also Autochrome plate; Brownie camera; Infrared photography; Instant photography.

FURTHER READING

Collins, Douglas. *The Story of Kodak.* New York: Harry N. Abrams, 1990.

Glendinning, Peter. *Color Photography: History, Theory, and Darkroom Technique.* Englewood Cliffs, N.J.: Prentice-Hall, 1985.
Wood, John. *The Art of the Autochrome: The Birth of Color Photography.* Iowa City: University of Iowa Press, 1993.

Color television

The invention: System for broadcasting full-color images over the airwaves.

The people behind the invention:
Peter Carl Goldmark (1906-1977), the head of the CBS research and development laboratory
William S. Paley (1901-1990), the businessman who took over CBS
David Sarnoff (1891-1971), the founder of RCA

The Race for Standardization

Although by 1928 color television had already been demonstrated in Scotland, two events in 1940 mark that year as the beginning of color television. First, on February 12, 1940, the Radio Corporation of America (RCA) demonstrated its color television system privately to a group that included members of the Federal Communications Commission (FCC), an administrative body that had the authority to set standards for an electronic color system. The demonstration did not go well; indeed, David Sarnoff, the head of RCA, canceled a planned public demonstration and returned his engineers to the Princeton, New Jersey, headquarters of RCA's laboratories.

Next, on September 1, 1940, the Columbia Broadcasting System (CBS) took the first step to develop a color system that would become the standard for the United States. On that day, CBS demonstrated color television to the public, based on the research of an engineer, Peter Carl Goldmark. Goldmark placed a set of spinning filters in front of the black-and-white television images, breaking them down into three primary colors and producing color television. The audience saw what was called "additive color."

Although Goldmark had been a researcher at CBS since January, 1936, he did not attempt to develop a color television system until March, 1940, after watching the Technicolor motion picture *Gone with the Wind* (1939). Inspired, Goldmark began to tinker in his tiny

CBS laboratory in the headquarters building in New York City.

If a decision had been made in 1940, the CBS color standard would have been accepted as the national standard. The FCC was, at that time, more concerned with trying to establish a black-and-white standard for television. Color television seemed decades away. In 1941, the FCC decided to adopt standards for black-and-white television only, leaving the issue of color unresolved—and the doors to the future of color broadcasting wide open. Control of a potentially lucrative market as well as personal rivalry threw William S. Paley, the head of CBS, and Sarnoff into a race for the control of color television. Both companies would pay dearly in terms of money and time, but it would take until the 1960's before the United States would become a nation of color television watchers.

RCA was at the time the acknowledged leader in the development of black-and-white television. CBS engineers soon discovered, however, that their company's color system would not work when combined with RCA black-and-white televisions. In other words, customers would need one set for black-and-white and one for color. Moreover, since the color system of CBS needed more broadcast frequency space than the black-and-white system in use, CBS was forced to ask the FCC to allocate new channel space in the ultrahigh frequency (UHF) band, which was then not being used. In contrast, RCA scientists labored to make a compatible color system that required no additional frequency space.

No Time to Wait

Following the end of World War II, in 1945, the suburbanites who populated new communities in America's cities wanted television sets right away; they did not want to wait for the government to decide on a color standard and then wait again while manufacturers redesigned assembly lines to make color sets. Rich with savings accumulated during the prosperity of the war years, Americans wanted to spend their money. After the war, the FCC saw no reason to open up proceedings regarding color systems. Black-and-white was operational; customers were waiting in line for the new electronic marvel. To give its engineers time to create a compatible color system, RCA skillfully lobbied the members of the FCC to take no action.

There were other problems with the CBS mechanical color television. It was noisy and large, and its color balance was hard to maintain. CBS claimed that through further engineering work, it would improve the actual sets. Yet RCA was able to convince other manufacturers to support it in preference to CBS principally because of its proven manufacturing track record.

In 1946, RCA demonstrated a new electronic color receiver with three picture tubes, one for each of the primary colors. Color reproduction was fairly true; although any movement on the screen caused color blurring, there was little flicker. It worked, however, and thus ended the invention phase of color television begun in 1940. The race for standardization would require seven more years of corporate struggle before the RCA system would finally win adoption as the national standard in 1953.

IMPACT

Through the 1950's, black-and-white television remained the order of the day. Through the later years of the decade, only the National Broadcasting Company (NBC) television network was regularly airing programs in color. Full production and presentation of shows in color during prime time did not come until the mid-1960's; most industry observers date 1972 as the true arrival of color television.

By 1972, color sets were found in more than half the homes in the United States. At that point, since color was so widespread, *TV Guide* stopped tagging color program listings with a special symbol and instead tagged only black-and-white shows, as it does to this day. Gradually, only cheap, portable sets were made for black-and-white viewing, while color sets came in all varieties from tiny handheld pocket televisions to mammoth projection televisions.

See also Autochrome plate; Community antenna television; Communications satellite; Fiber-optics; FM radio; Radio; Television; Transistor; Videocassette recorder.

FURTHER READING

Burns, R. W. *Television: An International History of the Formative Years.* London: Institution of Electrical Engineers in association with the Science Museum, 1998.

Fisher, David E., and Marshall Fisher. *Tube: The Invention of Television.* Washington, D.C.: Counterpoint, 1996.

Lewis, Tom. *Empire of the Air: The Men Who Made Radio.* New York: HarperPerennial, 1993.

Lyons, Eugene. *David Sarnoff: A Biography.* New York: Harper and Row, 1967.

Colossus computer

THE INVENTION: The first all-electronic calculating device, the Colossus computer was built to decipher German military codes during World War II.

THE PEOPLE BEHIND THE INVENTION:
Thomas H. Flowers, an electronics expert
Max H. A. Newman (1897-1984), a mathematician
Alan Mathison Turing (1912-1954), a mathematician
C. E. Wynn-Williams, a member of the Telecommunications Research Establishment

An Undercover Operation

In 1939, during World War II (1939-1945), a team of scientists, mathematicians, and engineers met at Bletchley Park, outside London, to discuss the development of machines that would break the secret code used in Nazi military communications. The Germans were using a machine called "Enigma" to communicate in code between headquarters and field units. Polish scientists, however, had been able to examine a German Enigma and between 1928 and 1938 were able to break the codes by using electromechanical codebreaking machines called "bombas." In 1938, the Germans made the Enigma more complicated, and the Polish were no longer able to break the codes. In 1939, the Polish machines and codebreaking knowledge passed to the British.

Alan Mathison Turing was one of the mathematicians gathered at Bletchley Park to work on codebreaking machines. Turing was one of the first people to conceive of the universality of digital computers. He first mentioned the "Turing machine" in 1936 in an article published in the Proceedings of the London Mathematical Society. The Turing machine, a hypothetical device that can solve any problem that involves mathematical computation, is not restricted to only one task—hence the universality feature.

Turing suggested an improvement to the Bletchley codebreaking machine, the "Bombe," which had been modeled on the Polish bomba. This improvement increased the computing power of the machine. The new codebreaking machine replaced the tedious method of decoding by hand, which in addition to being slow, was ineffective in dealing with complicated encryptions that were changed daily.

BUILDING A BETTER MOUSETRAP

The Bombe was very useful. In 1942, when the Germans started using a more sophisticated cipher machine known as the "Fish," Max H. A. Newman, who was in charge of one subunit at Bletchley Park, believed that an automated device could be designed to break the codes produced by the Fish. Thomas H. Flowers, who was in charge of a switching group at the Post Office Research Station at Dollis Hill, had been approached to build a special-purpose electromechanical device for Bletchley Park in 1941. The device was not useful, and Flowers was assigned to other problems.

Flowers began to work closely with Turing, Newman, and C. E. Wynn-Williams of the Telecommunications Research Establishment (TRE) to develop a machine that could break the Fish codes. The Dollis Hill team worked on the tape driving and reading problems, and Wynn-Williams's team at TRE worked on electronic counters and the necessary circuitry. Their efforts produced the "Heath Robinson," which could read two thousand characters per second. The Heath Robinson used vacuum tubes, an uncommon component in the early 1940's. The vacuum tubes performed more reliably and rapidly than the relays that had been used for counters. Heath Robinson and the companion machines proved that high-speed electronic devices could successfully do cryptoanalytic work (solve decoding problems).

Entirely automatic in operation once started, the Heath Robinson was put together at Bletchley Park in the spring of 1943. The Heath Robinson became obsolete for codebreaking shortly after it was put into use, so work began on a bigger, faster, and more powerful machine: the Colossus.

Flowers led the team that designed and built the Colossus in eleven months at Dollis Hill. The first Colossus (Mark I) was a bigger, faster version of the Heath Robinson and read about five thousand characters per second. Colossus had approximately fifteen hundred vacuum tubes, which was the largest number that had ever been used at that time. Although Turing and Wynn-Williams were not directly involved with the design of the Colossus, their previous work on the Heath Robinson was crucial to the project, since the first Colossus was based on the Heath Robinson.

Colossus became operational at Bletchley Park in December, 1943, and Flowers made arrangements for the manufacture of its components in case other machines were required. The request for additional machines came in March, 1944. The second Colossus, the Mark II, was extensively redesigned and was able to read twenty-five thousand characters per second because it was capable of performing parallel operations (carrying out several different operations at once, instead of one at a time); it also had a short-term memory. The Mark II went into operation on June 1, 1944. More machines were made, each with further modifications, until there were ten. The Colossus machines were special-purpose, program-controlled electronic digital computers, the only known electronic programmable computers in existence in 1944. The use of electronics allowed for a tremendous increase in the internal speed of the machine.

IMPACT

The Colossus machines gave Britain the best codebreaking machines of World War II and provided information that was crucial for the Allied victory. The information decoded by Colossus, the actual messages, and their influence on military decisions would remain classified for decades after the war.

The later work of several of the people involved with the Bletchley Park projects was important in British computer development after the war. Newman's and Turing's postwar careers were closely tied to emerging computer advances. Newman, who was interested in the impact of computers on mathematics, received a grant from the Royal Society in 1946 to establish a calculating machine laboratory at Manchester University. He was also involved with postwar computer growth in Britain.

Several other members of the Bletchley Park team, including Turing, joined Newman at Manchester in 1948. Before going to Manchester University, however, Turing joined Britain's National Physical Laboratory (NPL). At NPL, Turing worked on an advanced computer known as the Pilot Automatic Computing Engine (Pilot ACE). While at NPL, Turing proposed the concept of a stored program, which was a controversial but extremely important idea in computing. A "stored" program is one that remains in residence inside the computer, making it possible for a particular program and data to be fed through an input device simultaneously. (The Heath Robinson and Colossus machines were limited by utilizing separate input tapes, one for the program and one for the data to be analyzed.) Turing was among the first to explain the stored-program concept in print. He was also among the first to imagine how subroutines could be included in a program. (A subroutine allows separate tasks within a large program to be done in distinct modules; in effect, it is a detour within a program. After the completion of the subroutine, the main program takes control again.)

See also Apple II computer; Differential analyzer; ENIAC computer; IBM Model 1401 computer; Personal computer; Supercomputer; UNIVAC computer.

FURTHER READING

Carter, Frank. *Codebreaking with the Colossus Computer: Finding the K-Wheel Patterns—An Account of Some of the Techniques Used.* Milton Keynes, England: Bletchley Park Trust, 1997.

Gray, Paul. "Computer Scientist: Alan Turing." *Time* 153, no. 12 (March 29, 1999).

Hodges, Andrew. *Alan Turing: The Enigma.* New York: Walker, 2000.

Sale, Tony. *The Colossus Computer, 1943-1996: And How It Helped to Break the German Lorenz Cipher in World War II.* Cleobury Mortimer: M&M Baldwin, 1998.

COMMUNICATIONS SATELLITE

THE INVENTION: Telstar I, the world's first commercial communications satellite, opened the age of live, worldwide television by connecting the United States and Europe.

THE PEOPLE BEHIND THE INVENTION:
   Arthur C. Clarke (1917-   ), a British science-fiction writer who in 1945 first proposed the idea of using satellites as communications relays
   John R. Pierce (1910-   ), an American engineer who worked on the Echo and Telstar satellite communications projects

SCIENCE FICTION?

In 1945, Arthur C. Clarke suggested that a satellite orbiting high above the earth could relay television signals between different stations on the ground, making for a much wider range of transmission than that of the usual ground-based systems. Writing in the February, 1945, issue of Wireless World, Clarke said that satellites "could give television and microwave coverage to the entire planet."

In 1956, John R. Pierce at the Bell Telephone Laboratories of the American Telephone & Telegraph Company (AT&T) began to urge the development of communications satellites. He saw these satellites as a replacement for the ocean-bottom cables then being used to carry transatlantic telephone calls. In 1950, about one-and-a-half million transatlantic calls were made, and that number was expected to grow to three million by 1960, straining the capacity of the existing cables; in 1970, twenty-one million calls were made.

Communications satellites offered a good, cost-effective alternative to building more transatlantic telephone cables. On January 19, 1961, the Federal Communications Commission (FCC) gave permission for AT&T to begin Project Telstar, the first commercial communications satellite bridging the Atlantic Ocean. AT&T reached an agreement with the National Aeronautics and Space Administration (NASA) in July, 1961, in which AT&T would pay $3 million for each Telstar launch. The Telstar project involved about four hundred scientists, engineers, and technicians at the Bell Telephone Laboratories, twenty more technical personnel at AT&T headquarters, and the efforts of more than eight hundred other companies that provided equipment or services.

Telstar 1 was shaped like a faceted sphere, was 88 centimeters in diameter, and weighed 80 kilograms. Most of its exterior surface (sixty of the seventy-four facets) was covered by 3,600 solar cells to convert sunlight into 15 watts of electricity to power the satellite. Each solar cell was covered with artificial sapphire to reduce the damage caused by radiation. The main instrument was a two-way radio able to handle six hundred telephone calls at a time or one television channel.

The signal that the radio would send back to Earth was very weak—less than one-thirtieth the energy used by a household light bulb. Large ground antennas were needed to receive Telstar's faint signal. The main ground station was built by AT&T in Andover, Maine, on a hilltop informally called "Space Hill." A horn-shaped antenna, weighing 380 tons, with a length of 54 meters and an open end with an area of 1,097 square meters, was mounted so that it could rotate to track Telstar across the sky. To protect it from wind and weather, the antenna was built inside an inflated dome, 64 meters in diameter and 49 meters tall. It was, at the time, the largest inflatable structure ever built. A second, smaller horn antenna in Holmdel, New Jersey, was also used.

INTERNATIONAL COOPERATION

In February, 1961, the governments of the United States and England agreed to let the British Post Office and NASA work together to test experimental communications satellites. The British Post Office built a 26-meter-diameter steerable dish antenna of its own design at Goonhilly Downs, near Cornwall, England. Under a similar agreement, the French National Center for Telecommunications Studies constructed a ground station, almost identical to the Andover station, at Pleumeur-Bodou, Brittany, France.

After testing, Telstar 1 was moved to Cape Canaveral, Florida, and attached to the Thor-Delta launch vehicle built by the Douglas

Aircraft Company. The Thor-Delta was launched at 3:35 A.M. eastern standard time (EST) on July 10, 1962. Once in orbit, Telstar 1 took 157.8 minutes to circle the globe. The satellite came within range of the Andover station on its sixth orbit, and a television test pattern was transmitted to the satellite at 6:26 P.M. EST. At 6:30 P.M. EST, a tape-recorded black-and-white image of the American flag with the Andover station in the background, transmitted from Andover to Holmdel, opened the first television show ever broadcast by satellite. Live pictures of U.S. vice president Lyndon B. Johnson and other officials gathered at Carnegie Institution in Washington, D.C., followed on the AT&T program carried live on all three American networks.

Up to the moment of launch, it was uncertain if the French station would be completed in time to participate in the initial test. At 6:47 P.M. EST, however, Telstar's signal was picked up by the station in Pleumeur-Bodou, and Johnson's image became the first television transmission to cross the Atlantic. Pictures received at the French station were reported to be so clear that they looked like they had been sent from only forty kilometers away. Because of technical difficulties, the English station was unable to receive a clear signal.

The first formal exchange of programming between the United States and Europe occurred on July 23, 1962. This special eighteen-minute program, produced by the European Broadcasting Union, consisted of live scenes from major cities throughout Europe and was transmitted from Goonhilly Downs, where the technical difficulties had been corrected, to Andover via Telstar.

On the previous orbit, a program entitled "America, July 23, 1962," showing scenes from fifty television cameras around the United States, was beamed from Andover to Pleumeur-Bodou and seen by an estimated one hundred million viewers throughout Europe.

CONSEQUENCES

Telstar 1 and the communications satellites that followed it revolutionized the television news and sports industries. Before, television networks had to ship film across the oceans, meaning delays of hours or days between the time an event occurred and the broadcast of pictures of that event on television on another continent. Now, news of major significance, as well as sporting events, can be viewed live around the world. The impact on international relations also was significant, with world opinion becoming able to influence the actions of governments and individuals, since those actions could be seen around the world as the events were still in progress.

More powerful launch vehicles allowed new satellites to be placed in geosynchronous orbits, circling the earth at a speed the same as the earth's rotation rate. When viewed from the ground, these satellites appeared to remain stationary in the sky. This allowed continuous communications and greatly simplified the ground antenna system. By the late 1970's, private individuals had built small antennas in their backyards to receive television signals directly from the satellites.

See also Artificial satellite; Cruise missile; Rocket; Weather satellite.

FURTHER READING

McAleer, Neil. *Odyssey: The Authorised Biography of Arthur C. Clarke.* London: Victor Gollancz, 1992.
Pierce, John Robinson. *The Beginnings of Satellite Communications.* San Francisco: San Francisco Press, 1968.
_____. *Science, Art, and Communication.* New York: C. N. Potter, 1968.

COMMUNITY ANTENNA TELEVISION

THE INVENTION: A system for connecting households in isolated areas to common antennas to improve television reception, community antenna television was a forerunner of modern cable-television systems.

THE PEOPLE BEHIND THE INVENTION:
Robert J. Tarlton, the founder of CATV in eastern Pennsylvania
Ed Parsons, the founder of CATV in Oregon
Ted Turner (1938-   ), founder of the first cable superstation, WTBS

GROWING DEMAND FOR TELEVISION

Television broadcasting in the United States began in the late 1930's. After delays resulting from World War II, it exploded into the American public's consciousness. The new medium relied primarily on existing broadcasting stations that quickly converted from radio to television formats. Consequently, the reception of television signals was centralized in large cities. The demand for television quickly swept across the country. Ownership of television receivers increased dramatically, and those who could not afford their own flocked to businesses, usually taverns, or to the homes of friends with sets. People in urban areas had more opportunities to view the new medium and had the advantage of more broadcasts within the range of reception. Those in outlying regions were not so fortunate, as they struggled to see fuzzy pictures and were, in some cases, unable to receive a signal at all.

The situation for outlying areas worsened in 1948, when the Federal Communications Commission (FCC) implemented a ban on all new television stations while it considered how to expand the television market and how to deal with a controversy over color reception. This left areas without nearby stations in limbo, while people in areas with established stations reaped the benefits of new programming. The ban would remain in effect until 1952, when new stations came under construction across the country.

Poor reception in some areas and the FCC ban on new station construction together set the stage for the development of Community Antenna Television (CATV). CATV did not have a glamorous beginning. Late in 1949, two different men, frustrated by the slow movement of television to outlying areas, set up what would become the foundation of the multimillion-dollar cable industry.

Robert J. Tarlton was a radio salesman in Lansford, Pennsylvania, about sixty-five miles from Philadelphia. He wanted to move into television sales but lived in an area with poor reception. Together with friends, he founded Panther Valley Television and set up a master antenna in a mountain range that blocked the reception of Philadelphia-based broadcasting. For an installation fee of $125 and a fee of $3 per month, Panther Valley Television offered residents clear reception of the three Philadelphia stations via a coaxial cable wired to their homes. At the same time, Ed Parsons, of KAST radio in Astoria, Oregon, linked homes via coaxial cables to a master antenna set up to receive remote broadcasts. Both systems offered three channels, the major network affiliates, to subscribers. By 1952, when the FCC ban was lifted, some seventy CATV systems provided small and rural communities with the wonders of television. That same year, the National Cable Television Association was formed to represent the interests of the young industry.

Early systems could carry only one to three channels. In 1953, CATV began to use microwave relays, which could import distant signals to add more variety and pushed system capability to twelve channels. A system of towers began sprouting up across the country. These towers could relay a television signal from a powerful originating station to each cable system's main antenna. This further opened the reception available to subscribers.

PAY TELEVISION

The notion of pay television also began at this time. In 1951, the FCC authorized a test of Zenith Radio Corporation's Phonevision in Chicago. Scrambled images could be sent as electronic impulses over telephone lines, then unscrambled by devices placed in subscribers' homes. Subscribers could order a film over the telephone for a minimal cost, usually $1. Advertisers for the system promoted the idea of films for the "sick, aged, and sitterless." This early test was a forerunner of the premium, or pay, channels of later decades.

Network opposition to CATV came in the late 1950's. RCA chairman David Sarnoff warned against a pay television system that could soon fall under government regulation, as in the case of utilities. In April, 1959, the FCC found no basis for asserting jurisdiction or authority over CATV. This left the industry open to tremendous growth.

By 1960, the industry included 640 systems with 700,000 subscribers. Ten years later, 2,490 systems were in operation, serving more than 4.5 million households. This accelerated growth came at a price. In April, 1965, the FCC reversed itself and asserted authority over microwave-fed CATV. A year later, the entire cable system came under FCC control. The FCC quickly restricted the use of distant signals in the largest hundred markets.

The FCC movement to control cable systems stemmed from the agency's desire to balance the television market. From the onset of television broadcasting, the FCC strived to maintain a balanced programming schedule. The goal was to create local markets in which local affiliate stations prospered from advertising and other community support and would not be unduly harmed by competition from larger metropolitan stations. In addition, growth of the industry ideally was to be uniform, with large and small cities receiving equal consideration. Cable systems, particularly those that could receive distant signals via microwave relay, upset the balance. For example, a small Ohio town could receive New York channels as well as Chicago channels via cable, as opposed to receiving only the channels from one city.

The balance was further upset with the creation of a new communications satellite, COMSAT, in 1963. This technology allowed a signal to be sent to the satellite, retransmitted back to Earth, and then picked up by a receiving station. This further increased the range of cable offerings and moved the transmission of television signals to a national scale, as microwave-relayed transmissions worked best in a regional scope. These two factors led the FCC to freeze the cable industry from new development and construction in December, 1968. After 1972, when the cable freeze was lifted, the greatest impact of CATV would be felt.

TED TURNER

"The whole idea of grand things always turned me on," Ted Turner said in a 1978 *Playboy* magazine interview. Irrepressible, tenacious, and flamboyant, Turner was groomed from childhood for grandness.

Born Robert Edward Turner III in 1938 in Cincinnati, Ohio, he was raised by a harsh, demanding father who sent him to military preparatory schools and insisted he study business at Brown University instead of attending the U.S. Naval Academy, as the son wanted. Known as "Terrible Ted" in school for his high-energy, maverick ways, he became an champion debater, expert sailor, and natural leader. When the Turner Advertising Company failed in 1960, and his father committed suicide, young Turner took it over and parlayed it into an empire, acquiring or creating television stations and revolutionizing how they were broadcast to Americans.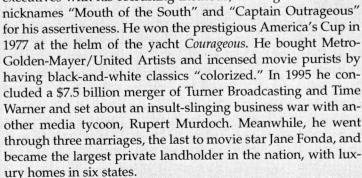

(George Bennett)

From then on he acquired, innovated, and, often, shocked. He bought the Atlanta Braves baseball team and Hawks basketball team, often angering sports executives with his recruiting methods, earning the nicknames "Mouth of the South" and "Captain Outrageous" for his assertiveness. He won the prestigious America's Cup in 1977 at the helm of the yacht *Courageous*. He bought Metro-Golden-Mayer/United Artists and incensed movie purists by having black-and-white classics "colorized." In 1995 he concluded a $7.5 billion merger of Turner Broadcasting and Time Warner and set about an insult-slinging business war with another media tycoon, Rupert Murdoch. Meanwhile, he went through three marriages, the last to movie star Jane Fonda, and became the largest private landholder in the nation, with luxury homes in six states.

However, Turner's life was not all acquisition. He started a charitable foundation and sponsored the Olympics-like Goodwill Games between the United States and the Soviet Union to improve relations, for which *Time* magazine named him its man of the year in 1991. However, Turner's grandest shocker came in 1997 when he promised to donate $1 billion—$100 million each year for a decade—to the United Nations to help in feeding the poor, resettling refugees, and eradicating land mines. And he publicly challenged other super-rich people to use their vast wealth similarly.

IMPACT

The founding of cable television had a two-tier effect on the American public. The immediate impact of CATV was the opening of television to areas cut off from network broadcasting as a result of distance or topographical obstructions. Cable brought television to those who would have otherwise missed the early years of the medium.

As technology furthered the capabilities of the industry, a second impact emerged. Along with the 1972 lifting of the ban on cable expansion, the FCC established strict guidelines for the advancement of the industry. Issuing a 500-page blueprint for the expansion of cable, the FCC included limits on the use of imported distant signals, required the blacking out of some specific programs (films and serials, for example), and limited pay cable to films that were more than two years old and to sports.

Another component of the guidelines required all systems that went into operation after March, 1972 (and all systems by March, 1977), to provide public access channels for education and local government. In addition, channels were to be made available for lease. These access channels opened information to subscribers that would not normally be available. Local governments and school boards began to broadcast meetings, and even high school athletics soon appeared via public access channels. These channels also provided space to local educational institutions for home-based courses in a variety of disciplines.

CABLE COMMUNICATIONS POLICY ACT

Further FCC involvement came in the 1984 Cable Communications Policy Act, which deregulated the industry and opened the door for more expansion. This act removed local control over cable service rates and virtually made monopolies out of local providers by limiting competition. The late 1980's brought a new technology, fiber optics, which promised to further advance the industry by increasing the quality of cable services and channel availability.

One area of the cable industry, pay television, took off in the 1970's and early 1980's. The first major pay channel was developed by the media giant Time-Life. It inaugurated Home Box Office (HBO) in 1975 as the first national satellite interconnected network. Early HBO programming primarily featured films but included no films less than two years old (meeting the 1972 FCC guidelines), no serials, and no advertisements. Other premium movie channels followed, including Showtime, Cinemax, and The Movie Channel. By the late 1970's, cable systems offered multiple premium channels to their subscribers.

Superstations were another component of the cable industry that boomed in the 1970's and 1980's. The first, WTBS, was owned and operated by Ted Turner and broadcast from Atlanta, Georgia. It emphasized films and reruns of old television series. Cable systems that broadcast WTBS were asked to allocate the signal to channel 17, thus creating uniformity across the country for the superstation. Chicago's WGN and New York City's WOR soon followed, gaining access to homes across the nation via cable. Both these superstations emphasized sporting events in the early years and expanded to include films and other entertainment in the 1980's.

Both pay channels and superstations transmitted via satellites (WTBS leased space from RCA, for example) and were picked up by cable systems across the country. Other stations with broadcasts intended solely for the cable industry opened in the 1980's. Ted Turner started the Cable News Network in 1980 and followed with the all-news network Headline News. He added another channel with the Turner Network Television (TNT) in 1988. Other 1980's additions included The Disney Channel, ESPN, The Entertainment Channel, The Discovery Channel, and Lifetime. The Cable-Satellite Public Affairs Network (C-SPAN) enhanced the cable industry's presence in Washington, D.C., by broadcasting sessions of the House of Representatives.

Specialized networks for particular audiences also developed. Music Television (MTV), featuring songs played along with video sequences, premiered in 1981. Nickelodeon, a children's channel, and VH-1, a music channel aimed at baby boomers rather than MTV's teenage audience, reflected the movement toward specialization. Other specialized channels, such as the Sci-Fi Channel and the Comedy Channel, went even further in targeting specific audiences.

CABLE AND THE PUBLIC

The impact on the American public was tremendous. Information and entertainment became available around the clock. Cable provided a new level of service, information, and entertainment unavailable to nonsubscribers. One phenomenon that exploded in the late 1980's was home shopping. Via The Home Shopping Club and QVC, two shopping channels offered through cable television, the American public could order a full range of products. Everything from jewelry to tools and home cleaning supplies to clothing and electronics was available to anyone with a credit card. Americans could now go shopping from home.

The cable industry was not without its competitors and critics. In the 1980's, the videocassette recorder (VCR) opened the viewing market. Prerecorded cassettes of recent film releases as well as classics were made available for purchase or for a small rental fee. National chains of video rental outlets, such as Blockbuster Video and Video Towne, offered thousands of titles for rent. Libraries also began to stock films. This created competition for the cable industry, in particular the premium movie channels. To combat this competition, channels began to offer original productions unavailable on videocassette. The combined effect of the cable industry and the videocassette market was devastating to the motion picture industry. The wide variety of programming available at home encouraged the American public, especially baby boomers with children, to stay home and watch cable or rented films instead of going to theaters.

Critics of the cable industry seized on the violence, sexual content, and graphic language found in some of cable's offerings. One parent responded by developing a lockout device that could make certain channels unavailable to children. Some premium channels developed an after-hours programming schedule that aired adult-theme programming only late at night. Another criticism stemmed from the repetition common on pay channels. As a result of the limited supply of and large demand for films, pay channels were forced to repeat programs several times within a month and to rebroadcast films that were several years old. This led consumers to question the value of the additional monthly fee paid for such channels. To combat the problem, premium channels increased efforts aimed at original production and added more films that had not been box-office hits.

By the early 1990's, as some eleven thousand cable systems were serving 56.2 million subscribers, a new cry for regulation began. Debates over services and increasingly high rates led the FCC and Congress to investigate the industry, opening the door for new guidelines on the cable industry. The non-cable networks—American Broadcasting Company (ABC), Columbia Broadcasting System (CBS), National Broadcasting Company (NBC), and newcomer Fox—stressed their concerns about the cable industry. These networks provided free programming, and cable systems profited from inclusion of network programming. Television industry representatives expressed the opinion that cable providers should pay for the privilege of retransmitting network broadcasts.

The impact on cable's subscribers, especially concerning monthly cable rates, came under heavy debate in public and government forums. The administration in Washington, D.C., expressed concern that cable rates had risen too quickly and for no obvious reason other than profit-seeking by what were essentially monopolistic local cable systems. What was clear was that the cable industry had transformed the television experience and was going to remain a powerful force within the medium. Regulators and television industry leaders were left to determine how to maintain an equitable coexistence within the medium.

See also Color television; Communications satellite; Fiber-optics; Telephone switching; Television.

FURTHER READING

Baldwin, Thomas F., and D. Stevens McVoy. *Cable Communication.* Englewood Cliffs, N.J.: Prentice-Hall, 1983.
Brenner, Daniel L., and Monroe E. Price. *Cable Television and Other Nonbroadcast Video: Law and Policy.* New York: Clark Boardman, 1986.
Burns, R. W. *Television: An International History of the Formative Years.* London: Institution of Electrical Engineers in Association with the Science Museum, 1998.

Coleman, Wim. *The Age of Broadcasting: Television.* Carlisle, Mass.: Discovery Enterprises, 1997.

Negrine, Ralph M., ed. *Cable Television and the Future of Broadcasting.* New York: St. Martin's Press, 1985.

Sconce, Jeffrey. *Haunted Media: Electronic Presence from Telegraphy to Television.* Durham, N.C.: Duke University Press, 2000.

Whittemore, Hank. *CNN: The Inside Story.* Boston: Little, Brown, 1990.

COMPACT DISC

THE INVENTION: A plastic disk on which digitized music or computer data is stored.

THE PEOPLE BEHIND THE INVENTION:

Akio Morita (1921-   ), a Japanese physicist and engineer who was a cofounder of Sony

Wisse Dekker (1924-   ), a Dutch businessman who led the Philips company

W. R. Bennett (1904-1983), an American engineer who was a pioneer in digital communications and who played an important part in the Bell Laboratories research program

DIGITAL RECORDING

The digital system of sound recording, like the analog methods that preceded it, was developed by the telephone companies to improve the quality and speed of telephone transmissions. The system of electrical recording introduced by Bell Laboratories in the 1920s was part of this effort. Even Edison's famous invention of the phonograph in 1877 was originally conceived as an accompaniment to the telephone. Although developed within the framework of telephone communications, these innovations found wide applications in the entertainment industry.

The basis of the digital recording system was a technique of sampling the electrical waveforms of sound called PCM, or pulse code modulation. PCM measures the characteristics of these waves and converts them into numbers. This technique was developed at Bell Laboratories in the 1930's to transmit speech. At the end of World War II, engineers of the Bell System began to adapt PCM technology for ordinary telephone communications.

The problem of turning sound waves into numbers was that of finding a method that could quickly and reliably manipulate millions of them. The answer to this problem was found in electronic computers, which used binary code to handle millions of computations in a few seconds. The rapid advance of computer technology and the semiconductor circuits that gave computers the power to handle complex calculations provided the means to bring digital sound technology into commercial use. In the 1960's, digital transmission and switching systems were introduced to the telephone network.

Pulse coded modulation of audio signals into digital code achieved standards of reproduction that exceeded even the best analog system, creating an enormous dynamic range of sounds with no distortion or background noise. The importance of digital recording went beyond the transmission of sound because it could be applied to all types of magnetic recording in which the source signal is transformed into an electric current. There were numerous commercial applications for such a system, and several companies began to explore the possibilities of digital recording in the 1970's.

Researchers at the Sony, Matsushita, and Mitsubishi electronics companies in Japan produced experimental digital recording systems. Each developed its own PCM processor, an integrated circuit that changes audio signals into digital code. It does not continuously transform sound but instead samples it by analyzing thousands of minute slices of it per second. Sony's PCM-F1 was the first analog-to-digital conversion chip to be produced. This gave Sony a lead in the research into and development of digital recording.

All three companies had strong interests in both audio and video electronics equipment and saw digital recording as a key technology because it could deal with both types of information simultaneously. They devised recorders for use in their manufacturing operations. After using PCM techniques to turn sound into digital code, they recorded this information onto tape, using not magnetic audio tape but the more advanced video tape, which could handle much more information. The experiments with digital recording occurred simultaneously with the accelerated development of video recording technology and owed much to the enhanced capabilities of video recorders. At this time, videocassette recorders were being developed in several corporate laboratories in Japan and Europe. The Sony Corporation was one of the companies developing video recorders at this time. Its U-matic machines were successfully used to record digitally. In 1972, the Nippon Columbia Company began to make its master recordings digitally on an Ampex video recording machine.

Links Among New Technologies

There were powerful links between the new sound recording systems and the emerging technologies of storing and retrieving video images. The television had proved to be the most widely used and profitable electronic product of the 1950's, but with the market for color television saturated by the end of the 1960's, manufacturers had to look for a replacement product. A machine to save and replay television images was seen as the ideal companion to the family TV set. The great consumer electronics companies—General Electric and RCA in the United States, Philips and Telefunken in Europe, and Sony and Matsushita in Japan—began experimental programs to find a way to save video images.

RCA's experimental teams took the lead in developing an optical videodisc system, called Selectavision, that used an electronic stylus to read changes in capacitance on the disc. The greatest challenge to them came from the Philips company of Holland. Its optical videodisc used a laser beam to read information on a revolving disc, in which a layer of plastic contained coded information. With the aid of the engineering department of the Deutsche Grammophon record company, Philips had an experimental laser disc in hand by 1964.

The Philips Laservision videodisc was not a commercial success, but it carried forward an important idea. The research and engineering work carried out in the laboratories at Eindhoven in Holland proved that the laser reader could do the job. More important, Philips engineers had found that this fragile device could be mass produced as a cheap and reliable component of a commercial product. The laser optical decoder was applied to reading the binary codes of digital sound. By the end of the 1970's, Philips engineers had produced a working system.

Ten years of experimental work on the Laservision system proved to be a valuable investment for the Philips corporation. Around 1979, it started to work on a digital audio disc (DAD) playback system. This involved more than the basic idea of converting the output of the PCM conversion chip onto a disc. The lines of pits on the compact disc carry a great amount of information: the left- and right-hand tracks of the stereo system are identified, and a sequence of pits also controls the motor speed and corrects any error in the laser reading of the binary codes.

This research was carried out jointly with the Sony Corporation of Japan, which had produced a superior method of encoding digital sound with its PCM chips. The binary codes that carried the information were manipulated by Sony's sixteen-bit microprocessor. Its PCM chip for analog-to-digital conversion was also employed. Together, Philips and Sony produced a commercial digital playback record that they named the compact disc. The name is significant, as it does more than indicate the size of the disc—it indicates family ties with the highly successful compact cassette. Philips and Sony had already worked to establish this standard in the magnetic tape format and aimed to make their compact disc the standard for digital sound reproduction.

Philips and Sony began to demonstrate their compact digital disc (CD) system to representatives of the audio industry in 1981. They were not alone in digital recording. The Japanese Victor Company, a subsidiary of Matsushita, had developed a version of digital recording from its VHD video disc design. It was called audio high density disc (AHD). Instead of the small CD disc, the AHD system used a ten-inch vinyl disc. Each digital recording system used a different PCM chip with a different rate of sampling the audio signal.

The recording and electronics industries' decision to standardize on the Philips/Sony CD system was therefore a major victory for these companies and an important event in the digital era of sound recording. Sony had found out the hard way that the technical performance of an innova-

Although not much larger than a 3.25-inch floppy disk, a compact disk can store more than five hundred times as much data. (PhotoDisc)

tion is irrelevant when compared with the politics of turning it into an industrywide standard. Although the pioneer in videocassette recorders, Sony had been beaten by its rival, Matsushita, in establishing the video recording standard. This mistake was not repeated in the digital standards negotiations, and many companies were persuaded to license the new technology. In 1982, the technology was announced to the public. The following year, the compact disc was on the market.

THE APEX OF SOUND TECHNOLOGY

The compact disc represented the apex of recorded sound technology. Simply put, here at last was a system of recording in which there was no extraneous noise—no surface noise of scratches and pops, no tape hiss, no background hum—and no damage was done to the recording as it was played. In principle, a digital recording will last forever, and each play will sound as pure as the first. The compact disc could also play much longer than the vinyl record or long-playing cassette tape.

Despite these obvious technical advantages, the commercial success of digital recording was not ensured. There had been several other advanced systems that had not fared well in the marketplace, and the conspicuous failure of quadrophonic sound in the 1970's had not been forgotten within the industry of recorded sound. Historically, there were two key factors in the rapid acceptance of a new system of sound recording and reproduction: a library of prerecorded music to tempt the listener into adopting the system and a continual decrease in the price of the playing units to bring them within the budgets of more buyers.

By 1984, there were about a thousand titles available on compact disc in the United States; that number had doubled by 1985. Although many of these selections were classical music—it was naturally assumed that audiophiles would be the first to buy digital equipment—popular music was well represented. The first CD available for purchase was an album by popular entertainer Billy Joel.

The first CD-playing units cost more than $1,000, but Akio Morita of Sony was determined that the company should reduce the price of players even if it meant selling them below cost. Sony's

Akio Morita

Akio Morita was born in Nagoya, Japan, in 1921 into a family owning one of the country's oldest, most prosperous sake breweries. As the eldest son, Morita was expected to take over its management from his father. However, business did not interest him as a child. Electronics did, especially radios. He made his own radio and phonograph and resolved to be a scientist. He succeeded, but in an ironic twist, he also became one of the twentieth century's most successful businessmen.

After taking a degree in physics from Osaka Imperial University in 1944, he worked at the Naval Research Center. There he met Masaru Ibuka. Although Ibuka was twelve years older and much more reserved in temperament, the two became fast friends. After World War II, they borrowed the equivalent of about $500 from Morita's father and opened the Tokyo Telecommunications Company, making voltmeters and, later, tape recorders.

To help along sluggish sales, Morita visited local schools to demonstrate the tape recorder's usefulness in teaching. He was so successful that a third of Japan's elementary schools bought them. From then on, Morita, as vice president of the company, was the lead man in marketing and sales strategy. He bought rights from West Electric Company to manufacture transistors in 1954, and soon the company was turning out transistor radios. Sales soared. They changed the name to Sony (based on the Latin word for sound, *sonus*) because it was more memorable.

Despite an American bias against Japanese products—which many Americans regarded as shoddy imitations—Morita launched Sony America in 1960. In 1963 Sony became the first Japanese company to sell its stock in America and in 1970 the first to be listed on the New York Stock Exchange, opening an American factory two years later. Morita became president of Sony Corporation in 1971 and board chairman in 1976.

In 1984 Sony earnings exceeded $5 billion, a ten-million percent increase in worth in less than forty years. As important for Japanese industry and national honor, Morita and Sony moved Japanese electronics into leading edge of technical sophistication and craftsmanship.

audio engineering department improved the performance of the players while reducing size and cost. By 1984, Sony had a small CD unit on the market for $300. Several of Sony's competitors, including Matsushita, had followed its lead into digital reproduction. There were several compact disc players available in 1985 that cost less than $500. Sony quickly applied digital technology to the popular personal stereo and to automobile sound systems. Sales of CD units increased roughly tenfold from 1983 to 1985.

IMPACT ON VINYL RECORDING

When the compact disc was announced in 1982, the vinyl record was the leading form of recorded sound, with 273 million units sold annually compared to 125 million prerecorded cassette tapes. The compact disc sold slowly, beginning with 800,000 units shipped in 1983 and rising to 53 million in 1986. By that time, the cassette tape had taken the lead, with slightly fewer than 350 million units. The vinyl record was in decline, with only about 110 million units shipped. Compact discs first outsold vinyl records in 1988. In the ten years from 1979 to 1988, the sales of vinyl records dropped nearly 80 percent. In 1989, CDs accounted for more than 286 million sales, but cassettes still led the field with total sales of 446 million. The compact disc finally passed the cassette in total sales in 1992, when more than 300 million CDs were shipped, an increase of 22 percent over the figure for 1991.

The introduction of digital recording had an invigorating effect on the industry of recorded sound, which had been unable to fully recover from the slump of the late 1970's. Sales of recorded music had stagnated in the early 1980's, and an industry accustomed to steady increases in output became eager to find a new product or style of music to boost its sales. The compact disc was the product to revitalize the market for both recordings and players. During the 1980's, worldwide sales of recorded music jumped from $12 billion to $22 billion, with about half of the sales volume accounted for by digital recordings by the end of the decade.

The success of digital recording served in the long run to undermine the commercial viability of the compact disc. This was a play-only technology, like the vinyl record before it. Once users had become accustomed to the pristine digital sound, they clamored for digital recording capability. The alliance of Sony and Philips broke down in the search for a digital tape technology for home use. Sony produced a digital tape system called DAT, while Philips responded with a digital version of its compact audio tape called DCC. Sony answered the challenge of DCC with its Mini Disc (MD) product, which can record and replay digitally.

The versatility of digital recording has opened up a wide range of consumer products. Compact disc technology has been incorporated into the computer, in which CD-ROM readers convert the digital code of the disc into sound and images. Many home computers have the capability to record and replay sound digitally. Digital recording is the basis for interactive audio/video computer programs in which the user can interface with recorded sound and images. Philips has established a strong foothold in interactive digital technology with its CD-I (compact disc interactive) system, which was introduced in 1990. This acts as a multimedia entertainer, providing sound, moving images, games, and interactive sound and image publications such as encyclopedias. The future of digital recording will be broad-based systems that can record and replay a wide variety of sounds and images and that can be manipulated by users of home computers.

See also Cassette recording; Dolby noise reduction; Electronic synthesizer; FM radio; Laser-diode recording process; Optical disk; Transistor; Videocassette recorder; Walkman cassette player.

FURTHER READING

Copeland, Peter. *Sound Recordings*. London: British Library, 1991.
Heerding, A. *A Company of Many Parts*. Cambridge: Cambridge University Press, 1998.
Marshall, David V. *Akio Morita and Sony*. Watford: Exley, 1995.
Morita, Akio, with Edwin M. Reingold, and Mitsuko Shimomura. *Made in Japan: Akio Morita and Sony*. London: HarperCollins, 1994.
Nathan, John. *Sony: The Private Life*. Boston, Mass.: Houghton Mifflin, 1999.
Schlender, Brenton R. "How Sony Keeps the Magic Going." *Fortune* 125 (February 24, 1992).

Compressed-Air-Accumulating Power Plant

The invention: Plants that can be used to store energy in the form of compressed air when electric power demand is low and use it to produce energy when power demand is high.

The organization behind the invention:
Nordwestdeutsche Kraftwerke, a Germany company

Power, Energy Storage, and Compressed Air

Energy, which can be defined as the capacity to do work, is essential to all aspects of modern life. One familiar kind of energy, which is produced in huge amounts by power companies, is electrical energy, or electricity. Most electricity is produced in a process that consists of two steps. First, a fossil fuel such as coal is burned and the resulting heat is used to make steam. Then, the steam is used to operate a turbine system that produces electricity. Electricity has myriad applications, including the operation of heaters, home appliances of many kinds, industrial machinery, computers, and artificial illumination systems.

An essential feature of electricity manufacture is the production of the particular amount of electricity that is needed at a given time. If moment-to-moment energy requirements are not met, the city or locality involved will experience a "blackout," the most obvious feature of which is the loss of electrical lighting. To prevent blackouts, it is essential to store extra electricity at times when power production exceeds power demands. Then, when power demands exceed the capacity to make energy by normal means, stored energy can be used to make up the difference.

One successful modern procedure for such storage is the compressed-air-accumulation process, pioneered by the Nordwestdeutsche Kraftwerke company's compressed-air-accumulating power plant, which opened in December, 1978. The plant, which is located in Huntorf, Germany (at the time, West Germany), makes compressed air during periods of low electricity demand, stores the air in an underground cavern, and uses it to produce extra electricity during periods of high demand.

PLANT OPERATION AND COMPONENTS

The German 300-megawatt compressed-air-accumulating power plant in Huntorf produces extra electricity from stored compressed air that will provide up to four hours per day of local peak electricity needs. The energy-storage process, which is vital to meeting very high peak electric power demands, is viable for electric power plants whose total usual electric outputs range from 25 megawatts to the 300 megawatts produced at Huntorf. It has been suggested, however, that the process is most suitable for 25- to 50-megawatt plants.

The energy-storage procedure used at Huntorf is quite simple. All the surplus electricity that is made in nonpeak-demand periods is utilized to drive an air compressor. The compressor pumps air from the surrounding atmosphere into an airtight underground storage cavern. When extra electricity is required, the stored compressed air is released and passed through a heating unit to be warmed, after which it is used to run gas-turbine systems that produce electricity. This sequence of events is the same as that used in any gas-turbine generating system; the only difference is that the compressed air can be stored for any desired period of time rather than having to be used immediately.

One requirement of any compressed-air-accumulating power plant is an underground storage chamber. The Huntorf plant utilizes a cavern that was hollowed out some 450 meters below the surface of the earth. The cavern was created by drilling a hole into an underground salt deposit and pumping in water. The water dissolved the salt, and the resultant saltwater solution (brine) was pumped out of the deposit. The process of pumping in water and removing brine was continued until the cavern reached the desired size. This type of storage cavern is virtually leak-free. The preparation of such underwater salt-dome caverns has been performed roughly since the middle of the twentieth century. Until the Huntorf endeavor, such caves were used to stockpile petroleum and natural gas for later use. It is also possible to use mined, hard-rock caverns

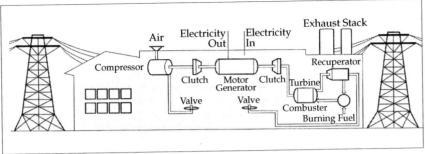

Schematic of a compressed-air-accumulating power plant.

for compressed-air accumulation when it is necessary to compress air to pressures higher than those that can be maintained effectively in a salt-dome cavern.

The essential machinery that must be added to conventional power plants to turn them into compressed-air-accumulating power plants are motor-driven air compressors and gas turbine generating systems. This equipment must be connected appropriately so that in the storage mode, the overall system will compress air for storage in the underground cavern, and in the power-production mode, the system will produce electricity from the stored compressed air.

Large compressed-air-accumulating power plants require specially constructed machinery. For example, the compressors that are used at Huntorf were developed specifically for that plant by Sulzer, a Swiss company. When the capacity of such plants is no higher than 50 megawatts, however, standard, readily available components can be used. This means that relatively small compressed-air-accumulating power plants can be constructed for a reasonable cost.

CONSEQUENCES

The development of compressed-air-accumulating power plants has had a significant impact on the electric power industry, adding to its capacity to store energy. The main storage methods available prior to the development of compressed-air-accumulation methodology were batteries and water that was pumped uphill (hydro-storage). Battery technology is expensive, and its capacity is insufficient for major, long-term power storage. Hydro-storage is a more viable technology.

Compressed-air energy-storage systems have several advantages over hydro-storage. First, they can be used in areas where flat terrain makes it impossible to use hydro-storage. Second, compressed-air storage is more efficient than hydro-storage. Finally, the fact that standard plant components can be used, along with several other factors, means that 25- to 50-megawatt compressed-air storage plants can be constructed much more quickly and cheaply than comparable hydro-storage plants.

The attractiveness of compressed-air-accumulating power plants has motivated efforts to develop hard-rock cavern construction techniques that cut costs and make it possible to use high-pressure air storage. In addition, aquifers (underground strata of porous rock that normally hold groundwater) have been used successfully for compressed-air storage. It is expected that compressed-air-accumulating power plants will be widely used in the future, which will help to decrease pollution and cut the use of fossil fuels.

See also Alkaline storage battery; Breeder reactor; Fuel cell; Geothermal power; Heat pump; Nuclear power plant; Tidal power plant.

FURTHER READING

"Compressed Air Stores Electricity." *Popular Science* 242, no. 5 (May, 1993).
Lee, Daehee. "Power to Spare: Compressed Air Energy Storage." *Mechanical Engineering* 113, no. 7 (July, 1991).
Shepard, Sam, and Septimus van der Linden. "Compressed Air Energy Storage Adapts Proven Technology to Address Market Opportunities." *Power Engineering* 105, no. 4 (April, 2001).
Zink, John C. "Who Says You Can't Store Electricity?" *Power Engineering* 101, no. 3 (March, 1997).

COMPUTER CHIPS

THE INVENTION: Also known as a microprocessor, a computer chip combines the basic logic circuits of a computer on a single silicon chip.

THE PEOPLE BEHIND THE INVENTION:

Robert Norton Noyce (1927-1990), an American physicist

William Shockley (1910-1989), an American coinventor of the transistor who was a cowinner of the 1956 Nobel Prize in Physics

Marcian Edward Hoff, Jr. (1937-   ), an American engineer

Jack St. Clair Kilby (1923-   ), an American researcher and assistant vice president of Texas Instruments

THE SHOCKLEY EIGHT

The microelectronics industry began shortly after World War II with the invention of the transistor. While radar was being developed during the war, it was discovered that certain crystalline substances, such as germanium and silicon, possess unique electrical properties that make them excellent signal detectors. This class of materials became known as "semiconductors," because they are neither conductors nor insulators of electricity.

Immediately after the war, scientists at Bell Telephone Laboratories began to conduct research on semiconductors in the hope that they might yield some benefits for communications. The Bell physicists learned to control the electrical properties of semiconductor crystals by "doping" (treating) them with minute impurities. When two thin wires for current were attached to this material, a crude device was obtained that could amplify the voice. The transistor, as this device was called, was developed late in 1947. The transistor duplicated many functions of vacuum tubes; it was also smaller, required less power, and generated less heat. The three Bell Laboratories scientists who guided its development—William Shockley, Walter H. Brattain, and John Bardeen—won the 1956 Nobel Prize in Physics for their work.

Shockley left Bell Laboratories and went to Palo Alto, California, where he formed his own company, Shockley Semiconductor Laboratories, which was a subsidiary of Beckman Instruments. Palo Alto is the home of Stanford University, which, in 1954, set aside 655 acres of land for a high-technology industrial area known as Stanford Research Park. One of the first small companies to lease a site there was Hewlett-Packard. Many others followed, and the surrounding area of Santa Clara County gave rise in the 1960's and 1970's to a booming community of electronics firms that became known as "Silicon Valley." On the strength of his prestige, Shockley recruited eight young scientists from the eastern United States to work for him. One was Robert Norton Noyce, an Iowa-bred physicist with a doctorate from the Massachusetts Institute of Technology. Noyce came to Shockley's company in 1956.

The "Shockley Eight," as they became known in the industry, soon found themselves at odds with their boss over issues of research and development. Seven of the dissenting scientists negotiated with industrialist Sherman Fairchild, and they convinced the remaining holdout, Noyce, to join them as their leader. The Shock-

Despite their tiny size, individual computer chips contain the basic logic circuits of entire computers. (PhotoDisc)

Jack St. Clair Kilby

Maybe the original, deepest inspiration for the integrated circuit chip was topographical: As a boy Jack Kilby (b.1923) often accompanied his father, an electrical engineer, on trips over the circuit of roads through his flat home state, Kansas.

In any case, he learned to love things electrical, and radios especially, from his father. Young Kilby had just started studying at the University of Illinois on his way to a degree in electrical engineering, when World War II started. He joined the Office of Strategic Services (OSS), which sent him into Japanese-occupied territory to train local freedom fighters. He found the radios given to him to be heavy and unreliable, so he got hold of components on his own and built better, smaller radios.

The "better, smaller" theme stayed with him. His first job out of college was with Centralab in Milwaukee, Wisconsin, where he designed ever smaller circuits. However, the bulky, hot vacuum tubes then in use limited miniaturization. In 1952, Centralab and Kilby eagerly incorporated the newly invented transistors into their designs. Kilby found, however, that all the electrical connections needed to hook up transistors and wires in a complex circuit also limited miniaturization.

He moved to Texas Instruments in 1958. The company was working on a modular approach to miniaturization with snap-together standardized parts. Kilby had a better idea: place everything for a specific circuit on a chip of silicon. Along with many other inventors, Kilby was soon looking for ways to put this new integrated circuit to work. He experimented with their use in computers and in generating solar power. He helped to develop the first hand-held calculator. Soon integrated circuits were in practically every electronic gadget, so that by the year 2000 his invention supported an electronic equipment industry that earned more than a trillion dollars a year.

Among his many awards, Kilby shared the 2000 Nobel Prize in Physics with Zhores I. Alferov and Herbert Kroemer, both of whom also miniaturized electronics.

ley Eight defected in 1957 to form a new company, Fairchild Semiconductor, in nearby Mountain View, California. Shockley's company, which never recovered from the loss of these scientists, soon went out of business.

INTEGRATING CIRCUITS

Research efforts at Fairchild Semiconductor and Texas Instruments, in Dallas, Texas, focused on putting several transistors on one piece, or "chip," of silicon. The first step involved making miniaturized electrical circuits. Jack St. Clair Kilby, a researcher at Texas Instruments, succeeded in making a circuit on a chip that consisted of tiny resistors, transistors, and capacitors, all of which were connected with gold wires. He and his company filed for a patent on this "integrated circuit" in February, 1959. Noyce and his associates at Fairchild Semiconductor followed in July of that year with an integrated circuit manufactured by means of a "planar process," which involved laying down several layers of semiconductor that were isolated by layers of insulating material. Although Kilby and Noyce are generally recognized as coinventors of the integrated circuit, Kilby alone received a membership in the National Inventors Hall of Fame for his efforts.

CONSEQUENCES

By 1968, Fairchild Semiconductor had grown to a point at which many of its key Silicon Valley managers had major philosophical differences with the East Coast management of their parent company. This led to a major exodus of top-level management and engineers. Many started their own companies. Noyce, Gordon E. Moore, and Andrew Grove left Fairchild to form a new company in Santa Clara called Intel with $2 million that had been provided by venture capitalist Arthur Rock. Intel's main business was the manufacture of computer memory integrated circuit chips. By 1970, Intel was able to develop and bring to market a random-access memory (RAM) chip that was subsequently purchased in large quantities by several major computer manufacturers, providing large profits for Intel.

In 1969, Marcian Edward Hoff, Jr., an Intel research and development engineer, met with engineers from Busicom, a Japanese firm. These engineers wanted Intel to design a set of integrated circuits for Busicom's desktop calculators, but Hoff told them their specifications were too complex. Nevertheless, Hoff began to think about the possi-

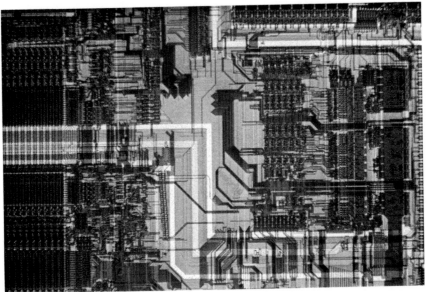

Circuitry of a typical computer chip. (PhotoDisc)

bility of incorporating all the logic circuits of a computer central processing unit (CPU) into one chip. He began to design a chip called a "microprocessor," which, when combined with a chip that would hold a program and one that would hold data, would become a small, general-purpose computer. Noyce encouraged Hoff and his associates to continue his work on the microprocessor, and Busicom contracted with Intel to produce the chip. Frederico Faggin, who was hired from Fairchild, did the chip layout and circuit drawings.

In January, 1971, the Intel team finished its first working microprocessor, the 4004. The following year, Intel made a higher-capacity microprocessor, the 8008, for Computer Terminals Corporation. That company contracted with Texas Instruments to produce a chip with the same specifications as the 8008, which was produced in June, 1972. Other manufacturers soon produced their own microprocessors.

The Intel microprocessor became the most widely used computer chip in the budding personal computer industry and may take significant credit for the PC "revolution" that soon followed. Microprocessors have become so common that people use them every day without realizing it. In addition to being used in computers, the microprocessor has found its way into automobiles, microwave ovens, wristwatches, telephones, and many other ordinary items.

See also Bubble memory; Floppy disk; Hard disk; Optical disk; Personal computer; Virtual machine.

FURTHER READING

Ceruzzi, Paul E. *A History of Modern Computing*. Cambridge, Mass.: MIT Press, 2000.

Reid, T. R. *The Chip: How Two Americans Invented the Microchip and Launched a Revolution*. New York: Random House, 2001.

Slater, Robert. *Portraits in Silicon*. Cambridge, Mass.: MIT Press, 1987.

CONTACT LENSES

THE INVENTION: Small plastic devices that fit under the eyelids, contact lenses, or "contacts," frequently replace the more familiar eyeglasses that many people wear to correct vision problems.

THE PEOPLE BEHIND THE INVENTION:

Leonardo da Vinci (1452-1519), an Italian artist and scientist
Adolf Eugen Fick (1829-1901), a German glassblower
Kevin Tuohy, an American optician
Otto Wichterle (1913-   ), a Czech chemist
William Feinbloom (1904-1985), an American optometrist

An Old Idea

There are two main types of contact lenses: hard and soft. Both types are made of synthetic polymers (plastics). The basic concept of the contact lens was conceived by Leonardo da Vinci in 1508. He proposed that vision could be improved if small glass ampules filled with water were placed in front of each eye. Nothing came of the idea until glass scleral lenses were invented by the German glassblower Adolf Fick. Fick's large, heavy lenses covered the pupil of the eye, its colored iris, and part of the sclera (the white of the eye). Fick's lenses were not useful, since they were painful to wear.

In the mid-1930's, however, plastic scleral lenses were developed by various organizations and people, including the German company I. G. Farben and the American optometrist William Feinbloom. These lenses were light and relatively comfortable; they could be worn for several hours at a time.

In 1945, the American optician Kevin Tuohy developed corneal lenses, which covered only the cornea of the eye. Reportedly, Tuohy's invention was inspired by the fact that his nearsighted wife could not bear scleral lenses but hated to wear eyeglasses. Tuohy's lenses were hard contact lenses made of rigid plastic, but they were much more comfortable than scleral lenses and could be worn for longer periods of time. Soon after, other people developed soft contact lenses, which cover both the cornea and the iris. At present, many kinds of contact lenses are available. Both hard and soft contact lenses have advantages for particular uses.

Eyes, Tears, and Contact Lenses

The camera-like human eye automatically focuses itself and adjusts to the prevailing light intensity. In addition, it never runs out of "film" and makes a continuous series of visual images. In the process of seeing, light enters the eye and passes through the clear, dome-shaped cornea, through the hole (the pupil) in the colored iris, and through the clear eye lens, which can change shape by means of muscle contraction. The lens focuses the light, which next passes across the jellylike "vitreous humor" and hits the retina. There, light-sensitive retinal cells send visual images to the optic nerve, which transmits them to the brain for interpretation.

Many people have 20/20 (normal) vision, which means that they can clearly see letters on a designated line of a standard eye chart placed 20 feet away. Nearsighted (myopic) people have vision of 20/40 or worse. This means that, 20 feet from the eye chart, they see clearly what people with 20/20 vision can see clearly at a greater distance.

Myopia (nearsightedness) is one of the four most common visual defects. The others are hyperopia, astigmatism, and presbyopia. All are called "refractive errors" and are corrected with appropriate eyeglasses or contact lenses. Myopia, which occurs in 30 percent of humans, occurs when the eyeball is too long for the lens's focusing ability and images of distant objects focus before they reach the retina, causing blurry vision. Hyperopia, or farsightedness, occurs when the eyeballs are too short. In hyperopia, the eye's lenses cannot focus images of nearby objects by the time those images reach the retina, resulting in blurry vision. A more common condition is astigmatism, in which incorrectly shaped corneas make all objects appear blurred. Finally, presbyopia, part of the aging process, causes the lens of the eye to lose its elasticity. It causes progressive difficulty in seeing nearby objects. In myopic, hyperopic, or astigmatic people, bifocal (two-lens) systems are used to correct presbyopia, whereas monofocal systems are used to correct presbyopia in people whose vision is otherwise normal.

WILLIAM FEINBLOOM

William Feinbloom started his career in eye care when he was only three, helping his father, an optometrist, in his practice. Born in Brooklyn, New York, in 1904, Feinbloom studied at the Columbia School of Optometry and graduated at nineteen. He later earned degrees in physics, mathematics, biophysics, and psychology, all of it to help him treat people who suffered visual impairments. His many achievements on the behalf of the partially sighted won him professional accolades as the "father of low vision."

In 1932, while working in a clinic, Feinbloom produced the first of his special vision-enhancing inventions. He ground three-power lenses, imitating the primary lens of a refracting telescope, and fit them in a frame for an elderly patient whose vision could not be treated. The patient was again able to see, and when news of this miracle later reached Pope Pius XI, he sent a special blessing to Feinbloom. He soon opened his own practice and during the next fifty years invented a series of new lenses for people with macular degeneration and other vision diseases, as well as making the first set of contact lenses in America.

In 1978 Feinbloom bequeathed his practice to the Pennsylvania College of Optometry, which named it the William Feinbloom Vision Rehabilitation Center. Every year the William Feinbloom Award honors a vision-care specialist who has improved the delivery and quality of optometric service. Feinbloom died in 1985.

Modern contact lenses, which many people prefer to eyeglasses, are used to correct all common eye defects as well as many others not mentioned here. The lenses float on the layer of tears that is made continuously to nourish the eye and keep it moist. They fit under the eyelids and either over the cornea or over both the cornea and the iris, and they correct visual errors by altering the eye's focal length enough to produce 20/20 vision. In addition to being more attractive than eyeglasses, contact lenses correct visual defects more effectively than eyeglasses can. Some soft contact lenses (all are made of flexible plastics) can be worn almost continuously. Hard lenses are made of more rigid plastic and last longer, though they can usually be worn only for six to nine hours at a time. The choice of hard or soft lenses must be made on an individual basis.

The disadvantages of contact lenses include the fact that they must be cleaned frequently to prevent eye irritation. Furthermore, people who do not produce adequate amounts of tears (a condition called "dry eyes") cannot wear them. Also, arthritis, many allergies, and poor manual dexterity caused by old age or physical problems make many people poor candidates for contact lenses.

IMPACT

The invention of Plexiglas hard scleral contact lenses set the stage for the development of the widely used corneal hard lenses by Tuohy. The development of soft contact lenses available to the general public began in Czechoslovakia in the 1960's. It led to the sale, starting in the

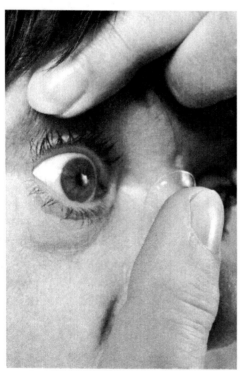

1970's, of the popular, soft contact lenses pioneered by Otto Wichterle. The Wichterle lenses, which cover both the cornea and the iris, are made of a plastic called HEMA (short for hydroxyethylmethylmethacrylate).

These very thin lenses have disadvantages that include the requirement of disinfection between uses, incomplete astigmatism correction, low durability, and the possibility of chemical combination with some medications, which can damage the eyes. Therefore, much research is being carried out to improve them. For this reason, and because of the continued

Contact lenses are placed directly on the surface of the eye. (Digital Stock)

popularity of hard lenses, new kinds of soft and hard lenses are continually coming on the market.

See also Artificial heart; Disposable razor; Hearing aid; Laser eye surgery; Pacemaker.

FURTHER READING

"The Contact Lens." *Newsweek* 130 (Winter, 1997/1998).
Hemphill, Clara. "A Quest for Better Vision: Spectacles over the Centuries." *New York Times* (August 8, 2000).
Koetting, Robert A. *History of the Contact Lens.* Irvine, Calif.: Allergan, 1978.
Lubick, Naomi. "The Hard and the Soft." *Scientific American* 283, no. 4 (October, 2000).

CORONARY ARTERY BYPASS SURGERY

THE INVENTION: The most widely used procedure of its type, coronary bypass surgery uses veins from legs to improve circulation to the heart.

THE PEOPLE BEHIND THE INVENTION:

Rene Favaloro (1923-2000), a heart surgeon

Donald B. Effler (1915-   ), a member of the surgical team that performed the first coronary artery bypass operation

F. Mason Sones (1918-   ), a physician who developed an improved technique of X-raying the heart's arteries

FIGHTING HEART DISEASE

In the mid-1960's, the leading cause of death in the United States was coronary artery disease, claiming nearly 250 deaths per 100,000 people. Because this number was so alarming, much research was being conducted on the heart. Most of the public's attention was focused on heart transplants performed separately by the famous surgeons Christiaan Barnard and Michael DeBakey. Yet other, less dramatic procedures were being developed and studied.

A major problem with coronary artery disease, besides the threat of death, is chest pain, or angina. Individuals whose arteries are clogged with fat and cholesterol are frequently unable to deliver enough oxygen to their heart muscles. This may result in angina, which causes enough pain to limit their physical activities. Some of the heart research in the mid-1960's was an attempt to find a surgical procedure that would eliminate angina in heart patients. The various surgical procedures had varying success rates.

In the late 1950's and early 1960's, a team of physicians in Cleveland was studying surgical procedures that would eliminate angina. The team was composed of Rene Favaloro, Donald B. Effler, F. Mason Sones, and Laurence Groves. They were working on the concept, proposed by Dr. Arthur M. Vineberg from McGill University in Montreal, of implanting a healthy artery from the chest into the heart. This bypass procedure would provide the heart with another

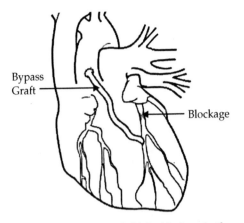

Bypass Graft

Blockage

Before bypass surgery (left) the blockage in the artery threatens to cut off bloodflow; after surgery to graft a piece of vein (right), the blood can flow around the blockage.

source of blood, resulting in enough oxygen to overcome the angina. Yet Vineberg's surgery was often ineffective because it was hard to determine exactly where to implant the new artery.

NEW TECHNIQUES

In order to make Vineberg's proposed operation successful, better diagnostic tools were needed. This was accomplished by the work of Sones. He developed a diagnostic procedure, called "arteriography," whereby a catheter was inserted into an artery in the arm, which he ran all the way into the heart. He then injected a dye into the coronary arteries and photographed them with a high-speed motion-picture camera. This provided an image of the heart, which made it easy to determine where the blockages were in the coronary arteries.

Using this tool, the team tried several new techniques. First, the surgeons tried to ream out the deposits found in the narrow portion of the artery. They found, however, that this actually reduced blood flow. Second, they tried slitting the length of the blocked area of the artery and suturing a strip of tissue that would increase the diameter of the opening. This was also ineffective because it often resulted in turbulent blood flow. Finally, the team attempted to reroute the flow of blood around the blockage by suturing in other tissue, such as a portion of a vein from the upper leg. This bypass procedure removed that part of the artery that was clogged and replaced it with a clear vessel, thereby restoring blood flow through the artery. This new method was introduced by Favaloro in 1967.

In order for Favaloro and other heart surgeons to perform coronary artery surgery successfully, several other medical techniques had to be developed. These included extracorporeal circulation and microsurgical techniques.

Extracorporeal circulation is the process of diverting the patient's blood flow from the heart and into a heart-lung machine. This procedure was developed in 1953 by U.S. surgeon John H. Gibbon, Jr. Since the blood does not flow through the heart, the heart can be temporarily stopped so that the surgeons can isolate the artery and perform the surgery on motionless tissue.

Microsurgery is necessary because some of the coronary arteries are less than 1.5 millimeters in diameter. Since these arteries had to be sutured, optical magnification and very delicate and sophisticated surgical tools were required. After performing this surgery on numerous patients, follow-up studies were able to determine the surgery's effectiveness. Only then was the value of coronary artery bypass surgery recognized as an effective procedure for reducing angina in heart patients.

CONSEQUENCES

According to the American Heart Association, approximately 332,000 bypass surgeries were performed in the United States in 1987, an increase of 48,000 from 1986. These figures show that the work by Favaloro and others has had a major impact on the health of United States citizens. The future outlook is also positive. It has been estimated that five million people had coronary artery disease in 1987. Of this group, an estimated 1.5 million had heart attacks and 500,000 died. Of those living, many experienced angina. Research has developed new surgical procedures and new drugs to help fight coronary artery disease. Yet coronary artery bypass surgery is still a major form of treatment.

See also Artificial blood; Artificial heart; Blood transfusion; Electrocardiogram; Heart-lung machine; Pacemaker.

FURTHER READING

Bing, Richard J. *Cardiology: The Evolution of the Science and the Art.* 2d ed. New Brunswick, N.J.: Rutgers University Press, 1999.

Faiola, Anthony. "Doctor's Suicide Strikes at Heart of Argentina's Health Care Crisis: Famed Cardiac Surgeon Championed the Poor." *Washington Post* (August 25, 2000).

Favaloro, René G. *The Challenging Dream of Heart Surgery: From the Pampas to Cleveland.* Boston: Little, Brown, 1994.

CRUISE MISSILE

THE INVENTION: Aircraft weapons system that makes it possible to attack both land and sea targets with extreme accuracy without endangering the lives of the pilots.

THE PERSON BEHIND THE INVENTION:

Rear Admiral Walter M. Locke (1930-   ), U.S. Navy project manager

FROM THE BUZZ BOMBS OF WORLD WAR II

During World War II, Germany developed and used two different types of missiles: ballistic missiles and cruise missiles. A ballistic missile is one that does not use aerodynamic lift in order to fly. It is fired into the air by powerful jet engines and reaches a high altitude; when its engines are out of fuel, it descends on its flight path toward its target. The German V-2 was the first ballistic missile. The United States and other countries subsequently developed a variety of highly sophisticated and accurate ballistic missiles.

The other missile used by Germany was a cruise missile called the V-1, which was also called the flying bomb or the buzz bomb. The V-1 used aerodynamic lift in order to fly, just as airplanes do. It flew relatively low and was slow; by the end of the war, the British, against whom it was used, had developed techniques for countering it, primarily by shooting it down.

After World War II, both the United States and the Soviet Union carried on the Germans' development of both ballistic and cruise missiles. The United States discontinued serious work on cruise missile technology during the 1950's: The development of ballistic missiles of great destructive capability had been very successful. Ballistic missiles armed with nuclear warheads had become the basis for the U.S. strategy of attempting to deter enemy attacks with the threat of a massive missile counterattack. In addition, aircraft carriers provided an air-attack capability similar to that of cruise missiles. Finally, cruise missiles were believed to be too vulnerable to being shot down by enemy aircraft or surface-to-air missiles.

While ballistic missiles are excellent for attacking large, fixed targets, they are not suitable for attacking moving targets. They can be very accurately aimed, but since they are not very maneuverable during their final descent, they are limited in their ability to change course to hit a moving target, such as a ship.

During the 1967 war, the Egyptians used a Soviet-built cruise missile to sink the Israeli ship *Elath*. The U.S. military, primarily the Navy and the Air Force, took note of the Egyptian success and within a few years initiated cruise missile development programs.

THE DEVELOPMENT OF CRUISE MISSILES

The United States probably could have developed cruise missiles similar to 1990's models as early as the 1960's, but it would have required a huge effort. The goal was to develop missiles that could be launched from ships and planes using existing launching equipment, could fly long distances at low altitudes at fairly high speeds, and could reach their targets with a very high degree of accuracy. If the missiles flew too slowly, they would be fairly easy to shoot down, like the German V-1's. If they flew at too high an altitude, they would be vulnerable to the same type of surface-based missiles that shot down Gary Powers, the pilot of the U.S. U2 spyplane, in 1960. If they were inaccurate, they would be of little use.

The early Soviet cruise missiles were designed to meet their performance goals without too much concern about how they would be launched. They were fairly large, and the ships that launched them required major modifications. The U.S. goal of being able to launch using existing equipment, without making major modifications to the ships and planes that would launch them, played a major part in their torpedo-like shape: Sea-Launched Cruise Missiles (SLCMs) had to fit in the submarine's torpedo tubes, and Air-Launched Cruise Missiles (ALCMs) were constrained to fit in rotary launchers. The size limitation also meant that small, efficient jet engines would be required that could fly the long distances required without needing too great a fuel load. Small, smart computers were needed to provide the required accuracy. The engine and computer technologies began to be available in the 1970's, and they blossomed in the 1980's.

The U.S. Navy initiated cruise missile development efforts in 1972; the Air Force followed in 1973. In 1977, the Joint Cruise Missile Project was established, with the Navy taking the lead. Rear Admiral Walter M. Locke was named project manager. The goal was to develop air-, sea-, and ground-launched cruise missiles. By coordinating activities, encouraging competition, and requiring the use of common components wherever possible, the cruise missile development program became a model for future weapon-system development efforts. The primary contractors included Boeing Aerospace Company, General Dynamics, and McDonnell Douglas.

In 1978, SLCMs were first launched from submarines. Over the next few years, increasingly demanding tests were passed by several versions of cruise missiles. By the mid-1980's, both antiship and antiland missiles were available. An antiland version could be guided to its target with extreme accuracy by comparing a map programmed into its computer to the picture taken by an on-board video camera.

The typical cruise missile is between 18 and 21 feet long, about 21 inches in diameter, and has a wingspan of between 8 and 12 feet. Cruise missiles travel slightly below the speed of sound and have a range of around 1,350 miles (antiland) or 250 miles (antiship). Both conventionally armed and nuclear versions have been fielded.

CONSEQUENCES

Cruise missiles have become an important part of the U.S. arsenal. They provide a means of attacking targets on land and water without having to put an aircraft pilot's life in danger. Their value was demonstrated in 1991 during the Persian Gulf War. One of their uses was to "soften up" defenses prior to sending in aircraft, thus reducing the risk to pilots. Overall estimates are that about 85 percent of cruise missiles used in the Persian Gulf War arrived on target, which is an outstanding record. It is believed that their extreme accuracy also helped to minimize noncombatant casualties.

See also Airplane; Atomic bomb; Hydrogen bomb; Rocket; Stealth aircraft; V-2 rocket.

FURTHER READING

Collyer, David G. *Buzz Bomb*. Deal, Kent, England: Kent Aviation Historical Research Society, 1994.

McDaid, Hugh, and David Oliver. *Robot Warriors: The Top Secret History of the Pilotless Plane*. London: Orion Media, 1997.

Macknight, Nigel. *Tomahawk Cruise Missile*. Osceola, Wis.: Motorbooks International, 1995.

Werrell, Kenneth P. *The Evolution of the Cruise Missile*. Maxwell Air Force Base, Ala.: Air University Press, 1997.

Cyclamate

THE INVENTION: An artificial sweetener introduced to the American market in 1950 under the tradename Sucaryl.

THE PERSON BEHIND THE INVENTION:
Michael Sveda (1912-1999), an American chemist

A FOOLHARDY EXPERIMENT

The first synthetic sugar substitute, saccharin, was developed in 1879. It became commercially available in 1907 but was banned for safety reasons in 1912. Sugar shortages during World War I (1914-1918) resulted in its reintroduction. Two other artificial sweeteners, Dulcin and P-4000, were introduced later but were banned in 1950 for causing cancer in laboratory animals.

In 1937, Michael Sveda was a young chemist working on his Ph.D. at the University of Illinois. A flood in the Ohio valley had ruined the local pipe-tobacco crop, and Sveda, a smoker, had been forced to purchase cigarettes. One day while in the laboratory, Sveda happened to brush some loose tobacco from his lips and noticed that his fingers tasted sweet. Having a curious, if rather foolhardy, nature, Sveda tasted the chemicals on his bench to find which one was responsible for the taste. The culprit was the forerunner of cyclohexylsulfamate, the material that came to be known as "cyclamate." Later, on reviewing his career, Sveda explained the serendipitous discovery with the comment: "God looks after . . . fools, children, and chemists."

Sveda joined E. I. Du Pont de Nemours and Company in 1939 and assigned the patent for cyclamate to his employer. In June of 1950, after a decade of testing on animals and humans, Abbott Laboratories announced that it was launching Sveda's artificial sweetener under the trade name Sucaryl. Du Pont followed with its sweetener product, Cyclan. A *Time* magazine article in 1950 announced the new product and noted that Abbott had warned that because the product was a sodium salt, individuals with kidney problems should consult their doctors before adding it to their food.

Cyclamate had no calories, but it was thirty to forty times sweeter than sugar. Unlike saccharin, cyclamate left no unpleasant aftertaste. The additive was also found to improve the flavor of some foods, such as meat, and was used extensively to preserve various foods. By 1969, about 250 food products contained cyclamates, including cakes, puddings, canned fruit, ice cream, salad dressings, and its most important use, carbonated beverages.

It was originally thought that cyclamates were harmless to the human body. In 1959, the chemical was added to the GRAS (generally recognized as safe) list. Materials on this list, such as sugar, salt, pepper, and vinegar, did not have to be rigorously tested before being added to food. In 1964, however, a report cited evidence that cyclamates and saccharin, taken together, were a health hazard. Its publication alarmed the scientific community. Numerous investigations followed.

SHOOTING THEMSELVES IN THE FOOT

Initially, the claims against cyclamate had been that it caused diarrhea or prevented drugs from doing their work in the body. By 1969, these claims had begun to include the threat of cancer. Ironically, the evidence that sealed the fate of the artificial sweetener was provided by Abbott itself.

A private Long Island company had been hired by Abbott to conduct an extensive toxicity study to determine the effects of long-term exposure to the cyclamate-saccharin mixtures often found in commercial products. The team of scientists fed rats daily doses of the mixture to study the effect on reproduction, unborn fetuses, and fertility. In each case, the rats were declared to be normal. When the rats were killed at the end of the study, however, those that had been exposed to the higher doses showed evidence of bladder tumors. Abbott shared the report with investigators from the National Cancer Institute and then with the U.S. Food and Drug Administration (FDA).

The doses required to produce the tumors were equivalent to an individual drinking 350 bottles of diet cola a day. That was more than one hundred times greater than that consumed even by those people who consumed a high amount of cyclamate. A six-person panel of scientists met to review the data and urged the ban of all cyclamates from foodstuffs. In October, 1969, amid enormous media coverage, the federal government announced that cyclamates were to be withdrawn from the market by the beginning of 1970.

In the years following the ban, the controversy continued. Doubt was cast on the results of the independent study linking sweetener use to tumors in rats, because the study was designed not to evaluate cancer risks but to explain the effects of cyclamate use over many years. Bladder parasites, known as "nematodes," found in the rats may have affected the outcome of the tests. In addition, an impurity found in some of the saccharin used in the study may have led to the problems observed. Extensive investigations such as the three-year project conducted at the National Cancer Research Center in Heidelberg, Germany, found no basis for the widespread ban.

In 1972, however, rats fed high doses of saccharin alone were found to have developed bladder tumors. At that time, the sweetener was removed from the GRAS list. An outright ban was averted by the mandatory use of labels alerting consumers that certain products contained saccharin.

IMPACT

The introduction of cyclamate heralded the start of a new industry. For individuals who had to restrict their sugar intake for health reasons, or for those who wished to lose weight, there was now an alternative to giving up sweet food.

The Pepsi-Cola company put a new diet drink formulation on the market almost as soon as the ban was instituted. In fact, it ran advertisements the day after the ban was announced showing the Diet Pepsi product boldly proclaiming "Sugar added—No Cyclamates."

Sveda, the discoverer of cyclamates, was not impressed with the FDA's decision on the sweetener and its handling of subsequent investigations. He accused the FDA of "a massive cover-up of elemental blunders" and claimed that the original ban was based on sugar politics and bad science.

For the manufacturers of cyclamate, meanwhile, the problem lay with the wording of the Delaney amendment, the legislation that regulates new food additives. The amendment states that the manufacturer must prove that its product is safe, rather than the FDA having to prove that it is unsafe. The onus was on Abbott Laboratories to deflect concerns about the safety of the product, and it remained unable to do so.

See also Aspartame; Genetically engineered insulin.

FURTHER READING

Kaufman, Leslie. "Michael Sveda, the Inventor of Cyclamates, Dies at Eighty Seven." *New York Times* (August 21, 1999).

Lawler, Philip F. *Sweet Talk: Media Coverage of Artificial Sweeteners.* Washington, D.C.: Media Institute, 1986.

Remington, Dennis W. *The Bitter Truth About Artificial Sweeteners.* Provo, Utah: Vitality House, 1987.

Whelan, Elizabeth M. "The Bitter Truth About a Sweetener Scare." *Wall Street Journal* (August 26, 1999).

CYCLOTRON

THE INVENTION: The first successful magnetic resonance accelerator for protons, the cyclotron gave rise to the modern era of particle accelerators, which are used by physicists to study the structure of atoms.

THE PEOPLE BEHIND THE INVENTION:

Ernest Orlando Lawrence (1901-1958), an American nuclear physicist who was awarded the 1939 Nobel Prize in Physics

M. Stanley Livingston (1905-1986), an American nuclear physicist

Niels Edlefsen (1893-1971), an American physicist

David Sloan (1905-   ), an American physicist and electrical engineer

THE BEGINNING OF AN ERA

The invention of the cyclotron by Ernest Orlando Lawrence marks the beginning of the modern era of high-energy physics. Although the energies of newer accelerators have increased steadily, the principles incorporated in the cyclotron have been fundamental to succeeding generations of accelerators, many of which were also developed in Lawrence's laboratory. The care and support for such machines have also given rise to "big science": the massing of scientists, money, and machines in support of experiments to discover the nature of the atom and its constituents.

At the University of California, Lawrence took an interest in the new physics of the atomic nucleus, which had been developed by the British physicist Ernest Rutherford and his followers in England, and which was attracting more attention as the development of quantum mechanics seemed to offer solutions to problems that had long preoccupied physicists. In order to explore the nucleus of the atom, however, suitable probes were required. An artificial means of accelerating ions to high energies was also needed.

During the late 1920's, various means of accelerating alpha particles, protons (hydrogen ions), and electrons had been tried, but none had been successful in causing a nuclear transformation when Lawrence entered the field. The high voltages required exceeded the resources available to physicists. It was believed that more than a million volts would be required to accelerate an ion to sufficient energies to penetrate even the lightest atomic nuclei. At such voltages, insulators broke down, releasing sparks across great distances. European researchers even attempted to harness lightning to accomplish the task, with fatal results.

Early in April, 1929, Lawrence discovered an article by a German electrical engineer that described a linear accelerator of ions that worked by passing an ion through two sets of electrodes, each of which carried the same voltage and increased the energy of the ions correspondingly. By spacing the electrodes appropriately and using an alternating electrical field, this "resonance acceleration" of ions could speed subatomic particles to many times the energy applied in each step, overcoming the problems presented when one tried to apply a single charge to an ion all at once. Unfortunately, the spacing of the electrodes would have to be increased as the ions were accelerated, since they would travel farther between each alternation of the phase of the accelerating charge, making an accelerator impractically long in those days of small-scale physics.

FAST-MOVING STREAMS OF IONS

Lawrence knew that a magnetic field would cause the ions to be deflected and form a curved path. If the electrodes were placed across the diameter of the circle formed by the ions' path, they should spiral out as they were accelerated, staying in phase with the accelerating charge until they reached the periphery of the magnetic field. This, it seemed to him, afforded a means of producing indefinitely high voltages without using high voltages by recycling the accelerated ions through the same electrodes. Many scientists doubted that such a method would be effective. No mechanism was known that would keep the circulating ions in sufficiently tight orbits to avoid collisions with the walls of the accelerating chamber. Others tried unsuccessfully to use resonance acceleration.

A graduate student, M. Stanley Livingston, continued Lawrence's work. For his dissertation project, he used a brass cylinder 10 centi-

ERNEST ORLANDO LAWRENCE

A man of great energy and gusty temper, Ernest Orlando Lawrence danced for joy when one of his cyclotrons accelerated a particle to more than the one million electron volts. That amount of power was important, according to contemporary theorists, because it was enough to penetrate the nucleus of a target atom. For giving physicists a tool with which to examine the subatomic realm, Lawrence received the 1939 Nobel Prize in Physics, among many other honors.

The grandson of Norwegian immigrants, Lawrence was born in Canton, South Dakota, in 1901. After high school, he went to St. Olaf's College, the University of South Dakota, the University of Minnesota, and Yale University, where he completed a doctorate in physics in 1925. After post-graduate fellowships at Yale, he became a professor at the University of California, Berkeley, the youngest on campus. In 1936 the university made him director of its radiation laboratory. Now named the Lawrence-Livermore National Laboratory, it stayed at the forefront of physics and high-technology research ever since.

Before World War II Lawrence and his brother, Dr. John Lawrence, also at the university, worked together to find practical biological and medical applications for the radioisotopes made in Lawrence's particle accelerators. During the war Lawrence participated in the Manhattan Project, which made the atomic bomb. He was a passionate anticommunist and after the war argued before Congress for funds to develop death rays and radiation bombs from research with his cyclotrons; however, he was also an American delegate to the Geneva Conference in 1958, which sought a ban on atomic bomb tests.

Lawrence helped solve the mystery of cosmic particles, invented a method for measuring ultra-small time intervals, and calculated with high precision the ratio of the charge of an electron to its mass, a fundamental constant of nature. Lawrence died in 1958 in Palo Alto, California.

meters in diameter sealed with wax to hold a vacuum, a half-pillbox of copper mounted on an insulated stem to serve as the electrode, and a Hartley radio frequency oscillator producing 10 watts. The hydrogen molecular ions were produced by a thermionic cathode (mounted near the center of the apparatus) from hydrogen gas admitted through an aperture in the side of the cylinder after a vacuum had been produced by a pump. Once formed, the oscillating electrical field drew out the ions and accelerated them as they passed through the cylinder. The accelerated ions spiraled out in a magnetic field produced by a 10-centimeter electromagnet to a collector. By November, 1930, Livingston had observed peaks in the collector current as he tuned the magnetic field through the value calculated to produce acceleration.

Borrowing a stronger magnet and tuning his radio frequency oscillator appropriately, Livingston produced 80,000-electronvolt ions at his collector on January 2, 1931, thus demonstrating the principle of magnetic resonance acceleration.

IMPACT

Demonstration of the principle led to the construction of a succession of large cyclotrons, beginning with a 25-centimeter cyclotron developed in the spring and summer of 1931 that produced one-million-electronvolt protons. With the support of the Research Corporation, Lawrence secured a large electromagnet that had been developed for radio transmission and an unused laboratory to house it: the Radiation Laboratory.

The 69-centimeter cyclotron built with the magnet was used to explore nuclear physics. It accelerated deuterons, ions of heavy water or deuterium that contain, in addition to the proton, the neutron, which was discovered by Sir James Chadwick in 1932. The accelerated deuteron, which injected neutrons into target atoms, was used to produce a wide variety of artificial radioisotopes. Many of these, such as technetium and carbon 14, were discovered with the cyclotron and were later used in medicine.

By 1939, Lawrence had built a 152-centimeter cyclotron for medical applications, including therapy with neutron beams. In that year, he won the Nobel Prize in Physics for the invention of the cyclotron and the production of radioisotopes. During World War II, Lawrence and the members of his Radiation Laboratory developed electromagnetic separation of uranium ions to produce the uranium 235 required for the atomic bomb. After the war, the 467-centimeter cyclotron was completed as a synchrocyclotron, which modulated the frequency of the accelerating fields to compensate for the increasing mass of ions as they approached the speed of light. The principle of synchronous acceleration, invented by Lawrence's associate, the American physicist Edwin Mattison McMillan, became fundamental to proton and electron synchrotrons.

The cyclotron and the Radiation Laboratory were the center of accelerator physics throughout the 1930's and well into the postwar era. The invention of the cyclotron not only provided a new tool for probing the nucleus but also gave rise to new forms of organizing scientific work and to applications in nuclear medicine and nuclear chemistry. Cyclotrons were built in many laboratories in the United States, Europe, and Japan, and they became a standard tool of nuclear physics.

See also Atomic bomb; Electron microscope; Field ion microscope; Geiger counter; Hydrogen bomb; Mass spectrograph; Neutrino detector; Scanning tunneling microscope; Synchrocyclotron; Tevatron accelerator.

FURTHER READING

Childs, Herbert. *An American Genius: The Life of Ernest Orlando Lawrence*. New York: Dutton, 1968.
Close, F. E., Michael Marten, and Christine Sutton. *The Particle Explosion*. New York: Oxford University Press, 1994.
Pais, Abraham. *Inward Bound: Of Matter and Forces in the Physical World*. New York: Clarendon Press, 1988.
Wilson, Elizabeth K. "Fifty Years of Heavy Chemistry." *Chemical and Engineering News* 78, no. 13 (March 27, 2000).

DIESEL LOCOMOTIVE

THE INVENTION: An internal combustion engine in which ignition is achieved by the use of high-temperature compressed air, rather than a spark plug.

THE PEOPLE BEHIND THE INVENTION:

Rudolf Diesel (1858-1913), a German engineer and inventor
Sir Dugold Clark (1854-1932), a British engineer
Gottlieb Daimler (1834-1900), a German engineer
Henry Ford (1863-1947), an American automobile magnate
Nikolaus Otto (1832-1891), a German engineer and Daimler's teacher

A BEGINNING IN WINTERTHUR

By the beginning of the twentieth century, new means of providing society with power were needed. The steam engines that were used to run factories and railways were no longer sufficient, since they were too heavy and inefficient. At that time, Rudolf Diesel, a German mechanical engineer, invented a new engine. His diesel engine was much more efficient than previous power sources. It also appeared that it would be able to run on a wide variety of fuels, ranging from oil to coal dust. Diesel first showed that his engine was practical by building a diesel-driven locomotive that was tested in 1912.

In the 1912 test runs, the first diesel-powered locomotive was operated on the track of the Winterthur-Romanston rail line in Switzerland. The locomotive was built by a German company, Gesellschaft für Thermo-Lokomotiven, which was owned by Diesel and his colleagues. Immediately after the test runs at Winterthur proved its efficiency, the locomotive—which had been designed to pull express trains on Germany's Berlin-Magdeburg rail line—was moved to Berlin and put into service. It worked so well that many additional diesel locomotives were built. In time, diesel engines were also widely used to power many other machines, including those that ran factories, motor vehicles, and ships.

RUDOLF DIESEL

Unbending, suspicious of others, but also exceptionally intelligent, Rudolf Christian Karl Diesel led a troubled life and came to a mysterious end. His parents, expatriate Germans, lived in Paris when he was born, 1858, and he spent his early childhood there. In 1870, just as he was starting his formal education, his family fled to England on the outbreak of the Franco-Prussian War, which turned the French against Germans. In England, Diesel spent much of his spare time in museums, educating himself. His father, a leather craftsman, was unable to support his family, so as a teenager Diesel was packed off to Augsburg, Germany, where he was largely on his own. Although these experiences made him fluent in English, French, and German, his was not a stable or happy childhood.

He threw himself into his studies, finishing his high school education three years ahead of schedule, and entered the Technical College of Munich, where he was the star student. Once, during his school years, he saw a demonstration of a Chinese firestick. The firestick was a tube with a plunger. When a small piece of flammable material was put in one end and the plunger pushed down rapidly toward it, the heat of the compressed air in the tube ignited the material. The demonstration later inspired Diesel to adapt the principle to an engine.

His was the first engine to run successfully with compressed air fuel ignition, but it was not the first design. So although he received the patent for the diesel engine, he had to fight challenges in court from other inventors over licensing rights. He always won, but the strain of litigation worsened his tendency to stubborn self-reliance, and this led him into difficulties. The first compression engines were unreliable and unwieldy, but Diesel rebuffed all suggestions for modifications, requiring that builders follow his original design. His attitude led to delays in development of the engine and lost him financial support.

In 1913, while crossing the English Channel aboard a ship, Diesel disappeared. His body was never found, and although the authorities concluded that Diesel committed suicide, no one knows what happened.

DIESELS, DIESELS EVERYWHERE

In the 1890's, the best engines available were steam engines that were able to convert only 5 to 10 percent of input heat energy to useful work. The burgeoning industrial society and a widespread network of railroads needed better, more efficient engines to help businesses make profits and to speed up the rate of transportation available for moving both goods and people, since the maximum speed was only about 48 kilometers per hour. In 1894, Rudolf Diesel, then thirty-five years old, appeared in Augsburg, Germany, with a new engine that he believed would demonstrate great efficiency.

The diesel engine demonstrated at Augsburg ran for only a short time. It was, however, more efficient than other existing engines. In addition, Diesel predicted that his engines would move trains faster than could be done by existing engines and that they would run on a wide variety of fuels. Experimentation proved the truth of his claims; even the first working motive diesel engine (the one used in the Winterthur test) was capable of pulling heavy freight and passenger trains at maximum speeds of up to 160 kilometers per hour.

By 1912, Diesel, a millionaire, saw the wide use of diesel locomotives in Europe and the United States and the conversion of hundreds of ships to diesel power. Rudolf Diesel's role in the story ends here, a result of his mysterious death in 1913—believed to be a suicide by the authorities—while crossing the English Channel on the steamer *Dresden*. Others involved in the continuing saga of diesel engines were the Britisher Sir Dugold Clerk, who improved diesel design, and the American Adolphus Busch (of beer-brewing fame), who bought the North American rights to the diesel engine.

The diesel engine is related to automobile engines invented by Nikolaus Otto and Gottlieb Daimler. The standard Otto-Daimler (or Otto) engine was first widely commercialized by American auto magnate Henry Ford. The diesel and Otto engines are internal-combustion engines. This means that they do work when a fuel is burned and causes a piston to move in a tight-fitting cylinder. In diesel engines, unlike Otto engines, the fuel is not ignited by a spark from a spark plug. Instead, ignition is accomplished by the use of high-temperature compressed air.

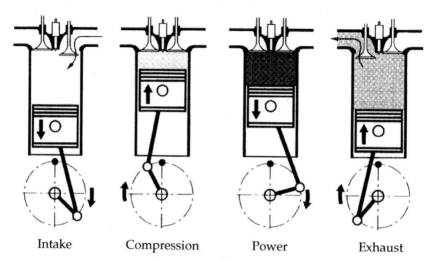

| Intake | Compression | Power | Exhaust |

The four strokes of a diesel engine. (Robert Bosch Corporation)

In common "two-stroke" diesel engines, pioneered by Sir Dugold Clerk, a starter causes the engine to make its first stroke. This draws in air and compresses the air sufficiently to raise its temperature to 900 to 1,000 degrees Fahrenheit. At this point, fuel (usually oil) is sprayed into the cylinder, ignites, and causes the piston to make its second, power-producing stroke. At the end of that stroke, more air enters as waste gases leave the cylinder; air compression occurs again; and the power-producing stroke repeats itself. This process then occurs continuously, without restarting.

IMPACT

Proof of the functionality of the first diesel locomotive set the stage for the use of diesel engines to power many machines. Although Rudolf Diesel did not live to see it, diesel engines were widely used within fifteen years after his death. At first, their main applications were in locomotives and ships. Then, because diesel engines are more efficient and more powerful than Otto engines, they were modified for use in cars, trucks, and buses.

At present, motor vehicle diesel engines are most often used in buses and long-haul trucks. In contrast, diesel engines are not as popular in automobiles as Otto engines, although European automakers make much wider use of diesel engines than American automakers do. Many enthusiasts, however, view diesel automobiles as the wave of the future. This optimism is based on the durability of the engine, its great power, and the wide range and economical nature of the fuels that can be used to run it. The drawbacks of diesels include the unpleasant odor and high pollutant content of their emissions.

Modern diesel engines are widely used in farm and earth-moving equipment, including balers, threshers, harvesters, bulldozers,rock crushers, and road graders. Construction of the Alaskan oil pipeline relied heavily on equipment driven by diesel engines. Diesel engines are also commonly used in sawmills, breweries, coal mines, and electric power plants.

Diesel's brainchild has become a widely used power source, just as he predicted. It is likely that the use of diesel engines will continue and will expand, as the demands of energy conservation require more efficient engines and as moves toward fuel diversification require engines that can be used with various fuels.

See also Bullet train; Gas-electric car; Internal combustion engine.

FURTHER READING

Cummins, C. Lyle. *Diesel's Engine*. Wilsonville, Oreg.: Carnot Press, 1993.
Diesel, Eugen. *From Engines to Autos: Five Pioneers in Engine Development and Their Contributions to the Automotive Industry*. Chicago: H. Regnery, 1960.
Nitske, Robert W., and Charles Morrow Wilson. *Rudolf Diesel: Pioneer of the Age of Power*. Norman: University of Oklahoma Press, 1965.

Differential Analyzer

The invention: An electromechanical device capable of solving differential equations.

The people behind the invention:
Vannevar Bush (1890-1974), an American electrical engineer
Harold L. Hazen (1901-1980), an American electrical engineer

Electrical Engineering Problems Become More Complex

After World War I, electrical engineers encountered increasingly difficult differential equations as they worked on vacuum-tube circuitry, telephone lines, and, particularly, long-distance power transmission lines. These calculations were lengthy and tedious. Two of the many steps required to solve them were to draw a graph manually and then to determine the area under the curve (essentially, accomplishing the mathematical procedure called integration).

In 1925, Vannevar Bush, a faculty member in the Electrical Engineering Department at the Massachusetts Institute of Technology (MIT), suggested that one of his graduate students devise a machine to determine the area under the curve. They first considered a mechanical device but later decided to seek an electrical solution. Realizing that a watt-hour meter such as that used to measure electricity in most homes was very similar to the device they needed, Bush and his student refined the meter and linked it to a pen that automatically recorded the curve.

They called this machine the Product Integraph, and MIT students began using it immediately. In 1927, Harold L. Hazen, another MIT faculty member, modified it in order to solve the more complex second-order differential equations (it originally solved only first-order equations).

The Differential Analyzer

The original Product Integraph had solved problems electrically, and Hazen's modification had added a mechanical integrator. Although the revised Product Integraph was useful in solving the types of problems mentioned above, Bush thought the machine could be improved by making it an entirely mechanical integrator, rather than a hybrid electrical and mechanical device.

In late 1928, Bush received funding from MIT to develop an entirely mechanical integrator, and he completed the resulting Differential Analyzer in 1930. This machine consisted of numerous interconnected shafts on a long, tablelike framework, with drawing boards flanking one side and six wheel-and-disk integrators on the other. Some of the drawing boards were configured to allow an operator to trace a curve with a pen that was linked to the Analyzer, thus providing input to the machine. The other drawing boards were configured to receive output from the Analyzer via a pen that drew a curve on paper fastened to the drawing board.

The wheel-and-disk integrator, which Hazen had first used in the revised Product Integraph, was the key to the operation of the Differential Analyzer. The rotational speed of the horizontal disk was the input to the integrator, and it represented one of the variables in the equation. The smaller wheel rolled on the top surface of the disk, and its speed, which was different from that of the disk, represented the integrator's output. The distance from the wheel to the center of the disk could be changed to accommodate the equation being solved, and the resulting geometry caused the two shafts to turn so that the output was the integral of the input. The integrators were linked mechanically to other devices that could add, subtract, multiply, and divide. Thus, the Differential Analyzer could solve complex equations involving many different mathematical operations. Because all the linkages and calculating devices were mechanical, the Differential Analyzer actually acted out each calculation. Computers of this type, which create an analogy to the physical world, are called analog computers.

The Differential Analyzer fulfilled Bush's expectations, and students and researchers found it very useful. Although each different problem required Bush's team to set up a new series of mechanical linkages, the researchers using the calculations viewed this as a minor inconvenience. Students at MIT used the Differential Analyzer in research for doctoral dissertations, master's theses, and bachelor's theses. Other researchers worked on a wide range of problems

Vannevar Bush

One of the most politically powerful scientists of the twentieth century, Vannevar Bush was born in 1890 in Everett, Massachusetts. He studied at Tufts College in Boston, not only earning two degrees in engineering but also registering his first patent while still an undergraduate. He worked for General Electric Company briefly after college and then conducted research on submarine detection for the U.S. Navy during World War I.

After the war he became a professor of electrical power transmission (and later dean of the engineering school) at the Massachusetts Institute of Technology (MIT). He also acted as a consultant for industry and started companies of his own, including (with two others) Raytheon Corporation. While at MIT he developed the Product Integraph and Differential Analyzer to aid in solving problems related to electrical power transmission.

Starting in 1939, Bush became a key science administrator. He was president of the Carnegie Foundation from 1939 until 1955, chaired the National Advisory Committee for Aeronautics from 1939 until 1941, in 1940 was appointed chairman of the President's National Defense Research Committee, and from 1941 until 1946 was director of the Office of Scientific Research and Development. This meant he was President Franklin Roosevelt's science adviser during World War II and oversaw wartime military research, including involvement in the Manhattan Project that build the first atomic bombs. After the war he worked for peaceful application of atomic power and was instrumental in inaugurating the National Science Foundation, which he directed, in 1950. Between 1957 and 1959 he served as chairman of MIT Corporation, retaining an honorary chairmanship thereafter.

All these political and administrative roles meant he exercised enormous influence in deciding which scientific projects were supported financially. Having received many honorary degrees and awards, including the National Medal of Science (1964), Bush died in 1974.

with the Differential Analyzer, mostly in electrical engineering, but also in atomic physics, astrophysics, and seismology. An English researcher, Douglas Hartree, visited Bush's laboratory in 1933 to learn about the Differential Analyzer and to use it in his own work on the atomic field of mercury. When he returned to England, he built several analyzers based on his knowledge of MIT's machine. The U.S. Army also built a copy in order to carry out the complex calculations required to create artillery firing tables (which specified the proper barrel angle to achieve the desired range). Other analyzers were built by industry and universities around the world.

IMPACT

As successful as the Differential Analyzer had been, Bush wanted to make another, better analyzer that would be more precise, more convenient to use, and more mathematically flexible. In 1932, Bush began seeking money for his new machine, but because of the Depression it was not until 1936 that he received adequate funding for the Rockefeller Analyzer, as it came to be known. Bush left MIT in 1938, but work on the Rockefeller Analyzer continued. It was first demonstrated in 1941, and by 1942, it was being used in the war effort to calculate firing tables and design radar antenna profiles. At the end of the war, it was the most important computer in existence.

All the analyzers, which were mechanical computers, faced serious limitations in speed because of the momentum of the machinery, and in precision because of slippage and wear. The digital computers that were being developed after World War II (even at MIT) were faster, more precise, and capable of executing more powerful operations because they were electrical computers. As a result, during the 1950's, they eclipsed differential analyzers such as those built by Bush. Descendants of the Differential Analyzer remained in use as late as the 1990's, but they played only a minor role.

See also Colossus computer; ENIAC computer; Mark I calculator; Personal computer; SAINT; UNIVAC computer.

FURTHER READING

Bush, Vannevar. *Pieces of the Action*. New York: Morrow, 1970.

Marcus, Alan I., and Howard P. Segal. *Technology in America*. Fort Worth, Tex.: Harcourt Brace College, 1999.

Spencer, Donald D. *Great Men and Women of Computing*. Ormond Beach, Fla.: Camelot Publishing, 1999.

Zachary, G. Pascal. *Endless Frontier: Vannevar Bush, Engineer of the American Century*. Cambridge, Mass.: MIT Press, 1999.

DIRIGIBLE

THE INVENTION: A rigid lighter-than-air aircraft that played a major role in World War I and in international air traffic until a disastrous accident destroyed the industry.

THE PEOPLE BEHIND THE INVENTION:
Ferdinand von Zeppelin (1838-1917), a retired German general
Theodor Kober (1865-1930), Zeppelin's private engineer

EARLY COMPETITION

When the Montgolfier brothers launched the first hot-air balloon in 1783, engineers—especially those in France—began working on ways to use machines to control the speed and direction of balloons. They thought of everything: rowing through the air with silk-covered oars; building movable wings; using a rotating fan, an airscrew, or a propeller powered by a steam engine (1852) or an electric motor (1882). At the end of the nineteenth century, the internal combustion engine was invented. It promised higher speeds and more power. Up to this point, however, the balloons were not rigid.

A rigid airship could be much larger than a balloon and could fly farther. In 1890, a rigid airship designed by David Schwarz of Dalmatia was tested in St. Petersburg, Russia. The test failed because there were problems with inflating the dirigible. A second test, in Berlin in 1897, was only slightly more successful, since the hull leaked and the flight ended in a crash.

Schwarz's airship was made of an entirely rigid aluminum cylinder. Ferdinand von Zeppelin had a different idea: His design was based on a rigid frame. Zeppelin knew about balloons from having fought in two wars in which they were used: the American Civil War of 1861-1865 and the Franco-Prussian War of 1870-1871. He wrote down his first "thoughts about an airship" in his diary on March 25, 1874, inspired by an article about flying and international mail. Zeppelin soon lost interest in this idea of civilian uses for an airship and concentrated instead on the idea that dirigible balloons might become an important part of modern warfare. He asked the

German government to fund his research, pointing out that France had a better military air force than Germany did. Zeppelin's patriotism was what kept him trying, in spite of money problems and technical difficulties.

In 1893, in order to get more money, Zeppelin tried to persuade the German military and engineering experts that his invention was practical. Even though a government committee decided that his work was worth a small amount of funding, the army was not sure that Zeppelin's dirigible was worth the cost. Finally, the committee chose Schwarz's design. In 1896, however, Zeppelin won the support of the powerful Union of German Engineers, which in May, 1898, gave him 800,000 marks to form a stock company called the Association for the Promotion of Airship Flights. In 1899, Zeppelin began building his dirigible in Manzell at Lake Constance. In July, 1900, the airship was finished and ready for its first test flight.

SEVERAL ATTEMPTS

Zeppelin, together with his engineer, Theodor Kober, had worked on the design since May, 1892, shortly after Zeppelin's retirement from the army. They had finished the rough draft by 1894, and though they made some changes later, this was the basic design of the Zeppelin. An improved version was patented in December, 1897.

In the final prototype, called the LZ 1, the engineers tried to make the airship as light as possible. They used a light internal combustion engine and designed a frame made of the light metal aluminum. The airship was 128 meters long and had a diameter of 11.7 meters when inflated. Twenty-four zinc-aluminum girders ran the length of the ship, being drawn together at each end. Sixteen rings held the body together. The engineers stretched an envelope of smooth cotton over the framework to reduce wind resistance and to protect the gas bags from the sun's rays. Seventeen gas bags made of rubberized cloth were placed inside the framework. Together they held more than 120,000 cubic meters of hydrogen gas, which would lift 11,090 kilograms. Two motor gondolas were attached to the sides, each with a 16-horsepower gasoline engine, spinning four propellers.

COUNT FERDINAND VON ZEPPELIN

The Zeppelin, the first lighter-than-air craft that was powered and steerable, began as a retirement project.

Count Ferdinand von Zeppelin was born near Lake Constance in southern Germany in 1838 and grew up in a family long used to aristocratic privilege and government service. After studying engineering at the University of Tübingen, he was commissioned as a lieutenant of engineers. In 1863 he traveled to the United States and, armed with a letter of introduction from President Abraham Lincoln, toured the Union emplacements. The observation balloons then used to see behind enemy lines impressed him. He learned all he could about them and even flew up in one to seven hundred feet.

His enthusiasm for airships stayed with him throughout his career, but he was not really able to apply himself to the problem until he retired (as a brigadier general) in 1890. Then he concentrated on the struggle to line up financing and attract talented help. He found investors for 90 percent of the money he needed and got the rest from his wife's inheritance. The first LZ's (*Luftschiff Zeppelin*) had troubles, but setbacks did not stop him. He was a stubborn, determined man. By the time he died in 1917 near Berlin he had seen ninety-two airships built. And because his design was so thoroughly associated with lighter-than-air vessels in the mind of the German public, they have ever after been known as zeppelins. However, he had already recognized their vulnerability as military aircraft, his main interest, and so he had turned his attention to designs for large airplanes as bombers.

The test flight did not go well. The two main questions—whether the craft was strong enough and fast enough—could not be answered because little things kept going wrong; for example, a crankshaft broke and a rudder jammed. The first flight lasted no more than eighteen minutes, with a maximum speed of 13.7 kilometers per hour. During all three test flights, the airship was in the air for a total of only two hours, going no faster than 28.2 kilometers per hour.

Zeppelin had to drop the project for some years because he ran out of money, and his company was dissolved. The LZ 1 was wrecked in the spring of 1901. A second airship was tested in November, 1905, and January, 1906. Both tests were unsuccessful, and in the end the ship was destroyed during a storm.

By 1906, however, the German government was convinced of the military usefulness of the airship, though it would not give money to Zeppelin unless he agreed to design one that could stay in the air for at least twenty-four hours. The third Zeppelin failed this test in the autumn of 1907. Finally, in the summer of 1908, the LZ 4 not only proved itself to the military but also attracted great publicity. It flew for more than twenty-four hours and reached a speed of more than 60 kilometers per hour. Caught in a storm at the end of this flight, the airship was forced to land and exploded, but money came from all over Germany to build another.

IMPACT

Most rigid airships were designed and flown in Germany. Of the 161 that were built between 1900 and 1938, 139 were made in Germany, and 119 were based on the Zeppelin design.

More than 80 percent of the airships were built for the military. The Germans used more than one hundred for gathering information and for bombing during World War I (1914-1918). Starting in May, 1915, airships bombed Warsaw, Poland; Bucharest, Romania; Salonika, Greece; and London, England. This was mostly a fear tactic, since the attacks did not cause great damage, and the English antiaircraft defense improved quickly. By 1916, the German army had lost so many airships that it stopped using them, though the navy continued.

Airships were first used for passenger flights in 1910. By 1914, the Delag (German Aeronautic Stock Company) used seven passenger airships for sightseeing trips around German cities. There were still problems with engine power and weather forecasting, and it was difficult to move the airships on the ground. After World War I, the Zeppelins that were left were given to the Allies as payment, and the Germans were not allowed to build airships for their own use until 1925.

In the 1920's and 1930's, it became cheaper to use airplanes for short flights, so airships were useful mostly for long-distance flight. A British airship made the first transatlantic flight in 1919. The British hoped to connect their empire by means of airships starting in 1924, but the 1930 crash of the R-101, in which most of the leading English aeronauts were killed, brought that hope to an end.

The United States Navy built the *Akron* (1931) and the *Macon* (1933) for long-range naval reconnaissance, but both airships crashed. Only the Germans continued to use airships on a regular basis. In 1929, the world tour of the *Graf Zeppelin* was a success. Regular flights between Germany and South America started in 1932, and in 1936, German airships bearing Nazi swastikas flew to Lakehurst, New Jersey. The tragic explosion of the hydrogen-filled *Hindenburg* in 1937, however, brought the era of the rigid airship to a close. The U.S. secretary of the interior vetoed the sale of nonflammable helium, fearing that the Nazis would use it for military purposes, and the German government had to stop transatlantic flights for safety reasons. In 1940, the last two remaining rigid airships were destroyed.

See also Airplane; Gyrocompass; Stealth aircraft; Supersonic passenger plane; Turbojet.

FURTHER READING

Brooks, Peter. *Zeppelin: Rigid Airships, 1893-1940.* London: Putman, 1992.

Chant, Christopher. *The Zeppelin: The History of German Airships from 1900-1937.* New York: Barnes and Noble Books, 2000.

Griehl, Manfred, and Joachim Dressel. *Zeppelin! The German Airship Story.* New York: Sterling Publishing, 1990.

Syon, Guillaume de. *Zeppelin!: Germany and the Airship, 1900-1939.* Baltimore: John Hopkins University Press, 2001.

DISPOSABLE RAZOR

THE INVENTION: An inexpensive shaving blade that replaced the traditional straight-edged razor and transformed shaving razors into a frequent household purchase item.

THE PEOPLE BEHIND THE INVENTION:

King Camp Gillette (1855-1932), inventor of the disposable razor

Steven Porter, the machinist who created the first three disposable razors for King Camp Gillette

William Emery Nickerson (1853-1930), an expert machine inventor who created the machines necessary for mass production

Jacob Heilborn, an industrial promoter who helped Gillette start his company and became a partner

Edward J. Stewart, a friend and financial backer of Gillette

Henry Sachs, an investor in the Gillette Safety Razor Company

John Joyce, an investor in the Gillette Safety Razor Company

William Painter (1838-1906), an inventor who inspired Gillette

George Gillette, an inventor, King Camp Gillette's father

A NEATER WAY TO SHAVE

In 1895, King Camp Gillette thought of the idea of a disposable razor blade. Gillette spent years drawing different models, and finally Steven Porter, a machinist and Gillette's associate, created from those drawings the first three disposable razors that worked. Gillette soon founded the Gillette Safety Razor Company, which became the leading seller of disposable razor blades in the United States.

George Gillette, King Camp Gillette's father, had been a newspaper editor, a patent agent, and an inventor. He never invented a very successful product, but he loved to experiment. He encouraged all of his sons to figure out how things work and how to improve on them. King was always inventing something new and had many patents, but he was unsuccessful in turning them into profitable businesses.

Gillette worked as a traveling salesperson for Crown Cork and

Seal Company. William Painter, one of Gillette's friends and the inventor of the crown cork, presented Gillette with a formula for making a fortune: Invent something that would constantly need to be replaced. Painter's crown cork was used to cap beer and soda bottles. It was a tin cap covered with cork, used to form a tight seal over a bottle. Soda and beer companies could use a crown cork only once and needed a steady supply.

King took Painter's advice and began thinking of everyday items that needed to be replaced often. After owning a Star safety razor for some time, King realized that the razor blade had not been improved for a long time. He studied all the razors on the market and found that both the common straight razor and the safety razor featured a heavy V-shaped piece of steel, sharpened on one side. King reasoned that a thin piece of steel sharpened on both sides would create a better shave and could be thrown away once it became dull. The idea of the disposable razor had been born.

Gillette made several drawings of disposable razors. He then made a wooden model of the razor to better explain his idea. Gillette's first attempt to construct a working model was unsuccessful, as the steel was too flimsy. Steven Porter, a Boston machinist, decided to try to make Gillette's razor from his drawings. He produced three razors, and in the summer of 1899 King was the first man to shave with a disposable razor.

CHANGING CONSUMER OPINION

In the early 1900's, most people considered a razor to be a once-in-a-lifetime purchase. Many fathers handed down their razors to their sons. Straight razors needed constant and careful attention to keep them sharp. The thought of throwing a razor in the garbage after several uses was contrary to the general public's idea of a razor. If Gillette's razor had not provided a much less painful and faster shave, it is unlikely that the disposable would have been a success. Even with its advantages, public opinion against the product was still difficult to overcome.

Financing a company to produce the razor proved to be a major obstacle. King did not have the money himself, and potential investors were skeptical. Skepticism arose both because of public perceptions of the product and because of its manufacturing process. Mass production appeared to be impossible, but the disposable razor would never be profitable if produced using the methods used to manufacture its predecessor.

William Emery Nickerson, an expert machine inventor, had looked at Gillette's razor and said it was impossible to create a machine to produce it. He was convinced to reexamine the idea and finally created a machine that would create a workable blade. In the process, Nickerson changed Gillette's original model. He improved the handle and frame so that it would better support the thin steel blade.

In the meantime, Gillette was busy getting his patent assigned to the newly formed American Safety Razor Company, owned by Gillette, Jacob Heilborn, Edward J. Stewart, and Nickerson. Gillette owned considerably more shares than anyone else. Henry Sachs provided additional capital, buying shares from Gillette.

The stockholders decided to rename the company the Gillette Safety Razor Company. It soon spent most of its money on machinery and lacked the capital it needed to produce and advertise its product. The only offer the company had received was from a group of New York investors who were willing to give $125,000 in exchange for 51 percent of the company. None of the directors wanted to lose control of the company, so they rejected the offer.

John Joyce, a friend of Gillette, rescued the financially insecure new company. He agreed to buy $100,000 worth of bonds from the company for sixty cents on the dollar, purchasing the bonds gradually as the company needed money. He also received an equivalent amount of company stock. After an investment of $30,000, Joyce had the option of backing out. This deal enabled the company to start manufacturing and advertising.

IMPACT

The company used $18,000 to perfect the machinery to produce the disposable razor blades and razors. Originally the directors wanted to sell each razor with twenty blades for three dollars. Joyce insisted on a price of five dollars. In 1903, five dollars was about one-third of the average American's weekly salary, and a high-quality straight razor could be purchased for about half that price.

The other directors were skeptical, but Joyce threatened to buy up all the razors for three dollars and sell them himself for five dollars. Joyce had the financial backing to make this promise good, so the directors agreed to the higher price.

The Gillette Safety Razor Company contracted with Townsend & Hunt for exclusive sales. The contract stated that Townsend & Hunt would buy 50,000 razors with twenty blades each during a period of slightly more than a year and would purchase 100,000 sets per year for the following four years. The first advertisement for the product appeared in *System Magazine* in early fall of 1903, offering the razors by mail order. By the end of 1903, only fifty-one razors had been sold.

Since Gillette and most of the directors of the company were not salaried, Gillette had needed to keep his job as salesman with Crown Cork and Seal. At the end of 1903, he received a promotion that meant relocation from Boston to London. Gillette did not want to go and pleaded with the other directors, but they insisted that the company could not afford to put him on salary. The company decided to reduce the number of blades in a set from twenty to twelve in an effort to increase profits without noticeably raising the cost of a set. Gillette resigned the title of company president and left for England.

Shortly thereafter, Townsend & Hunt changed its name to the Gillette Sales Company, and three years later the sales company sold out to the parent company for $300,000. Sales of the new type of razor were increasing rapidly in the United States, and Joyce wanted to sell patent rights to European companies for a small percentage of sales. Gillette thought that that would be a horrible mistake and quickly traveled back to Boston. He had two goals: to stop the sale of patent rights, based on his conviction that the foreign market would eventually be very lucrative, and to become salaried by the company. Gillette accomplished both these goals and soon moved back to Boston.

Despite the fact that Joyce and Gillette had been good friends for a long time, their business views often differed. Gillette set up a holding company in an effort to gain back controlling interest in the Gillette Safety Razor Company. He borrowed money and convinced his allies in the company to invest in the holding company, eventually regaining control. He was reinstated as president of the company. One clear disagreement was that Gillette wanted to relocate the company to Newark, New Jersey, and Joyce thought that that would be a waste of money. Gillette authorized company funds to be invested in a Newark site. The idea was later dropped, costing the company a large amount of capital. Gillette was not a very wise businessman

KING CAMP GILLETTE

At age sixteen, King Camp Gillette (1855-1932) saw all of his family's belongings consumed in the Great Chicago Fire. He had to drop out of school because of it and earn his own living. The catastrophe and the sudden loss of security that followed shaped his ambitions. He was not about to risk destitution ever again.

He made himself a successful traveling salesman but still felt he was earning too little. So he turned his mind to inventions, hoping to get rich quick. The disposable razor was his only venture, but it was enough. After its long preparation for marketing Gillette's invention and some subsequent turmoil among its board of directors, the Gillette Safety Razor Company was a phenomenal success and a bonanza for Gillette. He became wealthy. He retired in 1913, just ten years after the company opened, his security assured.

His mother had written cookbooks, one of which was a bestseller. As an adult, Gillette got the writing bug himself and wrote four books, but his theme was far loftier than cooking—social theory and security for the masses. Like Karl Marx he argued that economic competition squanders human resources and leads to deprivation, which in turn leads to crime. So, he reasoned, getting rid of economic competition will end misery and crime. He recommended that a centralized agency plan production and oversee distribution, a recommendation that America resoundingly ignored. However, other ideas of his eventually found acceptance, such as air conditioning for workers and government assistance for the unemployed.

In 1922 Gillette moved to Los Angeles, California, and devoted himself to raising oranges and collecting his share of the company profits. However, he seldom felt free enough with his money to donate it to charity or finance social reform.

and made many costly mistakes. Joyce even accused him of deliberately trying to keep the stock price low so that Gillette could purchase more stock. Joyce eventually bought out Gillette, who retained his title as president but had little say about company business.

With Gillette out of a management position, the company became more stable and more profitable. The biggest problem the company faced was that it would soon lose its patent rights. After the patent expired, the company would have competition. The company decided that it could either cut prices (and therefore profits) to compete with the lower-priced disposables that would inevitably enter the market, or it could create a new line of even better razors. The company opted for the latter strategy. Weeks before the patent expired, the Gillette Safety Razor Company introduced a new line of razors.

Both World War I and World War II were big boosts to the company, which contracted with the government to supply razors to almost all the troops. This transaction created a huge increase in sales and introduced thousands of young men to the Gillette razor. Many of them continued to use Gillettes after returning from the war.

Aside from the shaky start of the company, its worst financial difficulties were during the Great Depression. Most Americans simply could not afford Gillette blades, and many used a blade for an extended time and then resharpened it rather than throwing it away. If it had not been for the company's foreign markets, the company would not have shown a profit during the Great Depression. Gillette's obstinacy about not selling patent rights to foreign investors proved to be an excellent decision.

The company advertised through sponsoring sporting events, including the World Series. Gillette had many celebrity endorsements from well-known baseball players. Before it became too expensive for one company to sponsor an entire event, Gillette had exclusive advertising during the World Series, various boxing matches, the Kentucky Derby, and football bowl games. Sponsoring these events was costly, but sports spectators were the typical Gillette customers.

The Gillette Company created many products that complemented razors and blades, including shaving cream, women's razors, and electric razors. The company expanded into new products including women's cosmetics, writing utensils, deodorant, and wigs. One of the main reasons for obtaining a more diverse product line was that a one-product company is less stable, especially in a volatile market. The Gillette Company had learned that lesson in the Great Depression. Gillette continued to thrive by following the principles the company had used from the start. The majority of Gillette's profits came from foreign markets, and its employees looked to improve products and find opportunities in other departments as well as their own.

See also Contact lenses; Memory metal; Steelmaking process.

FURTHER READING

Adams, Russell B., Jr. *King C. Gillette: The Man and His Wonderful Shaving Device*. Boston: Little, Brown, 1978.
Dowling, Tim. *Inventor of the Disposable Culture: King Camp Gillette, 1855-1932*. London: Short, 2001.
"Gillette: Blade-runner." *The Economist* 327 (April 10, 1993).
Killgren, Lucy. "Nicking Gillette." *Marketing Week* 22 (June 17, 1999).
McKibben, Gordon. *Cutting Edge: Gillette's Journey to Global Leadership*. Boston, Mass.: Harvard Business School Press, 1998.
Thomas, Robert J. *New Product Success Stories: Lessons from Leading Innovators*. New York: John Wiley, 1995.
Zeien, Alfred M. *The Gillette Company*. New York: Newcomen Society of the United States, 1999.

DOLBY NOISE REDUCTION

THE INVENTION: Electronic device that reduces the signal-to-noise ratio of sound recordings and greatly improves the sound quality of recorded music.

THE PEOPLE BEHIND THE INVENTION:
Emil Berliner (1851-1929), a German inventor
Ray Milton Dolby (1933-   ), an American inventor
Thomas Alva Edison (1847-1931), an American inventor

PHONOGRAPHS, TAPES, AND NOISE REDUCTION

The main use of record, tape, and compact disc players is to listen to music, although they are also used to listen to recorded speeches, messages, and various forms of instruction. Thomas Alva Edison invented the first sound-reproducing machine, which he called the "phonograph," and patented it in 1877. Ten years later, a practical phonograph (the "gramophone") was marketed by a German, Emil Berliner. Phonographs recorded sound by using diaphragms that vibrated in response to sound waves and controlled needles that cut grooves representing those vibrations into the first phonograph records, which in Edison's machine were metal cylinders and in Berliner's were flat discs. The recordings were then played by reversing the recording process: Placing a needle in the groove in the recorded cylinder or disk caused the diaphragm to vibrate, re-creating the original sound that had been recorded.

In the 1920's, electrical recording methods developed that produced higher-quality recordings, and then, in the 1930's, stereophonic recording was developed by various companies, including the British company Electrical and Musical Industries (EMI). Almost simultaneously, the technology of tape recording was developed. By the 1940's, long-playing stereo records and tapes were widely available. As recording techniques improved further, tapes became very popular, and by the 1960's, they had evolved into both studio master recording tapes and the audio cassettes used by consumers.

Hisses and other noises associated with sound recording and its environment greatly diminished the quality of recorded music. In 1967, Ray Dolby invented a noise reducer, later named "Dolby A," that could be used by recording studios to reduce tape signal-to-noise ratios. Several years later, his "Dolby B" system, designed for home use, became standard equipment in all types of play-back machines. Later, Dolby and others designed improved noise-suppression systems.

RECORDING AND TAPE NOISE

Sound is made up of vibrations of varying frequencies—sound waves—that sound recorders can convert into grooves on plastic re-cords, varying magnetic arrangements on plastic tapes covered with iron particles, or tiny pits on compact discs. The following dis-cussion will focus on tape recordings, for which the original Dolby noise reducers were designed.

Tape recordings are made by a process that converts sound waves into electrical impulses that cause the iron particles in a tape to reorganize themselves into particular magnetic arrangements. The process is reversed when the tape is played back. In this pro-cess, the particle arrangements are translated first into electrical im-pulses and then into sound that is produced by loudspeakers. Erasing a tape causes the iron particles to move back into their origi-nal spatial arrangement.

Whenever a recording is made, undesired sounds such as hisses, hums, pops, and clicks can mask the nuances of recorded sound, an-noying and fatiguing listeners. The first attempts to do away with undesired sounds (noise) involved making tapes, recording de-vices, and recording studios quieter. Such efforts did not, however, remove all undesired sounds.

Furthermore, advances in recording technology increased the problem of noise by producing better instruments that "heard" and transmitted to recordings increased levels of noise. Such noise is of-ten caused by the components of the recording system; tape hiss is an example of such noise. This type of noise is most discernible in quiet passages of recordings, because loud recorded sounds often mask it.

RAY DOLBY

Ray Dolby, born in Portland, Oregon, in 1933, became an electronics engineer while still in high school in 1952. That is when he began working part time for Ampex Corporation, helping develop the first videotape recorder. He was responsible for the electronics in the Ampex VTR, which was marketed in 1956. The next year he finished a bachelor of science degree at Stanford University, won a Marshall Scholarship and National Science Foundation grant, and went to Cambridge University in England for graduate studies. He received a Ph.D. in 1961 and a fellowship to Pembroke College, during which he also consulted for the United Kingdom Atomic Energy Authority.

After two years in India as a United Nations adviser, he set up Dolby Laboratories in London. It was there that he produced the sound suppression equipment that made him famous to audiophiles and movie goers, particularly in the 1970's for the Dolby stereo ("surround sound") that enlivened such blockbusters as *Star Wars*. In 1976 he moved to San Francisco and opened new offices for his company. The holder of more than fifty patents, Dolby published monographs on videotape recording, long wavelength X-ray analysis, and noise reduction. He is among the most honored scientists in the recording industry. Among many other awards, he received an Oscar, Emmy, Samuel L. Warner Memorial Award, gold and silver medals from the Audio Engineering Society, and the National Medal of Technology. England made him an honorary Officer of the Most Excellent Order of the British Empire, and Cambridge University and York University awarded him honorary doctorates.

Because of the problem of noise in quiet passages of recorded sound, one early attempt at noise suppression involved the reduction of noise levels by using "dynaural" noise suppressors. These devices did not alter the loud portions of a recording; instead, they reduced the very high and very low frequencies in the quiet passages in which noise became most audible. The problem with such devices was, however, that removing the high and low frequencies could also affect the desirable portions of the recorded sound. These suppressors could not distinguish desirable from undesirable sounds. As recording techniques improved, dynaural noise suppressors caused more and more problems, and their use was finally discontinued.

Another approach to noise suppression is sound compression during the recording process. This compression is based on the fact that most noise remains at a constant level throughout a recording, regardless of the sound level of a desired signal (such as music). To carry out sound compression, the lowest-level signals in a recording are electronically elevated above the sound level of all noise. Musical nuances can be lost when the process is carried too far, because the maximum sound level is not increased by devices that use sound compression. To return the music or other recorded sound to its normal sound range for listening, devices that "expand" the recorded music on playback are used. Two potential problems associated with the use of sound compression and expansion are the difficulty of matching the two processes and the introduction into the recording of noise created by the compression devices themselves.

In 1967, Ray Dolby developed Dolby A to solve these problems as they related to tape noise (but not to microphone signals) in the recording and playing back of studio master tapes. The system operated by carrying out ten-decibel compression during recording and then restoring (noiselessly) the range of the music on playback. This was accomplished by expanding the sound exactly to its original range. Dolby A was very expensive and was thus limited to use in recording studios. In the early 1970's, however, Dolby invented the less expensive Dolby B system, which was intended for consumers.

CONSEQUENCES

The development of Dolby A and Dolby B noise-reduction systems is one of the most important contributions to the high-quality recording and reproduction of sound. For this reason, Dolby A quickly became standard in the recording industry. In similar fashion, Dolby B was soon incorporated into virtually every high-fidelity stereo cassette deck to be manufactured.

Dolby's discoveries spurred advances in the field of noise reduction. For example, the German company Telefunken and the Japanese companies Sanyo and Toshiba, among others, developed their own noise-reduction systems. Dolby Laboratories countered by producing an improved system: Dolby C. The competition in the area of noise reduction continues, and it will continue as long as changes in recording technology produce new, more sensitive recording equipment.

See also Cassette recording; Compact disc; Electronic synthesizer; FM radio; Radio; Transistor; Transistor radio; Walkman cassette player.

FURTHER READING

Alkin, E. G. M. *Sound Recording and Reproduction.* 3d ed. Boston: Focal Press, 1996.
Baldwin, Neil. *Edison: Inventing the Century.* Chicago: University of Chicago Press, 2001.
Wile, Frederic William. *Emile Berliner, Maker of the Microphone.* New York: Arno Press, 1974.

ELECTRIC CLOCK

THE INVENTION: Electrically powered time-keeping device with a quartz resonator that has led to the development of extremely accurate, relatively inexpensive electric clocks that are used in computers and microprocessors.

THE PERSON BEHIND THE INVENTION:
Warren Alvin Marrison (1896-1980), an American scientist

FROM COMPLEX MECHANISMS TO QUARTZ CRYSTALS

William Alvin Marrison's fabrication of the electric clock began a new era in time-keeping. Electric clocks are more accurate and more reliable than mechanical clocks, since they have fewer moving parts and are less likely to malfunction.

An electric clock is a device that generates a string of electric pulses. The most frequently used electric clocks are called "free running" and "periodic," which means that they generate a continuous sequence of electric pulses that are equally spaced. There are various kinds of electronic "oscillators" (materials that vibrate) that can be used to manufacture electric clocks.

The material most commonly used as an oscillator in electric clocks is crystalline quartz. Because quartz (silicon dioxide) is a completely oxidized compound (which means that it does not deteriorate readily) and is virtually insoluble in water, it is chemically stable and resists chemical processes that would break down other materials. Quartz is a "piezoelectric" material, which means that it is capable of generating electricity when it is subjected to pressure or stress of some kind. In addition, quartz has the advantage of generating electricity at a very stable frequency, with little variation. For these reasons, quartz is an ideal material to use as an oscillator.

THE QUARTZ CLOCK

A quartz clock is an electric clock that makes use of the piezoelectric properties of a quartz crystal. When a quartz crystal vibrates, a

Early electric clock. (PhotoDisc)

difference of electric potential is produced between two of its faces. The crystal has a natural frequency (rate) of vibration that is determined by its size and shape. If the crystal is placed in an oscillating electric circuit that has a frequency that is nearly the same as that of the crystal, it will vibrate at its natural frequency and will cause the frequency of the entire circuit to match its own frequency.

Piezoelectricity is electricity, or "electric polarity," that is caused by the application of mechanical pressure on a "dielectric" material (one that does not conduct electricity), such as a quartz crystal. The process also works in reverse; if an electric charge is applied to the dielectric material, the material will experience a mechanical distortion. This reciprocal relationship is called "the piezoelectric effect." The phenomenon of electricity being generated by the application of mechanical pressure is called the direct piezoelectric effect, and the phenomenon of mechanical stress being produced as a result of the application of electricity is called the converse piezoelectric effect.

When a quartz crystal is used to create an oscillator, the natural frequency of the crystal can be used to produce other frequencies that can power clocks. The natural frequency of a quartz crystal is nearly constant if precautions are taken when it is cut and polished and if it is maintained at a nearly constant temperature and pressure. After a quartz crystal has been used for some time, its fre-

WARREN ALVIN MARRISON

Born in Invenary, Canada, in 1896, Warren Alvin Marrison completed high school at Kingston Collegiate Institute in Ontario and attended Queen's University in Kingston, where he studied science. World War I interrupted his studies, and while serving in the Royal Flying Corps as an electronics researcher, he began his life-long interest in radio. He graduated from university with a degree in engineering physics in 1920, transferred to Harvard University in 1921, and earned a master's degree.

After his studies, he worked for the Western Electric Company in New York, helping to develop a method to record sound on film. He moved to the company's Bell Laboratory in 1925 and studied how to produce frequency standards for radio transmissions. This research led him to use quartz crystals as oscillators, and he was able to step down the frequency enough that it could power a motor. Because the motor revolved at the same rate as the crystal's frequency, he could determine the number of vibrations per time unit of the crystal and set a frequency standard. However, because the vibrations were constant over time, the crystal also measured time, and a new type of clock was born.

For his work, Marrison received the British Horological Institute's Gold Medal in 1947 and the Clockmakers' Company's Tompion Medal in 1955. He died in California in 1980.

quency usually varies slowly as a result of physical changes. If allowances are made for such changes, quartz-crystal clocks such as those used in laboratories can be manufactured that will accumulate errors of only a few thousandths of a second per month. The quartz crystals that are typically used in watches, however, may accumulate errors of tens of seconds per year.

There are other materials that can be used to manufacture accurate electric clocks. For example, clocks that use the element rubidium typically would accumulate errors no larger than a few ten-thousandths of a second per year, and those that use the element cesium would experience errors of only a few millionths of a second per year. Quartz is much less expensive than rarer materials such as rubidium and cesium, and it is easy to use in such common applications as computers. Thus, despite their relative inaccuracy, electric quartz clocks are extremely useful and popular, particularly for applications that require accurate timekeeping over a relatively short period of time. In such applications, quartz clocks may be adjusted periodically to correct for accumulated errors.

IMPACT

The electric quartz clock has contributed significantly to the development of computers and microprocessors. The computer's control unit controls and synchronizes all data transfers and transformations in the computer system and is the key subsystem in the computer itself. Every action that the computer performs is implemented by the control unit.

The computer's control unit uses inputs from a quartz clock to derive timing and control signals that regulate the actions in the system that are associated with each computer instruction. The control unit also accepts, as input, control signals generated by other devices in the computer system.

The other primary impact of the quartz clock is in making the construction of multiphase clocks a simple task. A multiphase clock is a clock that has several outputs that oscillate at the same frequency. These outputs may generate electric waveforms of different shapes or of the same shape, which makes them useful for various applications. It is common for a computer to incorporate a single-phase quartz clock that is used to generate a two-phase clock.

See also Atomic clock; Carbon dating; Electric refrigerator; Fluorescent lighting; Microwave cooking; Television; Vacuum cleaner; Washing machine.

FURTHER READING

Barnett, Jo Ellen. *Time's Pendulum: From Sundials to Atomic Clocks, the Fascinating History of Time Keeping and How Our Discoveries Changed the World.* San Diego: Harcourt Brace, 1999.

Dennis, Maggie, and Carlene Stephens. "Engineering Time: Inventing the Electronic Wristwatch." *British Journal for the History of Science* 33, no. 119 (December, 2000).

Ganeri, Anita. *From Candle to Quartz Clock: The Story of Time and Timekeeping*. London: Evna Brothers, 1996.

Thurber, Karl. "All the Time in the World." *Popular Electronics* 14, no. 10 (October, 1997).

Electric refrigerator

The invention: An electrically powered and hermetically sealed food-storage appliance that replaced iceboxes, improved production, and lowered food-storage costs.

The people behind the invention:
Marcel Audiffren, a French monk
Christian Steenstrup (1873-1955), an American engineer
Fred Wolf, an American engineer

Ice Preserves America's Food

Before the development of refrigeration in the United States, a relatively warm climate made it difficult to preserve food. Meat spoiled within a day and milk could spoil within an hour after milking. In early America, ice was stored below ground in icehouses that had roofs at ground level. George Washington had a large icehouse at his Mount Vernon estate. By 1876, America was consuming more than 2 million tons of ice each year, which required 4,000 horses and 10,000 men to deliver.

Several related inventions were needed before mechanical refrigeration was developed. James Watt invented the condenser, an important refrigeration system component, in 1769. In 1805, Oliver Evans presented the idea of continuous circulation of a refrigerant in a closed cycle. In this closed cooling cycle, a liquid refrigerant evaporates to a gas at low temperature, absorbing heat from its environment and thereby producing "cold," which is circulated around an enclosed cabinet. To maintain this cooling cycle, the refrigerant gas must be returned to liquid form through condensation by compression. The first closed-cycle vapor-compression refrigerator, which was patented by Jacob Perkins in 1834, used ether as a refrigerant.

Iceboxes were used in homes before refrigerators were developed. Ice was cut from lakes and rivers in the northern United States or produced by ice machines in the southern United States. An ice machine using air was patented by John Gorrie at New Orleans in 1851. Ferdinand Carre introduced the first successful commercial ice machine, which used ammonia as a refrigerant, in 1862, but it was too large for home use and produced only a pound of ice per hour. Ice machinery became very dependable after 1890 but was plagued by low efficiency. Very warm summers in 1890 and 1891 cut natural ice production dramatically and increased demand for mechanical ice production. Ice consumption continued to increase after 1890; by 1914, 21 million tons of ice were used annually. The high prices charged for ice and the extremely low efficiency of home iceboxes gradually led the public to demand a substitute for ice refrigeration.

REFRIGERATION FOR THE HOME

Domestic refrigeration required a compact unit with a built-in electric motor that did not require supervision or maintenance. Marcel Audiffren, a French monk, conceived the idea of an electric refrigerator for home use around 1910. The first electric refrigerator, which was invented by Fred Wolf in 1913, was called the Domelre, which stood for domestic electric refrigerator. This machine used condensation equipment that was housed in the home's basement. In 1915, Alfred Mellowes built the first refrigerator to contain all of its components; this machine was known as Guardian's Frigerator. General Motors acquired Guardian in 1918 and began to mass produce refrigerators. Guardian was renamed Frigidaire in 1919. In 1918, the Kelvinator Company, run by Edmund Copeland, built the first refrigerator with automatic controls, the most important of which was the thermostatic switch. Despite these advances, by 1920 only a few thousand homes had refrigerators, which cost about $1,000 each.

The General Electric Company (GE) purchased the rights to the General Motors refrigerator, which was based on an improved design submitted by one of its engineers, Christian Steenstrup. Steenstrup's innovative design included a motor and reciprocating compressor that were hermetically sealed with the refrigerant. This unit, known as the GE Monitor Top, was first produced in 1927. A patent on this machine was filed for in 1926 and granted to Steenstrup in 1930. Steenstrup became chief engineer of GE's electric refrigeration department and accumulated thirty-nine additional patents in refrigeration over the following years. By 1936, he had more than one hundred patents to his credit in refrigeration and other areas.

Further refinement of the refrigerator evolved with the development of Freon, a nonexplosive, nontoxic, and noncorrosive refrigerant discovered by Thomas Midgely, Jr., in 1928. Freon used lower pressures than ammonia did, which meant that lighter materials and lower temperatures could be used in refrigeration.

During the years following the introduction of the Monitor Top, the cost of refrigerators dropped from $1,000 in 1918 to $400 in 1926, and then to $170 in 1935. Sales of units increased from 200,000 in 1926 to 1.5 million in 1935.

Initially, refrigerators were sold separately from their cabinets, which commonly were used wooden iceboxes. Frigidaire began making its own cabinets in 1923, and by 1930, refrigerators that combined machinery and cabinet were sold.

Throughout the 1930's, refrigerators were well-insulated, hermetically sealed steel units that used evaporator coils to cool the food compartment. The refrigeration system was transferred from on top of to below the food storage area, which made it possible to raise the food storage area to a more convenient level. Special light bulbs that produced radiation to kill taste- and odor-bearing bacteria were used in refrigerators. Other developments included sliding shelves, shelves in doors, rounded and styled cabinet corners, ice cube trays, and even a built-in radio.

The freezing capacity of early refrigerators was inadequate. Only a package or two of food could be kept cool at a time, ice cubes melted, and only a minimal amount of food could be kept frozen. The two-temperature refrigerator consisting of one compartment providing normal cooling and a separate compartment for freezing was developed by GE in 1939. Evaporator coils for cooling were placed within the refrigerator walls, providing more cooling capacity and more space for food storage. Frigidaire introduced a Cold Wall compartment, while White-Westinghouse introduced a Colder Cold system. After World War II, GE introduced the refrigerator-freezer combination.

IMPACT

Audiffren, Wolf, Steenstrup, and others combined the earlier inventions of Watt, Perkins, and Carre with the development of electric motors to produce the electric refrigerator. The development of domestic electric refrigeration had a tremendous effect on the quality of home life. Reliable, affordable refrigeration allowed consumers a wider selection of food and increased flexibility in their daily consumption. The domestic refrigerator with increased freezer capacity spawned the growth of the frozen food industry. Without the electric refrigerator, households would still depend on unreliable supplies of ice.

See also Fluorescent lighting; Food freezing; Freeze-drying; Microwave cooking; Refrigerant gas; Robot (household); Tupperware; Vacuum cleaner; Washing machine.

FURTHER READING

Anderson, Oscar Edward. *Refrigeration in America: A History of a New Technology and Its Impact*. Princeton: Princeton University Press, 1953.

Donaldson, Barry, Bernard Nagengast, and Gershon Meckler. *Heat and Cold: Mastering the Great Indoors: A Selective History of Heating, Ventilation, Air-Conditioning and Refrigeration from the Ancients to the 1930's*. Atlanta, Ga.: American Society of Heating, Refrigerating and Air-Conditioning Engineers, 1994.

Woolrich, Willis Raymond. *The Men Who Created Cold: A History of Refrigeration*. New York: Exposition Press, 1967.

Electrocardiogram

The invention: Device for analyzing the electrical currents of the human heart.

The people behind the invention:
Willem Einthoven (1860-1927), a Dutch physiologist and winner of the 1924 Nobel Prize in Physiology or Medicine
Augustus D. Waller (1856-1922), a German physician and researcher
Sir Thomas Lewis (1881-1945), an English physiologist

Horse Vibrations

In the late 1800's, there was substantial research interest in the electrical activity that took place in the human body. Researchers studied many organs and systems in the body, including the nerves, eyes, lungs, muscles, and heart. Because of a lack of available technology, this research was tedious and frequently inaccurate. Therefore, the development of the appropriate instrumentation was as important as the research itself.

The initial work on the electrical activity of the heart (detected from the surface of the body) was conducted by Augustus D. Waller and published in 1887. Many credit him with the development of the first electrocardiogram. Waller used a Lippmann's capillary electrometer (named for its inventor, the French physicist Gabriel-Jonas Lippmann) to determine the electrical charges in the heart and called his recording a "cardiograph." The recording was made by placing a series of small tubes on the surface of the body. The tubes contained mercury and sulfuric acid. As an electrical current passed through the tubes, the mercury would expand and contract. The resulting images were projected onto photographic paper to produce the first cardiograph. Yet Waller had only limited sucess with the device and eventually abandoned it.

In the early 1890's, Willem Einthoven, who became a good friend of Waller, began using the same type of capillary tube to study the electrical currents of the heart. Einthoven also had a difficult time working with the instrument. His laboratory was located in an old wooden building near a cobblestone street. Teams of horses pulling heavy wagons would pass by and cause his laboratory to vibrate. This vibration affected the capillary tube, causing the cardiograph to be unclear. In his frustration, Einthoven began to modify his laboratory. He removed the floorboards and dug a hole some ten to fifteen feet deep. He lined the walls with large rocks to stabilize his instrument. When this failed to solve the problem, Einthoven, too, abandoned the Lippmann's capillary tube. Yet Einthoven did not abandon the idea, and he began to experiment with other instruments.

ELECTROCARDIOGRAPHS OVER THE PHONE

In order to continue his research on the electrical currents of the heart, Einthoven began to work with a new device, the d'Arsonval galvanometer (named for its inventor, the French biophysicist Arsène d'Arsonval). This instrument had a heavy coil of wire suspended between the poles of a horseshoe magnet. Changes in electrical activity would cause the coil to move; however, Einthoven found that the coil was too heavy to record the small electrical changes found in the heart. Therefore, he modified the instrument by replacing the coil with a silver-coated quartz thread (string). The movements could be recorded by transmitting the deflections through a microscope and projecting them on photographic film. Einthoven called the new instrument the "string galvanometer."

In developing his string galvanomter, Einthoven was influenced by the work of one of his teachers, Johannes Bosscha. In the 1850's, Bosscha had published a study describing the technical complexities of measuring very small amounts of electricity. He proposed the idea that a galvanometer modified with a needle hanging from a silk thread would be more sensitive in measuring the tiny electric currents of the heart.

By 1905, Einthoven had improved the string galvanometer to the point that he could begin using it for clinical studies. In 1906, he had his laboratory connected to the hospital in Leiden by a telephone wire. With this arrangement, Einthoven was able to study in his laboratory electrocardiograms derived from patients in the

WILLEM EINTHOVEN

Willem Einthoven was born in 1860 on the Island of Java, now part of Indonesia. His father was a Dutch army medical officer, and his mother was the daughter of the Finance Director for the Dutch East Indies. When his father died in 1870, his mother moved with her six children to Utrecht, Holland.

Einthoven entered the University of Utrecht in 1878 intending to become a physician like his father, but physics and physiology attracted him more. During his education two research projects that he conducted brought him notoriety. The first involved the articulation of the elbow, which he undertook after a sports injury of his own elbow. (He remained an avid participant in sports his whole life.) The second, which earned him his doctorate in 1885, examined stereoscopy and color variation. Because of the keen investigative abilities these studies displayed, he was at once appointed professor of physiology at the University of Leiden. He took up the position the next year, after qualifying as a general practitioner.

Einthoven conducted research into asthma and the optics and electrical activity of vision before turning his attention to the heart. He developed the electrocardiogram in order to measure the heart's electrical activity accurately and tested its applications and capacities with many students and visiting scientists, helping thereby to widen interest in it as a diagnostic tool. For this work he received the 1924 Nobel Prize in Physiology or Medicine.

In his later years, Einthoven studied problems in acoustics and the electrical activity of the sympathetic nervous system. He died in Leiden in 1927.

hospital, which was located a mile away. With this source of subjects, Einthoven was able to use his galvanometer to study many heart problems. As a result of these studies, Einthoven identified the following heart problems: blocks in the electrical conduction system of the heart; premature beats of the heart, including two premature beats in a row; and enlargements of the various chambers of the heart. He was also able to study how the heart behaved during the administration of cardiac drugs.

A major researcher who communicated with Einthoven about the electrocardiogram was Sir Thomas Lewis, who is credited with developing the electrocardiogram into a useful clinical tool. One of Lewis's important accomplishments was his identification of atrial fibrillation, the overactive state of the upper chambers of the heart. During World War I, Lewis was involved with studying soldiers' hearts. He designed a series of graded exercises, which he used to test the soldiers' ability to perform work. From this study, Lewis was able to use similar tests to diagnose heart disease and to screen recruits who had heart problems.

IMPACT

As Einthoven published additional studies on the string galvanometer in 1903, 1906, and 1908, greater interest in his instrument was generated around the world. In 1910, the instrument, now called the "electrocardiograph," was installed in the United States. It was the foundation of a new laboratory for the study of heart disease at Johns Hopkins University.

As time passed, the use of the electrocardiogram—or "EKG," as it is familiarly known—increased substantially. The major advantage of the EKG is that it can be used to diagnose problems in the heart without incisions or the use of needles. It is relatively painless for the patient; in comparison with other diagnostic techniques, moreover, it is relatively inexpensive.

Recent developments in the use of the EKG have been in the area of stress testing. Since many heart problems are more evident during exercise, when the heart is working harder, EKGs are often given to patients as they exercise, generally on a treadmill. The clinician gradually increases the intensity of work the patient is doing while monitoring the patient's heart. The use of stress testing has helped to make the EKG an even more valuable diagnostic tool.

See also Amniocentesis; Artificial heart; Blood transfusion; CAT scanner; Coronary artery bypass surgery; Electroencephalogram; Heart-lung machine; Mammography; Nuclear magnetic resonance; Pacemaker; Ultrasound; X-ray image intensifier.

Further Reading

Cline, Barbara Lovett. *Men Who Made a New Physics: Physicists and the Quantum Theory.* Chicago: University of Chicago Press, 1987.

Hollman, Arthur. *Sir Thomas Lewis: Pioneer Cardiologist and Clinical Scientist.* New York: Springer, 1997.

Lewis, Thomas. *Collected Works on Heart Disease.* 1912. Reprint. New York: Classics of Cardiology Library, 1991.

Snellen, H. A. *Two Pioneers of Electrocardiography: The Correspondence Between Einthoven and Lewis from 1908-1926.* Rotterdam: Donker Academic Publications, 1983.

_____. *Willem Einthoven, 1860-1927, Father of Electrocardiography: Life and Work, Ancestors and Contemporaries.* Boston: Kluwer Academic Publishers, 1995.

ELECTROENCEPHALOGRAM

THE INVENTION: A system of electrodes that measures brain wave patterns in humans, making possible a new era of neurophysiology.

THE PEOPLE BEHIND THE INVENTION:
Hans Berger (1873-1941), a German psychiatrist and research scientist
Richard Caton (1842-1926), an English physiologist and surgeon

THE ELECTRICAL ACTIVITY OF THE BRAIN

Hans Berger's search for the human electroencephalograph (English physiologist Richard Caton had described the electroencephalogram, or "brain wave," in rabbits and monkeys in 1875) was motivated by his desire to find a physiological method that might be applied successfully to the study of the long-standing problem of the relationship between the mind and the brain. His scientific career, therefore, was directed toward revealing the psychophysical relationship in terms of principles that would be rooted firmly in the natural sciences and would not have to rely upon vague philosophical or mystical ideas.

During his early career, Berger attempted to study psychophysical relationships by making plethysmographic measurements of changes in the brain circulation of patients with skull defects. In plethysmography, an instrument is used to indicate and record by tracings the variations in size of an organ or part of the body. Later, Berger investigated temperature changes occurring in the human brain during mental activity and the action of psychoactive drugs. He became disillusioned, however, by the lack of psychophysical understanding generated by these investigations.

Next, Berger turned to the study of the electrical activity of the brain, and in the 1920's he set out to search for the human electroencephalogram. He believed that the electroencephalogram would finally provide him with a physiological method capable of furnishing insight into mental functions and their disturbances.

Berger made his first unsuccessful attempt at recording the electrical activity of the brain in 1920, using the scalp of a bald medical student. He then attempted to stimulate the cortex of patients with skull defects by using a set of electrodes to apply an electrical current to the skin covering the defect. The main purpose of these stimulation experiments was to elicit subjective sensations. Berger hoped that eliciting these sensations might give him some clue about the nature of the relationship between the physiochemical events produced by the electrical stimulus and the mental processes revealed by the patients' subjective experience. The availability of many patients with skull defects—in whom the pulsating surface of the brain was separated from the stimulating electrodes by only a few millimeters of tissue—reactivated Berger's interest in recording the brain's electrical activity.

HANS BERGER

Hans Berger, the father of electroencephalography, was born in Neuses bei Coburn, Germany, in 1873. He entered the University of Jena in 1892 as a medical student and became an assistant in the psychiatric clinic in 1897. In 1912 he was appointed the clinic's chief doctor and then its director and a university professor of psychiatry. In 1919 he was chosen as rector of the university.

Berger hoped to settle the long-standing philosophical question about the brain and the mind by finding observable physical processes that correlated with thought and feelings. He started off by studying the blood circulation in the head and brain temperature. Even though this work founded psychophysiology, he failed to find objective evidence of subjective states until he started examining fluctuations in the electrical potential of the brain in 1924. His 1929 paper describing the electroencephalograph later provided medicine with a basic diagnostic tool, but the instrument proved to be a very confusing probe of the human psyche for him. His colleagues in psychiatry and medicine did not accept his relationships of physical phenomena and mental states.

Berger retired as professor emeritus in 1938 and died three years later in Jena.

SMALL, TREMULOUS MOVEMENTS

Berger used several different instruments in trying to detect brain waves, but all of them used a similar method of recording. Electrical oscillations deflected a mirror upon which a light beam was projected. The deflections of the light beam were proportional to the magnitude of the electrical signals. The movement of the spot of the light beam was recorded on photographic paper moving at a speed no greater than 3 centimeters per second.

In July, 1924, Berger observed small, tremulous movements of the instrument while recording from the skin overlying a bone defect in a seventeen-year-old patient. In his first paper on the electroencephalogram, Berger described this case briefly as his first successful recording of an electroencephalogram. At the time of these early studies, Berger already had used the term "electroencephalogram" in his diary. Yet for several years he had doubts about the origin of the electrical signals he recorded. As late as 1928, he almost abandoned his electrical recording studies.

The publication of Berger's first paper on the human encephalogram in 1929 had little impact on the scientific world. It was either ignored or regarded with open disbelief. At this time, even when Berger himself was not completely free of doubts about the validity of his findings, he managed to continue his work. He published additional contributions to the study of the electroencephalogram in a series of fourteen papers. As his research progressed, Berger became increasingly confident and convinced of the significance of his discovery.

IMPACT

The long-range impact of Berger's work is incontestable. When Berger published his last paper on the human encephalogram in 1938, the new approach to the study of brain function that he inaugurated in 1929 had gathered momentum in many centers, both in Europe and in the United States. As a result of his pioneering work, a new diagnostic method had been introduced into medicine. Physiology had acquired a new investigative tool. Clinical neurophysiology had been liberated from its dependence upon the functional anatomical approach, and electrophysiological exploration of complex functions of the central nervous system had begun in earnest. Berger's work had finally received its well-deserved recognition. Many of those who undertook the study of the electroencephalogram were able to bring a far greater technical knowledge of neurophysiology to bear upon the problems of the electrical activity of the brain. Yet the community of neurological scientists has not ceased to look with respect to the founder of electroencephalography, who, despite overwhelming odds and isolation, opened a new area of neurophysiology.

See also Amniocentesis; CAT scanner; Electrocardiogram; Mammography; Nuclear magnetic resonance; Ultrasound; X-ray image intensifier.

FURTHER READING

Barlow, John S. *The Electroencephalogram: Its Patterns and Origins.* Cambridge, Mass.: MIT Press, 1993.
Berger, Hans. *Hans Berger on the Electroencephalogram of Man.* New York: Elsevier, 1969.

Electron Microscope

The invention: A device for viewing extremely small objects that uses electron beams and "electron lenses" instead of the light rays and optical lenses used by ordinary microscopes.

The people behind the invention:
Ernst Ruska (1906-1988), a German engineer, researcher, and inventor who shared the 1986 Nobel Prize in Physics
Hans Busch (1884-1973), a German physicist
Max Knoll (1897-1969), a German engineer and professor
Louis de Broglie (1892-1987), a French physicist who won the 1929 Nobel Prize in Physics

Reaching the Limit

The first electron microscope was constructed by Ernst Ruska and Max Knoll in 1931. Scientists who look into the microscopic world always demand microscopes of higher and higher resolution (resolution is the ability of an optical instrument to distinguish closely spaced objects). As early as 1834, George Airy, the eminent British astronomer, theorized that there should be a natural limit to the resolution of optical microscopes. In 1873, two Germans, Ernst Abbe, cofounder of the Karl Zeiss Optical Works at Jena, and Hermann von Helmholtz, the famous physicist and philosopher, independently published papers on this issue. Both arrived at the same conclusion as Airy: Light is limited by the size of its wavelength. Specifically, light cannot resolve smaller than one-half the height of its wavelength.

One solution to this limitation was to experiment with light, or electromagnetic radiation, or shorter and shorter wavelengths. At the beginning of the twentieth century, Joseph Edwin Barnard experimented on microscopes using ultraviolet light. Such instruments, however, only modestly improved the resolution. In 1912, German physicist Max von Laue considered using X rays. At the time, however, it was hard to turn "X-ray microscopy" into a physical reality. The wavelengths of X rays are exceedingly short, but for the most part they are used to penetrate matter, not to illuminate objects. It appeared that microscopes had reached their limit.

MATTER WAVES

In a new microscopy, then, light—even electromagnetic radiation in general—as the medium that traditionally carried image information, had to be replaced by a new medium. In 1924, French theoretical physicist Louis de Broglie advanced a startling hypothesis: Matter on the scale of subatomic particles possesses wave characteristics. De Broglie also concluded that the speed of low-mass subatomic particles, such as electrons, is related to wavelength. Specifically, higher speeds correspond to shorter wavelengths.

When Knoll and Ruska built the first electron microscope in 1931, they had never heard about de Broglie's "matter wave." Ruska recollected that when, in 1932, he and Knoll first learned about de Broglie's idea, he realized that those matter waves would have to be many times shorter in wavelength than light waves.

The core component of the new instrument was the electron beam, or "cathode ray," as it was usually called then. The cathode-ray tube was invented in 1857 and was the source of a number of discoveries, including X rays. In 1896, Olaf Kristian Birkeland, a Norwegian scientist, after experimenting with the effect of parallel magnetic fields on the electron beam of the cathode-ray tube, concluded that cathode rays that are concentrated on a focal point by means of a magnet are as effective as parallel light rays that are concentrated by means of a lens.

From around 1910, German physicist Hans Busch was the leading researcher in the field. In 1926, he published his theory on the trajectories of electrons in magnetic fields. His conclusions confirmed and expanded upon those of Birkeland. As a result, Busch has been recognized as the founder of a new field later known as "electron optics." His theoretical study showed, among other things, that the analogy between light and lenses on the one hand, and electron beams and electromagnetic lenses, on the other hand, was accurate.

Ernst Ruska

Ernst August Friedrich Ruska was born in 1906 in Heidelberg to Professor Julius Ruska and his wife, Elisabeth. In 1925 he left home for the Technical College of Munich, moving two years later to the Technical College of Berlin and gaining practical training at nearby Siemens and Halsk Limited. During his university days he became interested in vacuum tube technology and worked at the Institute of High Voltage, participating in the development of a high performance cathode ray oscilloscope.

His interests also lay with the theory and application of electron optics. In 1929, as part of his graduate work, Ruska published a proof of Hans Busch's theory explaining possible lens-like effects of a magnetic field on an electron stream, which led to the invention of the polschuh lens. It formed the core of the electron microscope that Ruska built with his mentor, Max Kroll, in 1931.

Ruska completed his doctoral studies in 1934, but he had already found work in industry, believing that further technical development of electron microscopes was beyond the means of university laboratories. He worked for Fernseh Limited from 1933 to 1937 and for Siemens from 1937 to 1955. Following World War II he helped set up the Institute of Electron Optics and worked in the Faculty of Medicine and Biology of the German Academy of Sciences. He joined the Fritz Haber Institute of the Max Planck Society in Berlin in 1949 and took over as director of its Institute for Electron Microscopy in 1955, keeping the position until he retired in 1974.

His life-long work with electron microscopy earned Ruska half of the 1986 Nobel Prize in Physics. He died two years later. To honor his memory, European manufacturers of electron microscopes instituted the Ernst Ruska Prizes, one for researchers of materials and optics and one for biomedical researchers.

Beginning in 1928, Ruska, as a graduate student at the Berlin Institute of Technology, worked on refining Busch's work. He found that the energy of the electrons in the beam was not uniform. This nonuniformity meant that the images of microscopic objects would ultimately be fuzzy. Knoll and Ruska were able to work from the recognition of this problem to the design and materialization of a concentrated electron "writing spot" and to the actual construction of the electron microscope. By April, 1931, they had established a technological landmark with the "first constructional realization of an electron microscope."

IMPACT

The world's first electron microscope, which took its first photographic record on April 7, 1931, was rudimentary. Its two-stage total magnification was only sixteen times larger than the sample. Since Ruska and Knoll's creation, however, progress in electron microscopy has been spectacular. Such an achievement is one of the prominent examples that illustrate the historically unprecedented pace of science and technology in the twentieth century.

In 1935, for the first time, the electron microscope surpassed the optical microscope in resolution. The problem of damaging the specimen by the heating effects of the electron beam proved to be more difficult to resolve. In 1937, a team at the University of Toronto constructed the first generally usable electron microscope. In 1942, a group headed by James Hillier at the Radio Corporation of America produced commercial transmission electron microscopes. In 1939 and 1940, research papers on electron microscopes began to appear in Sweden, Canada, the United States, and Japan; from 1944 to 1947, papers appeared in Switzerland, France, the Soviet Union, The Netherlands, and England. Following research work in laboratories, commercial transmission electron microscopes using magnetic lenses with short focal lengths also appeared in these countries.

See also Cyclotron; Field ion microscope; Geiger counter; Mass spectrograph; Neutrino detector; Scanning tunneling microscope; Synchrocyclotron; Tevatron accelerator; Ultramicroscope.

FURTHER READING

Cline, Barbara Lovett. *Men Who Made a New Physics: Physicists and the Quantum Theory*. Chicago: University of Chicago Press, 1987.

Hawkes, P. W. *The Beginnings of Electron Microscopy.* Orlando: Academic Press, 1985.

Marton, Ladislaus. *Early History of the Electron Microscope.* 2d ed. San Francisco: San Francisco Press, 1994.

Rasmussen, Nicolas. *Picture Control: The Electron Microscope and the Transformation of Biology in America, 1940-1960.* Stanford, Calif.: Stanford University Press, 1997.

Electronic synthesizer

The invention: Portable electronic device that both simulates the
sounds of acoustic instruments and creates entirely new sounds.

The person behind the invention:
Robert A. Moog (1934-   ), an American physicist, engineer,
and inventor

From Harmonium to Synthesizer

The harmonium, or acoustic reed organ, is commonly viewed as
having evolved into the modern electronic synthesizer that can be
used to create many kinds of musical sounds, from the sounds of
single or combined acoustic musical instruments to entirely original
sounds. The first instrument to be called a synthesizer was patented
by the Frenchman J. A. Dereux in 1949. Dereux's synthesizer, which
amplified the acoustic properties of harmoniums, led to the development of the recording organ.

Next, several European and American inventors altered and
augmented the properties of such synthesizers. This stage of the
process was followed by the invention of electronic synthesizers, which initially used electronically generated sounds to imitate
acoustic instruments. It was not long, however, before such synthesizers were used to create sounds that could not be produced by any
other instrument. Among the early electronic synthesizers were
those made in Germany by Herbert Elmert and Robert Beyer in
1953, and the American Olsen-Belar synthesizers, which were developed in 1954. Continual research produced better and better versions of these large, complex electronic devices.

Portable synthesizers, which are often called "keyboards," were
then developed for concert and home use. These instruments became extremely popular, especially in rock music. In 1964, Robert A.
Moog, an electronics professor, created what are thought by many
to be the first portable synthesizers to be made available to the public. Several other well-known portable synthesizers, such as ARP
and Buchla synthesizers, were also introduced at about the same

time. Currently, many companies manufacture studio-quality synthesizers of various types.

Synthesizer Components and Operation

Modern synthesizers make music electronically by building up musical phrases via numerous electronic circuits and combining those phrases to create musical compositions. In addition to duplicating the sounds of many instruments, such synthesizers also enable their users to create virtually any imaginable sound. Many sounds have been created on synthesizers that could not have been created in any other way.

Synthesizers use sound-processing and sound-control equipment that controls "white noise" audio generators and oscillator circuits. This equipment can be manipulated to produce a huge variety of sound frequencies and frequency mixtures in the same way that a beam of white light can be manipulated to produce a particular color or mixture of colors.

Once the desired products of a synthesizer's noise generator and oscillators are produced, percussive sounds that contain all or many audio frequencies are mixed with many chosen individual sounds and altered by using various electronic processing components. The better the quality of the synthesizer, the more processing components it will possess. Among these components are sound amplifiers, sound mixers, sound filters, reverberators, and sound combination devices.

Sound amplifiers are voltage-controlled devices that change the dynamic characteristics of any given sound made by a synthesizer. Sound mixers make it possible to combine and blend two or more manufactured sounds while controlling their relative volumes. Sound filters affect the frequency content of sound mixtures by increasing or decreasing the amplitude of the sound frequencies within particular frequency ranges, which are called "bands."

Sound filters can be either band-pass filters or band-reject filters. They operate by increasing or decreasing the amplitudes of sound frequencies within given ranges (such as treble or bass). Reverberators (or "reverb" units) produce artificial echoes that can have significant musical effects. There are also many other varieties of sound-

Robert Moog

Robert Moog, born in 1934, grew up in the Queens borough of New York City, a tough area for a brainy kid. To avoid the bullies who picked on him because he was a nerd, Moog spent a lot of time helping his father with his hobby, electronics. At fourteen, he built his own theremin, an eerie-sounding forerunner of electric instruments.

Moog's mother, meanwhile, force-fed him piano lessons. He liked science better and majored in physics at Queens College and then Cornell University, but he did not forget the music. While in college, he designed a kit for making theremins and advertised it, selling enough of them to run up a sizable bankroll. Also while in college, Moog, acting on a suggestion from a composer, put together the first easy-to-play electronic synthesizer. Other music synthesizers already existed, but they were large, complex, and expensive—suitable only for recording studios. When Moog unveiled his synthesizer in 1965, it was portable, sold for one-tenth the price, and gave musicians virtually an orchestra at their fingertips. It became a stage instrument.

Walter Carlos used a Moog synthesizer in 1969 for his album *Switched-on Bach*, electronic renditions of Johann Sebastian Bach's concertos. It was a hit and won a Grammy award. The album made Moog and his new instrument famous. Its reputation grew when the Beatles used it for "Because" on *Abbey Road* and Carlos recorded the score for Stanley Kubrick's classic movie *A Clockwork Orange* on a Moog. With the introduction of the even more portable Minimoog, the popularity of synthesizers soared, especially among rock musicians but also in jazz and other styles.

Moog sold his company and moved to North Carolina in 1978. There he started another company, Big Briar, devoted to designing special instruments, such as a keyboard that can be played with as much expressive subtlety as a violin and an interactive piano.

processing elements, among them sound-envelope generators, spatial locators, and frequency shifters. Ultimately, the sound-combination devices put together the results of the various groups of audio generating and processing elements, shaping the sound that has been created into its final form.

A variety of control elements are used to integrate the operation of synthesizers. Most common is the keyboard, which provides the name most often used for portable electronic synthesizers. Portable synthesizer keyboards are most often pressure-sensitive devices (meaning that the harder one presses the key, the louder the resulting sound will be) that resemble the black-and-white keyboards of more conventional musical instruments such as the piano and the organ. These synthesizer keyboards produce two simultaneous outputs: control voltages that govern the pitches of oscillators, and timing pulses that sustain synthesizer responses for as long as a particular key is depressed.

Unseen but present are the integrated voltage controls that control overall signal generation and processing. In addition to voltage controls and keyboards, synthesizers contain buttons and other switches that can transpose their sound ranges and other qualities. Using the appropriate buttons or switches makes it possible for a single synthesizer to imitate different instruments—or groups of instruments—at different times. Other synthesizer control elements include sample-and-hold devices and random voltage sources that make it possible to sustain particular musical effects and to add various effects to the music that is being played, respectively.

Electronic synthesizers are complex and flexible instruments. The various types and models of synthesizers make it possible to produce many different kinds of music, and many musicians use a variety of keyboards to give them great flexibility in performing and recording.

IMPACT

The development and wide dissemination of studio and portable synthesizers has led to their frequent use to combine the sound properties of various musical instruments; a single musician can thus produce, inexpensively and with a single instrument, sound combinations that previously could have been produced only by a large number of musicians playing various instruments. (Understandably, many players of acoustic instruments have been upset by this development, since it means that they are hired to play less often than they were before synthesizers were developed.) Another consequence of synthesizer use has been the development of entirely original varieties of sound, although this area has been less thoroughly explored, for commercial reasons. The development of synthesizers has also led to the design of other new electronic music-making techniques and to the development of new electronic musical instruments.

Opinions about synthesizers vary from person to person—and, in the case of certain illustrious musicians, from time to time. One well-known musician initially proposed that electronic synthesizers would replace many or all conventional instruments, particularly pianos. Two decades later, though, this same musician noted that not even the best modern synthesizers could match the quality of sound produced by pianos made by manufacturers such as Steinway and Baldwin.

See also Broadcaster guitar; Cassette recording; Compact disc; Dolby noise reduction; Transistor.

FURTHER READING

Hopkin, Bart. *Gravikords, Whirlies and Pyrophones: Experimental Musical Instruments*. Roslyn, N.Y.: Ellipsis Arts, 1996.
Koener, Brendan I.. "Back to Music's Future." *U.S. News & World Report* 122, no. 8 (March 3, 1997).
Nunziata, Susan. "Moog Keyboard Offers Human Touch." *Billboard* 104, no. 7 (February 15, 1992).
Shapiro, Peter. *Modulations: A History of Electronic Music: Throbbing Words on Sound*. New York: Caipirinha Productions, 2000.

ENIAC COMPUTER

THE INVENTION: The first general-purpose electronic digital computer.

THE PEOPLE BEHIND THE INVENTION:
John Presper Eckert (1919-1995), an electrical engineer
John William Mauchly (1907-1980), a physicist, engineer, and professor
John von Neumann (1903-1957), a Hungarian American mathematician, physicist, and logician
Herman Heine Goldstine (1913-   ), an army mathematician
Arthur Walter Burks (1915-   ), a philosopher, engineer, and professor
John Vincent Atanasoff (1903-1995), a mathematician and physicist

A TECHNOLOGICAL REVOLUTION

The Electronic Numerical Integrator and Calculator (ENIAC) was the first general-purpose electronic digital computer. By demonstrating the feasibility and value of electronic digital computation, it initiated the computer revolution. The ENIAC was developed during World War II (1939-1945) at the Moore School of Electrical Engineering by a team headed by John William Mauchly and John Presper Eckert, who were working on behalf of the U.S. Ordnance Ballistic Research Laboratory (BRL) at the Aberdeen Proving Ground in Maryland. Early in the war, the BRL's need to generate ballistic firing tables already far outstripped the combined abilities of the available differential analyzers and teams of human computers.

In 1941, Mauchly had seen the special-purpose electronic computer developed by John Vincent Atanasoff to solve sets of linear equations. Atanasoff's computer was severely limited in scope and was never fully completed. The functioning prototype, however, helped convince Mauchly of the feasibility of electronic digital computation and so led to Mauchly's formal proposal in April, 1943, to develop the general-purpose ENIAC. The BRL, in desperate need of computational help, agreed to fund the project, with Lieutenant

Herman Heine Goldstine overseeing it for the U.S. Army.

This first substantial electronic computer was designed, built, and debugged within two and one-half years. Even given the highly talented team, it could be done only by taking as few design risks as possible. The ENIAC ended up as an electronic version of prior computers: Its functional organization was similar to that of the differential analyzer, while it was programmed via a plugboard (which was something like a telephone switchboard), much like the earlier electromechanical calculators made by the International Business Machines (IBM) Corporation. Another consequence was that the internal representation of numbers was decimal rather than the now-standard binary, since the familiar electromechanical computers used decimal digits.

Although the ENIAC was completed only after the end of the war, it was used primarily for military purposes. In fact, the first production run on the system was a two-month calculation needed for the design of the hydrogen bomb. John von Neumann, working as a consultant to both the Los Alamos Scientific Laboratory and the ENIAC project, arranged for the production run immediately prior to ENIAC's formal dedication in 1946.

A Very Fast Machine

The ENIAC was an impressive machine: It contained 18,000 vacuum tubes, weighed 27 metric tons, and occupied a large room. The final cost to the U.S. Army was about $486,000. For this price, the army received a machine that computed up to a thousand times faster than its electromechanical precursors; for example, addition and subtraction required only 200 microseconds (200 millionths of a second). At its dedication ceremony, the ENIAC was fast enough to calculate a fired shell's trajectory faster than the shell itself took to reach its target.

The machine also was much more complex than any predecessor and employed a risky new technology in vacuum tubes; this caused much concern about its potential reliability. In response to this concern, Eckert, the lead engineer, imposed strict safety factors on all components, requiring the design to use components at a level well below the manufacturers' specified limits. The result was a machine that ran for as long as three days without a hardware malfunction.

Programming the ENIAC was effected by setting switches and physically connecting accumulators, function tables (a kind of manually set read-only memory), and control units. Connections were made via cables running between plugboards. This was a laborious and error-prone process, often requiring a one-day set time.

The team recognized this problem, and in early 1945, Eckert, Mauchly, and Neumann worked on the design of a new machine. Their basic idea was to treat both program and data in the same way, and in particular to store them in the same high-speed memory; in other words, they planned to produce a stored-program computer. Neumann described and explained this design in his "First Draft of a Report on the EDVAC" (EDVAC is an acronym for Electronic Discrete Variable Automatic Computer). In his report, Neumann contributed new design techniques and provided the first general, comprehensive description of the stored-program architecture.

After the delivery of the ENIAC, Neumann suggested that it could be wired up so that a set of instructions would be permanently available and could be selected by entries in the function tables. Engineers implemented the idea, providing sixty instructions that could be invoked from the programs stored into the function tables. Despite slowing down the computer's calculations, this technique was so superior to plugboard programming that it was used exclusively thereafter. In this way, the ENIAC was converted into a kind of primitive stored-program computer.

Impact

The ENIAC's electronic speed and the stored-program design of the EDVAC posed a serious engineering challenge: to produce a computer memory that would be large, inexpensive, and fast. Without such fast memories, the electronic control logic would spend most of its time idling. Vacuum tubes themselves (used in the control) were not an effective answer because of their large power requirements and heat generation.

The EDVAC design draft proposed using mercury delay lines, which had been used earlier in radars. These delay lines converted an electronic signal into a slower acoustic signal in a mercury solution; for continuous storage, the signal picked up at the other end was regenerated and sent back into the mercury. Maurice Vincent Wilkes at the University of Cambridge was the first to complete such a system, in May, 1949. One month earlier, Frederick Calland Williams and Tom Kilburn at Manchester University had brought their prototype computer into operation, which used cathode-ray tubes (CRTs) for its main storage. Thus, England took an early lead in developing computing systems, largely because of a more immediate practical design approach.

In the meantime, Eckert and Mauchly formed the Electronic Control Company (later the Eckert-Mauchly Computer Corporation). They produced the Binary Automatic Computer (BINAC) in 1949 and the Universal Automatic Computer (UNIVAC) I in 1951; both machines used mercury storage.

The memory problem that the ENIAC introduced was finally resolved with the invention of the magnetic core in the early 1950's. Core memory was installed on the ENIAC and soon on all new machines. The ENIAC continued in operation until October, 1955, when parts of it were retired to the Smithsonian Institution. The ENIAC proved the viability of digital electronics and led directly to the development of stored-program computers. Its impact can be seen in every modern digital computer.

See also Apple II computer; BINAC computer; Colossus computer; IBM Model 1401 computer; Personal computer; Supercomputer; UNIVAC computer.

FURTHER READING

Burks, Alice R., and Arthur W. Burks. *The First Electronic Computer: The Atanasoff Story.* Ann Arbor: University of Michigan Press, 1990.

McCarney, Scott. *ENIAC: The Triumphs and Tragedies of the World's First Computer.* New York: Berkley Books, 2001.

Slater, Robert. *Portraits in Silicon.* Cambridge, Mass.: MIT Press, 1989.

Stern, Nancy B. *From ENIAC to UNIVAC: An Appraisal of the Eckert-Mauchly Computers.* Bedford, Mass.: Digital Press, 1981.

Fax Machine

The invention: Originally known as the "facsimile machine," a machine that converts written and printed images into electrical signals that can be sent via telephone, computer, or radio.

The person behind the invention:
Alexander Bain (1818-1903), a Scottish inventor

Sending Images

The invention of the telegraph and telephone during the latter half of the nineteenth century gave people the ability to send information quickly over long distances. With the invention of radio and television technologies, voices and moving pictures could be seen around the world as well. Oddly, however, the facsimile process—which involves the transmission of pictures, documents, or other physical data over distance—predates all these modern devices, since a simple facsimile apparatus (usually called a fax machine) was patented in 1843 by Alexander Bain. This early device used a pendulum to synchronize the transmitting and receiving units; it did not convert the image into an electrical format, however, and it was quite crude and impractical. Nevertheless, it reflected the desire to send images over long distances, which remained a technological goal for more than a century.

Facsimile machines developed in the period around 1930 enabled news services to provide newspapers around the world with pictures for publication. It was not until the 1970's, however, that technological advances made small fax machines available for everyday office use.

Scanning Images

Both the fax machines of the 1930's and those of today operate on the basis of the same principle: scanning. In early machines, an image (a document or a picture) was attached to a roller, placed in the fax machine, and rotated at a slow and fixed speed (which must be the same at each end of the link) in a bright light. Light from the image was reflected from the document in varying degrees, since dark areas reflect less light than lighter areas do. A lens moved across the page one line at a time, concentrating and directing the reflected light to a photoelectric tube. This tube would respond to the change in light level by varying its electric output, thus converting the image into an output signal whose intensity varied with the changing light and dark spots of the image. Much like the signal from a microphone or television camera, this modulated (varying) wave could then be broadcast by radio or sent over telephone lines to a receiver that performed a reverse function. At the receiving end, a light bulb was made to vary its intensity to match the varying intensity of the incoming signal. The output of the light bulb was concentrated through a lens onto photographically sensitive paper, thus re-creating the original image as the paper was rotated.

Early fax machines were bulky and often difficult to operate. Advances in semiconductor and computer technology in the 1970's, however, made the goal of creating an easy-to-use and inexpensive fax machine realistic. Instead of a photoelectric tube that consumes a relatively large amount of electrical power, a row of small photodiode semiconductors is used to measure light intensity. Instead of a power-consuming light source, low-power light-emitting diodes (LEDs) are used. Some 1,728 light-sensitive diodes are placed in a row, and the image to be scanned is passed over them one line at a time. Each diode registers either a dark or a light portion of the image. As each diode is checked in sequence, it produces a signal for one picture element, also known as a "pixel" or "pel." Because many diodes are used, there is no need for a focusing lens; the diode bar is as wide as the page being scanned, and each pixel represents a portion of a line on that page.

Since most fax transmissions take place over public telephone system lines, the signal from the photodiodes is transmitted by means of a built-in computer modem in much the same format that computers use to transmit data over telephone lines. The receiving fax uses its modem to convert the audible signal into a sequence that varies in intensity in proportion to the original signal. This varying signal is then sent in proper sequence to a row of 1,728 small wires over which a chemically treated paper is passed. As each wire receives a signal that represents a black portion of the scanned image, the wire heats and, in contact with the paper, produces a black dot that corresponds to the transmitted pixel. As the page is passed over these wires one line at a time, the original image is re-created.

Consequences

The fax machine has long been in use in many commercial and scientific fields. Weather data in the form of pictures are transmitted from orbiting satellites to ground stations; newspapers receive photographs from international news sources via fax; and, using a very expensive but very high-quality fax device, newspapers and magazines are able to transmit full-size proof copies of each edition to printers thousands of miles away so that a publication edited in one country can reach newsstands around the world quickly.

With the technological advances that have been made in recent years, however, fax transmission has become a part of everyday life, particularly in business and research environments. The ability to send quickly a copy of a letter, document, or report over thousands of miles means that information can be shared in a matter of minutes rather than in a matter of days. In fields such as advertising and architecture, it is often necessary to send pictures or drawings to remote sites. Indeed, the fax machine has played an important role in providing information to distant observers of political unrest when other sources of information (such as radio, television, and newspapers) are shut down.

In fact, there has been a natural coupling of computers, modems, and fax devices. Since modern faxes are sent as computer data over phone lines, specialized and inexpensive modems (which allow two computers to share data) have been developed that allow any computer user to send and receive faxes without bulky machines. For example, a document—including drawings, pictures, or graphics of some kind—is created in a computer and transmitted directly to another fax machine. That computer can also receive a fax transmission and either display it on the computer's screen or print it on the local printer. Since fax technology is now within the reach of almost anyone who is interested in using it, there is little doubt that it will continue to grow in popularity.

See also Communications satellite; Instant photography; Internet; Personal computer; Xerography.

FURTHER READING

Bain, Alexander, and Leslie William Davidson. *Autobiography.* New York: Longmans, Green, 1973.

Cullen, Scott. "Telecommunications in the Office." *Office Systems* 16, no. 12 (December, 1999).

Holtzmann, Gerald J. "Just the Fax." *Inc.* 20, no. 13 (September 15, 1998).

Hunkin, Tim. "Just Give Me the Fax." *New Scientist* 137, no. 1860 (February 13, 1993).

FIBER-OPTICS

THE INVENTION: The application of glass fibers to electronic communications and other fields to carry large volumes of information
quickly, smoothly, and cheaply over great distances.

THE PEOPLE BEHIND THE INVENTION:

Samuel F. B. Morse (1791-1872), the American artist and
inventor who developed the electromagnetic telegraph
system

Alexander Graham Bell (1847-1922), the Scottish American
inventor and educator who invented the telephone and the
photophone

Theodore H. Maiman (1927-   ), the American physicist and
engineer who invented the solid-state laser

Charles K. Kao (1933-   ), a Chinese-born electrical engineer

Zhores I. Alferov (1930-   ), a Russian physicist and
mathematician

THE SINGING SUN

In 1844, Samuel F. B. Morse, inventor of the telegraph, sent his famous message, "What hath God wrought?" by electrical impulses
traveling at the speed of light over a 66-kilometer telegraph wire
strung between Washington, D.C., and Baltimore. Ever since that
day, scientists have worked to find faster, less expensive, and more
efficient ways to convey information over great distances.

At first, the telegraph was used to report stock-market prices and
the results of political elections. The telegraph was quite important
in the American Civil War (1861-1865). The first transcontinental
telegraph message was sent by Stephen J. Field, chief justice of the
California Supreme Court, to U.S. president Abraham Lincoln on
October 24, 1861. The message declared that California would remain loyal to the Union. By 1866, telegraph lines had reached all
across the North American continent and a telegraph cable had
been laid beneath the Atlantic Ocean to link the Old World with the
New World.

ZHORES I. ALFEROV

To create a telephone system that transmitted with light, perfecting fiber-optic cables was only half the solution. There also had to be a small, reliable, energy-efficient light source. In the 1960's engineers realized that lasers were the best candidate. However, early gas lasers were bulky, and semiconductor lasers, while small, were temperamental and had to be cooled in liquid nitrogen. Nevertheless, the race was on to devise a semiconductor laser that produced a continuous beam and did not need to be cooled. The race was between a Bell Labs team in the United States and a Russian team led by Zhores I. Alferov, neither of which knew much about the other.

Alferov was born in 1930 in Vitebsk, Byelorussia, then part of the Soviet Union. He earned a degree in electronics from the V. I. Ulyanov (Lenin) Electrotechnical Institute in Leningrad (now St. Petersburg). As part of his graduate studies, he became a researcher at the A. F. Ioffe Physico-Technical Institute in the same city, receiving a doctorate in physics and mathematics in 1970. By then he was one of the world's leading experts in semiconductor lasers.

Alferov found that he could improve the laser's performance by sandwiching very thin layers of gallium arsenide and metal, insulated in silicon, in such a way that electrons flowed only along a 0.03 millimeter strip, producing light in the process. This double heterojunction narrow-stripe laser was the answer, producing a steady beam at room temperature. Alferov published his results a month before the American team came up with almost precisely the same solution.

The question of who was first was not settled until much later, during which time both Bell Labs and Alferov's institute went on to further refinements of the technology. Alferov rose to become a dean at the St. Petersburg Technical University and vice-president of the Russian Academy of Sciences. In 2000 he shared the Nobel Prize in Physics.

Another American inventor made the leap from the telegraph to the telephone. Alexander Graham Bell, a teacher of the deaf, was interested in the physical way speech works. In 1875, he started experimenting with ways to transmit sound vibrations electrically. He realized that an electrical current could be adjusted to resemble the vibrations of speech. Bell patented his invention on March 7, 1876. On July 9, 1877, he founded the Bell Telephone Company.

In 1880, Bell invented a device called the "photophone." He used it to demonstrate that speech could be transmitted on a beam of light. Light is a form of electromagnetic energy. It travels in a vibrating wave. When the amplitude (height) of the wave is adjusted, a light beam can be made to carry messages. Bell's invention included a thin mirrored disk that converted sound waves directly into a beam of light. At the receiving end, a selenium resistor connected to a headphone converted the light back into sound. "I have heard a ray of sun laugh and cough and sing," Bell wrote of his invention.

Although Bell proved that he could transmit speech over distances of several hundred meters with the photophone, the device was awkward and unreliable, and it never became popular as the telephone did. Not until one hundred years later did researchers find important practical uses for Bell's idea of talking on a beam of light.

Two other major discoveries needed to be made first: development of the laser and of high-purity glass. Theodore H. Maiman, an American physicist and electrical engineer at Hughes Research Laboratories in Malibu, California, built the first laser. The laser produces an intense, narrowly focused beam of light that can be adjusted to carry huge amounts of information. The word itself is an acronym for *l*ight *a*mplification by the *s*timulated *e*mission of *r*adiation.

It soon became clear, though, that even bright laser light can be broken up and absorbed by smog, fog, rain, and snow. So in 1966, Charles K. Kao, an electrical engineer at the Standard Telecommunications Laboratories in England, suggested that glass fibers could be used to transmit message-carrying beams of laser light without disruption from weather.

Fiber Optics Are Tested

Optical glass fiber is made from common materials, mostly silica, soda, and lime. The inside of a delicate silica glass tube is coated with a hundred or more layers of extremely thin glass. The tube is then heated to 2,000 degrees Celsius and collapsed into a thin glass rod, or preform. The preform is then pulled into thin strands of fiber. The fibers are coated with plastic to protect them from being nicked or scratched, and then they are covered in flexible cable.

Fiber optic strands. (PhotoDisc)

The earliest glass fibers contained many impurities and defects, so they did not carry light well. Signal repeaters were needed every few meters to energize (amplify) the fading pulses of light. In 1970, however, researchers at the Corning Glass Works in New York developed a fiber pure enough to carry light at least one kilometer without amplification.

The telephone industry quickly became involved in the new fiber-optics technology. Researchers believed that a bundle of optical fibers as thin as a pencil could carry several hundred telephone calls at the same time. Optical fibers were first tested by telephone companies in big cities, where the great volume of calls often overloaded standard underground phone lines.

On May 11, 1977, American Telephone & Telegraph Company (AT&T), along with Illinois Bell Telephone, Western Electric, and Bell Telephone Laboratories, began the first commercial test of fiber-optics telecommunications in downtown Chicago. The system consisted of a 2.4-kilometer cable laid beneath city streets. The cable, only 1.3 centimeters in diameter, linked an office building in the downtown business district with two telephone exchange centers. Voice and video signals were coded into pulses of laser light and transmitted through the hair-thin glass fibers. The tests showed that a single pair of fibers could carry nearly six hundred telephone conversations at once very reliably and at a reasonable cost.

Six years later, in October, 1983, Bell Laboratories succeeded in transmitting the equivalent of six thousand telephone signals through an optical fiber cable that was 161 kilometers long. Since that time, countries all over the world, from England to Indonesia, have developed optical communications systems.

CONSEQUENCES

Fiber optics has had a great impact on telecommunications. A single fiber can now carry thousands of conversations with no electrical interference. These fibers are less expensive, weigh less, and take up much less space than copper wire. As a result, people can carry on conversations over long distances without static and at a low cost.

One of the first uses of fiber optics and perhaps its best-known application is the fiberscope, a medical instrument that permits internal examination of the human body without surgery or X-ray techniques. The fiberscope, or endoscope, consists of two fiber bundles. One of the fiber bundles transmits bright light into the patient, while the other conveys a color image back to the eye of the physician. The fiberscope has been used to look for ulcers, cancer, and polyps in the stomach, intestine, and esophagus of humans. Medical instruments, such as forceps, can be attached to the fiberscope, allowing the physician to perform a range of medical procedures, such as clearing a blocked windpipe or cutting precancerous polyps from the colon.

See also Cell phone; Community antenna television; Communications satellite; FM radio; Laser; Long-distance radiotelephony; Long-distance telephone; Telephone switching.

FURTHER READING

Carey, John, and Neil Gross. "The Light Fantastic: Optoelectronics May Revolutionize Computers—and a Lot More." *Business Week* (May 10, 1993).
Free, John. "Fiber Optics Head for Home." *Popular Science* 238 (March, 1991).
Hecht, Jeff. *City of Light: The Story of Fiber Optics*. Oxford: Oxford University Press, 1999.
Paul, Noel C. "Laying Down the Line with Huge Projects to Circle the Globe in Fiber Optic Cable." *Christian Science Monitor* (March 29, 2001).
Shinal, John G., with Timothy J. Mullaney. "At the Speed of Light." *Business Week* (October 9, 2000).

FIELD ION MICROSCOPE

THE INVENTION: A microscope that uses ions formed in high-voltage electric fields to view atoms on metal surfaces.

THE PEOPLE BEHIND THE INVENTION:
Erwin Wilhelm Müller (1911-1977), a physicist, engineer, and research professor
J. Robert Oppenheimer (1904-1967), an American physicist

TO SEE BENEATH THE SURFACE

In the early twentieth century, developments in physics, especially quantum mechanics, paved the way for the application of new theoretical and experimental knowledge to the problem of viewing the atomic structure of metal surfaces. Of primary importance were American physicist George Gamow's 1928 theoretical explanation of the field emission of electrons by quantum mechanical means and J. Robert Oppenheimer's 1928 prediction of the quantum mechanical ionization of hydrogen in a strong electric field.

In 1936, Erwin Wilhelm Müller developed his field emission microscope, the first in a series of instruments that would exploit these developments. It was to be the first instrument to view atomic structures—although not the individual atoms themselves—directly. Müller's subsequent field ion microscope utilized the same basic concepts used in the field emission microscope yet proved to be a much more powerful and versatile instrument. By 1956, Müller's invention allowed him to view the crystal lattice structure of metals in atomic detail; it actually showed the constituent atoms.

The field emission and field ion microscopes make it possible to view the atomic surface structures of metals on fluorescent screens. The field ion microscope is the direct descendant of the field emission microscope. In the case of the field emission microscope, the images are projected by electrons emitted directly from the tip of a metal needle, which constitutes the specimen under investigation.

These electrons produce an image of the atomic lattice structure of the needle's surface. The needle serves as the electron-donating electrode in a vacuum tube, also known as the "cathode." A fluorescent screen that serves as the electron-receiving electrode, or "anode," is placed opposite the needle. When sufficient electrical voltage is applied across the cathode and anode, the needle tip emits electrons, which strike the screen. The image produced on the screen is a projection of the electron source—the needle surface's atomic lattice structure.

Müller studied the effect of needle shape on the performance of the microscope throughout much of 1937. When the needles had been properly shaped, Müller was able to realize magnifications of up to 1 million times. This magnification allowed Müller to view what he called "maps" of the atomic crystal structure of metals, since the needles were so small that they were often composed of only one simple crystal of the material. While the magnification may have been great, however, the resolution of the instrument was severely limited by the physics of emitted electrons, which caused the images Müller obtained to be blurred.

IMPROVING THE VIEW

In 1943, while working in Berlin, Müller realized that the resolution of the field emission microscope was limited by two factors. The electron velocity, a particle property, was extremely high and uncontrollably random, causing the micrographic images to be blurred. In addition, the electrons had an unsatisfactorily high wavelength. When Müller combined these two factors, he was able to determine that the field emission microscope could never depict single atoms; it was a physical impossibility for it to distinguish one atom from another.

By 1951, this limitation led him to develop the technology behind the field ion microscope. In 1952, Müller moved to the United States and founded the Pennsylvania State University Field Emission Laboratory. He perfected the field ion microscope between 1952 and 1956.

The field ion microscope utilized positive ions instead of electrons to create the atomic surface images on the fluorescent screen.

ERWIN MÜLLER

Erwin Müller's scientific goal was to see an individual atom, and to that purpose he invented ever more powerful microscopes. He was born in Berlin, Germany, in 1911 and attended the city's Technische Hochschule, earning a diploma in engineering in 1935 and a doctorate in physics in 1936. Following his studies he worked as an industrial researcher. Still a neophyte scientist, he discovered the principle of the field emission microscope and was able to produce an image of a structure only two nanometers in diameter on the surface of a cathode. In 1941 Müller discovered field desorption by reversing the polarity of the electron emitter at very low temperatures so that surface atoms evaporated in the electric field. In 1947 he left industry and began an academic career, teaching physical chemistry at the Altenburg Engineering School. The following year he was appointed a department head at the Fritz Haber Institute. While there, he found that by having a cathode absorb gas ions and then re-emit them he could produce greater magnification.

In 1952 Müller became a professor at Pennsylvania State University. Applying the new field-ion emission principle, he was able to achieve his goal, images of individual atoms, in 1956. Almost immediately chemists and physicists adopted the field-ion microscope to conduct basic research concerning the underlying behavior of field ionization and interactions among absorbed atoms. He further aided such research by coupling a field-ion microscope and mass spectrometer, calling the combination an atom-probe field-ion microscope; it could both magnify and chemically analyze atoms.

Müller died in 1977. He received the National Medal of Science posthumously, one of many honors for his contributions to microscopy.

When an easily ionized gas—at first hydrogen, but usually helium, neon, or argon—was introduced into the evacuated tube, the emitted electrons ionized the gas atoms, creating a stream of positively charged particles, much as Oppenheimer had predicted in 1928. Müller's use of positive ions circumvented one of the resolution problems inherent in the use of imaging electrons. Like the electrons, however, the positive ions traversed the tube with unpredictably random velocities. Müller eliminated this problem by cryogenically cooling the needle tip with a supercooled liquefied gas such as nitrogen or hydrogen.

By 1956, Müller had perfected the means of supplying imaging positive ions by filling the vacuum tube with an extremely small quantity of an inert gas such as helium, neon, or argon. By using such a gas, Müller was assured that no chemical reaction would occur between the needle tip and the gas; any such reaction would alter the surface atomic structure of the needle and thus alter the resulting microscopic image. The imaging ions allowed the field ion microscope to image the emitter surface to a resolution of between two and three angstroms, making it ten times more accurate than its close relative, the field emission microscope.

CONSEQUENCES

The immediate impact of the field ion microscope was its influence on the study of metallic surfaces. It is a well-known fact of materials science that the physical properties of metals are influenced by the imperfections in their constituent lattice structures. It was not possible to view the atomic structure of the lattice, and thus the finest detail of any imperfection, until the field ion microscope was developed. The field ion microscope is the only instrument powerful enough to view the structural flaws of metal specimens in atomic detail.

Although the instrument may be extremely powerful, the extremely large electrical fields required in the imaging process preclude the instrument's application to all but the heartiest of metallic specimens. The field strength of 500 million volts per centimeter exerts an average stress on metal specimens in the range of almost 1 ton per square millimeter. Metals such as iron and platinum can withstand this strain because of the shape of the needles into which they are formed. Yet this limitation of the instrument makes it extremely difficult to examine biological materials, which cannot withstand the amount of stress that metals can. A practical by-product in the study of field ionization—field evaporation—eventually permitted scientists to view large biological molecules.

Field evaporation also allowed surface scientists to view the atomic structures of biological molecules. By embedding molecules such as phthalocyanine within the metal needle, scientists have been able to view the atomic structures of large biological molecules by field evaporating much of the surrounding metal until the biological material remains at the needle's surface.

See also Cyclotron; Electron microscope; Mass spectrograph; Neutrino detector; Scanning tunneling microscope; Sonar; Synchrocyclotron; Tevatron accelerator; Ultramicroscope.

FURTHER READING

Gibson, J. M. "Tools for Probing 'Atomic' Action." *IEEE Spectrum* 22, no. 12 (December, 1985).
Kunetka, James W. *Oppenheimer: The Years of Risk*. Englewood Cliffs, N.J.: Prentice-Hall, 1982.
Schweber, Silvan S. *In the Shadow of the Bomb: Bethe, Oppenheimer, and the Moral Responsibility of the Scientist*. Princeton, N.J.: Princeton University Press, 2000.
Tsong, Tien Tzou. *Atom-Probe Field Ion Microscopy: Field Ion Emission and Surfaces and Interfaces at Atomic Resolution*. New York: Cambridge University Press, 1990.

FLOPPY DISK

THE INVENTION: Inexpensive magnetic medium for storing and moving computer data.

THE PEOPLE BEHIND THE INVENTION:

Andrew D. Booth (1918-   ), an English inventor who developed paper disks as a storage medium

Reynold B. Johnson (1906-1998), a design engineer at IBM's research facility who oversaw development of magnetic disk storage devices

Alan Shugart (1930-   ), an engineer at IBM's research laboratory who first developed the floppy disk as a means of mass storage for mainframe computers

FIRST TRIES

When the International Business Machines (IBM) Corporation decided to concentrate on the development of computers for business use in the 1950's, it faced a problem that had troubled the earliest computer designers: how to store data reliably and inexpensively. In the early days of computers (the early 1940's), a number of ideas were tried. The English inventor Andrew D. Booth produced spinning paper disks on which he stored data by means of punched holes, only to abandon the idea because of the insurmountable engineering problems he foresaw.

The next step was "punched" cards, an idea first used when the French inventor Joseph-Marie Jacquard invented an automatic weaving loom for which patterns were stored in pasteboard cards. The idea was refined by the English mathematician and inventor Charles Babbage for use in his "analytical engine," an attempt to build a kind of computing machine. Although it was simple and reliable, it was not fast enough, nor did it store enough data, to be truly practical.

The Ampex Corporation demonstrated its first magnetic audiotape recorder after World War II (1939-1945). Shortly after that, the *Binary Automatic Computer* (BINAC) was introduced with a storage device that appeared to be a large tape recorder. A more advanced machine, the *Univ*ersal *A*utomatic *C*omputer (UNIVAC), used metal tape instead of plastic (plastic was easily stretched or even broken). Unfortunately, metal tape was considerably heavier, and its edges were razor-sharp and thus dangerous. Improvements in plastic tape eventually produced sturdy media, and magnetic tape became (and remains) a practical medium for storage of computer data.

Still later designs combined Booth's spinning paper disks with magnetic technology to produce rapidly rotating "drums." Whereas a tape might have to be fast-forwarded nearly to its end to locate a specific piece of data, a drum rotating at speeds up to 12,500 revolutions per minute (rpm) could retrieve data very quickly and could store more than 1 million bits (or approximately 125 kilobytes) of data.

In May, 1955, these drums evolved, under the direction of Reynold B. Johnson, into IBM's hard disk unit. The hard disk unit consisted of fifty platters, each 2 feet in diameter, rotating at 1,200 rpm. Both sides of the disk could be used to store information. When the operator wished to access the disk, at his or her command a read/write head was moved to the right disk and to the side of the disk that held the desired data. The operator could then read data from or record data onto the disk. To speed things even more, the next version of the device, similar in design, employed one hundred read/write heads—one for each of its fifty double-sided disks. The only remaining disadvantage was its size, which earned IBM's first commercial unit the nickname "jukebox."

THE FIRST FLOPPY

The floppy disk drive developed directly from hard disk technology. It did not take shape until the late 1960's under the direction of Alan Shugart (it was announced by IBM as a ready product in 1970). First created to help restart the operating systems of mainframe computers that had gone dead, the floppy seemed in some ways to be a step back, for it operated more slowly than a hard disk drive and did not store as much data. Initially, it consisted of a single thin plastic disk eight inches in diameter and was developed without the protective envelope in which it is now universally encased. The addition of that jacket gave the floppy its single greatest advantage over the hard disk: portability with reliability.

Another advantage soon became apparent: The floppy is resilient to damage. In a hard disk drive, the read/write heads must hover thousandths of a centimeter over the disk surface in order to attain maximum performance. Should even a small particle of dust get in the way, or should the drive unit be bumped too hard, the head may "crash" into the surface of the disk and ruin its magnetic coating; the result is a permanent loss of data. Because the floppy operates with the read-write head in contact with the flexible plastic disk surface, individual particles of dust or other contaminants are not nearly as likely to cause disaster.

As a result of its advantages, the floppy disk was the logical choice for mass storage in personal computers (PCs), which were developed a few years after the floppy disk's introduction. The floppy is still an important storage device even though hard disk drives for PCs have become less expensive. Moreover, manufacturers continually are developing new floppy formats and new floppy disks that can hold more data.

Three-and-one-half-inch disks improved on the design of earlier floppies by protecting their magnetic media within hard plastic shells and using sliding metal flanges to protect the surfaces on which recording heads make contact. (PhotoDisc)

CONSEQUENCES

Personal computing would have developed very differently were it not for the availability of inexpensive floppy disk drives. When IBM introduced its PC in 1981, the machine provided as standard equipment a connection for a cassette tape recorder as a storage device; a floppy disk was only an option (though an option few did not take). The awkwardness of tape drives—their slow speed and sequential nature of storing data—presented clear obstacles to the acceptance of the personal computer as a basic information tool. By contrast, the floppy drive gives computer users relatively fast storage at low cost.

Floppy disks provided more than merely economical data storage. Since they are built to be removable (unlike hard drives), they represented a basic means of transferring data between machines. Indeed, prior to the popularization of *local area networks* (LANs), the floppy was known as a "sneaker" network: One merely carried the disk by foot to another computer.

Floppy disks were long the primary means of distributing new software to users. Even the very flexible floppy showed itself to be quite resilient to the wear and tear of postal delivery. Later, the 3.5-inch disk improved upon the design of the original 8-inch and 5.25-inch floppies by protecting the disk medium within a hard plastic shell and by using a sliding metal door to protect the area where the read/write heads contact the disk.

By the late 1990's, floppy disks were giving way to new data-storage media, particularly CD-ROMs—durable laser-encoded disks that hold more than 700 megabytes of data. As the price of blank CDs dropped dramatically, floppy disks tended to be used mainly for short-term storage of small amounts of data. Floppy disks were also being used less and less for data distribution and transfer, as computer users turned increasingly to sending files via e-mail on the Internet, and software providers made their products available for downloading on Web sites.

See also Bubble memory; Compact disc; Computer chips; Hard disk; Optical disk; Personal computer.

FURTHER READING

Brandel, Mary. "IBM Fashions the Floppy." *Computerworld* 33, no. 23 (June 7, 1999).

Chposky, James, and Ted Leonsis. *Blue Magic: The People, Power, and Politics Behind the IBM Personal Computer.* New York: Facts on File, 1988.

Freiberger, Paul, and Michael Swaine. *Fire in the Valley: The Making of the Personal Computer.* New York: McGraw-Hill, 2000.

Grossman. Wendy. *Remembering the Future: Interviews from Personal Computer World.* New York: Springer, 1997.

Fluorescent lighting

THE INVENTION: A form of electrical lighting that uses a glass tube coated with phosphor that gives off a cool bluish light and emits ultraviolet radiation.

THE PEOPLE BEHIND THE INVENTION:

Vincenzo Cascariolo (1571-1624), an Italian alchemist and shoemaker

Heinrich Geissler (1814-1879), a German glassblower

Peter Cooper Hewitt (1861-1921), an American electrical engineer

Celebrating the "Twelve Greatest Inventors"

On the night of November 23, 1936, more than one thousand industrialists, patent attorneys, and scientists assembled in the main ballroom of the Mayflower Hotel in Washington, D.C., to celebrate the one hundredth anniversary of the U.S. Patent Office. A transport liner over the city radioed the names chosen by the Patent Office as America's "Twelve Greatest Inventors," and, as the distinguished group strained to hear those names, "the room was flooded for a moment by the most brilliant light yet used to illuminate a space that size."

Thus did *The New York Times* summarize the commercial introduction of the fluorescent lamp. The twelve inventors present were Thomas Alva Edison, Robert Fulton, Charles Goodyear, Charles Hall, Elias Howe, Cyrus Hall McCormick, Ottmar Mergenthaler, Samuel F. B. Morse, George Westinghouse, Wilbur Wright, and Eli Whitney. There was, however, no name to bear the honor for inventing fluorescent lighting. That honor is shared by many who participated in a very long series of discoveries.

The fluorescent lamp operates as a low-pressure, electric discharge inside a glass tube that contains a droplet of mercury and a gas, commonly argon. The inside of the glass tube is coated with fine particles of phosphor. When electricity is applied to the gas, the mercury gives off a bluish light and emits ultraviolet radiation.

When bathed in the strong ultraviolet radiation emitted by the mercury, the phosphor fluoresces (emits light).

The setting for the introduction of the fluorescent lamp began at the beginning of the 1600's, when Vincenzo Cascariolo, an Italian shoemaker and alchemist, discovered a substance that gave off a bluish glow in the dark after exposure to strong sunlight. The fluorescent substance was apparently barium sulfide and was so unusual for that time and so valuable that its formulation was kept secret for a long time. Gradually, however, scholars became aware of the preparation secrets of the substance and studied it and other luminescent materials.

Further studies in fluorescent lighting were made by the German physicist Johann Wilhelm Ritter. He observed the luminescence of phosphors that were exposed to various "exciting" lights. In 1801, he noted that some phosphors shone brightly when illuminated by light that the eye could not see (ultraviolet light). Ritter thus discovered the ultraviolet region of the light spectrum. The use of phosphors to transform ultraviolet light into visible light was an important step in the continuing development of the fluorescent lamp.

Further studies in fluorescent lighting were made by the German physicist Johann Wilhelm Ritter. He observed the luminescence of phosphors that were exposed to various "exciting" lights. In 1801, he noted that some phosphors shone brightly when illuminated by light that the eye could not see (ultraviolet light). Ritter thus discovered the ultraviolet region of the light spectrum. The use of phosphors to transform ultraviolet light into visible light was an important step in the continuing development of the fluorescent lamp. The British mathematician and physicist Sir George Gabriel Stokes studied the phenomenon as well. It was he who, in 1852, termed the afterglow "fluorescence."

GEISSLER TUBES

While these advances were being made, other workers were trying to produce a practical form of electric light. In 1706, the English physicist Francis Hauksbee devised an electrostatic generator, which is used to accelerate charged particles to very high levels of electrical energy. He then connected the device to a glass "jar," used a vacuum pump to evacuate the jar to a low pressure, and tested his generator. In so doing, Hauksbee obtained the first human-made electrical glow discharge by "capturing lightning" in a jar.

In 1854, Heinrich Geissler, a glassblower and apparatus maker, opened his shop in Bonn, Germany, to make scientific instruments; in 1855, he produced a vacuum pump that used liquid mercury as an evacuation fluid. That same year, Geissler made the first gaseous conduction lamps while working in collaboration with the German scientist Julius Plücker. Plücker referred to these lamps as "Geissler tubes." Geissler was able to create red light with neon gas filling a lamp and light of nearly all colors by using certain types of gas within each of the lamps. Thus, both the neon sign business and the science of spectroscopy were born.

Geissler tubes were studied extensively by a variety of workers. At the beginning of the twentieth century, the practical American engineer Peter Cooper Hewitt put these studies to use by marketing the first low-pressure mercury vapor lamps. The lamps were quite successful, although they required high voltage for operation, emitted an eerie blue-green, and shone dimly by comparison with their eventual successor, the fluorescent lamp. At about the same time, systematic studies of phosphors had finally begun.

By the 1920's, a number of investigators had discovered that the low-pressure mercury vapor discharge marketed by Hewitt was an extremely efficient method for producing ultraviolet light, if the mercury and rare gas pressures were properly adjusted. With a phosphor to convert the ultraviolet light back to visible light, the Hewitt lamp made an excellent light source.

IMPACT

The introduction of fluorescent lighting in 1936 presented the public with a completely new form of lighting that had enormous advantages of high efficiency, long life, and relatively low cost.

By 1938, production of fluorescent lamps was well under way. By April, 1938, four sizes of fluorescent lamps in various colors had been offered to the public and more than two hundred thousand lamps had been sold.

During 1939 and 1940, two great expositions—the New York

World's Fair and the San Francisco International Exposition— helped popularize fluorescent lighting. Thousands of tubular fluorescent lamps formed a great spiral in the "motor display salon," the car showroom of the General Motors exhibit at the New York World's Fair. Fluorescent lamps lit the Polish Restaurant and hung in vertical clusters on the flagpoles along the Avenue of the Flags at the fair, while two-meter-long, upright fluorescent tubes illuminated buildings at the San Francisco International Exposition.

When the United States entered World War II (1939-1945), the demand for efficient factory lighting soared. In 1941, more than twenty-one million fluorescent lamps were sold. Technical advances continued to improve the fluorescent lamp. By the 1990's, this type of lamp supplied most of the world's artificial lighting.

See also Electric clock; Electric refrigerator; Microwave cooking; Television; Tungsten filament; Vacuum cleaner; Washing machine.

FURTHER READING

Bowers, B. "New Lamps for Old: The Story of Electric Lighting." *IEE Review* 41, no. 6 (November 16, 1995).
Dake, Henry Carl, and Jack De Ment. *Fluorescent Light and Its Applications, Including Location and Properties of Fluorescent Materials.* Brooklyn, N.Y.: Chemical Publishing, 1941.
"EPA Sees the Light on Fluorescent Bulbs." *Environmental Health Perspectives* 107, no. 12 (December, 1999).
Harris, J. B. "Electric Lamps, Past and Present." *Engineering Science and Education Journal* 2, no. 4 (August, 1993).
"How Fluorescent Lighting Became Smaller." *Consulting-Specifying Engineer* 23, no. 2 (February, 1998).

FM RADIO

THE INVENTION: A method of broadcasting radio signals by modulating the frequency, rather than the amplitude, of radio waves, FM radio greatly improved the quality of sound transmission.

THE PEOPLE BEHIND THE INVENTION:
Edwin H. Armstrong (1890-1954), the inventor of FM radio broadcasting
David Sarnoff (1891-1971), the founder of RCA

AN ENTIRELY NEW SYSTEM

Because early radio broadcasts used amplitude modulation (AM) to transmit their sounds, they were subject to a sizable amount of interference and static. Since good AM reception relies on the amount of energy transmitted, energy sources in the atmosphere between the station and the receiver can distort or weaken the original signal. This is particularly irritating for the transmission of music.

Edwin H. Armstrong provided a solution to this technological constraint. A graduate of Columbia University, Armstrong made a significant contribution to the development of radio with his basic inventions for circuits for AM receivers. (Indeed, the monies Armstrong received from his earlier inventions financed the development of the frequency modulation, or FM, system.) Armstrong was one among many contributors to AM radio. For FM broadcasting, however, Armstrong must be ranked as the most important inventor.

During the 1920's, Armstrong established his own research laboratory in Alpine, New Jersey, across the Hudson River from New York City. With a small staff of dedicated assistants, he carried out research on radio circuitry and systems for nearly three decades. At that time, Armstrong also began to teach electrical engineering at Columbia University.

From 1928 to 1933, Armstrong worked diligently at his private laboratory at Columbia University to construct a working model of an FM radio broadcasting system. With the primitive limitations then imposed on the state of vacuum tube technology, a number of

Armstrong's experimental circuits required as many as one hundred tubes. Between July, 1930, and January, 1933, Armstrong filed four basic FM patent applications. All were granted simultaneously on December 26, 1933.

Armstrong sought to perfect FM radio broadcasting, not to offer radio listeners better musical reception but to create an entirely new radio broadcasting system. On November 5, 1935, Armstrong made his first public demonstration of FM broadcasting in New York City to an audience of radio engineers. An amateur station based in suburban Yonkers, New York, transmitted these first signals. The scientific world began to consider the advantages and disadvantages of Armstrong's system; other laboratories began to craft their own FM systems.

Corporate Conniving

Because Armstrong had no desire to become a manufacturer or broadcaster, he approached David Sarnoff, head of the Radio Corporation of America (RCA). As the owner of the top manufacturer of radio sets and the top radio broadcasting network, Sarnoff was interested in all advances of radio technology. Armstrong first demonstrated FM radio broadcasting for Sarnoff in December, 1933. This was followed by visits from RCA engineers, who were sufficiently impressed to recommend to Sarnoff that the company conduct field tests of the Armstrong system.

In 1934, Armstrong, with the cooperation of RCA, set up a test transmitter at the top of the Empire State Building, sharing facilities with the experimental RCA television transmitter. From 1934 through 1935, tests were conducted using the Empire State facility, to mixed reactions of RCA's best engineers. AM radio broadcasting already had a performance record of nearly two decades. The engineers wondered if this new technology could replace something that had worked so well.

This less-than-enthusiastic evaluation fueled the skepticism of RCA lawyers and salespeople. RCA had too much invested in the AM system, both as a leading manufacturer and as the dominant owner of the major radio network of the time, the National Broadcasting Company (NBC). Sarnoff was in no rush to adopt FM. To change systems would risk the millions of dollars RCA was making as America emerged from the Great Depression.

In 1935, Sarnoff advised Armstrong that RCA would cease any further research and development activity in FM radio broadcasting. (Still, engineers at RCA laboratories continued to work on FM to protect the corporate patent position.) Sarnoff declared to the press that his company would push the frontiers of broadcasting by concentrating on research and development of radio with pictures, that is, television. As a tangible sign, Sarnoff ordered that Armstrong's FM radio broadcasting tower be removed from the top of the Empire State Building.

Armstrong was outraged. By the mid-1930's, the development of FM radio broadcasting had become a mission for Armstrong. For the remainder of his life, Armstrong devoted his considerable talents to the promotion of FM radio broadcasting.

IMPACT

After the break with Sarnoff, Armstrong proceeded with plans to develop his own FM operation. Allied with two of RCA's biggest manufacturing competitors, Zenith and General Electric, Armstrong pressed ahead. In June of 1936, at a Federal Communications Commission (FCC) hearing, Armstrong proclaimed that FM broadcasting was the only static-free, noise-free, and uniform system—both day and night—available. He argued, correctly, that AM radio broadcasting had none of these qualities.

During World War II (1939-1945), Armstrong gave the military permission to use FM with no compensation. That patriotic gesture cost Armstrong millions of dollars when the military soon became all FM. It did, however, expand interest in FM radio broadcasting. World War II had provided a field test of equipment and use.

By the 1970's, FM radio broadcasting had grown tremendously. By 1972, one in three radio listeners tuned into an FM station some time during the day. Advertisers began to use FM radio stations to reach the young and affluent audiences that were turning to FM stations in greater numbers.

By the late 1970's, FM radio stations were outnumbering AM stations. By 1980, nearly half of radio listeners tuned into FM stations on a regular basis. A decade later, FM radio listening accounted for more than two-thirds of audience time. Armstrong's predictions that listeners would prefer the clear, static-free sounds offered by FM radio broadcasting had come to pass by the mid-1980's, nearly fifty years after Armstrong had commenced his struggle to make FM radio broadcasting a part of commercial radio.

See also Community antenna television; Communications satellite; Dolby noise reduction; Fiber-optics; Radio; Radio crystal sets; Television; Transistor radio.

FURTHER READING

Lewis, Tom. *Empire of the Air: The Men Who Made Radio.* New York: HarperPerennial, 1993.
Sobel, Robert. *RCA.* New York: Stein and Day, 1986.
Streissguth, Thomas. *Communications: Sending the Message.* Minneapolis, Minn.: Oliver Press, 1997.

Food freezing

The invention: It was long known that low temperatures helped to protect food against spoiling; the invention that made frozen food practical was a method of freezing items *quickly*. Clarence Birdseye's quick-freezing technique made possible a revolution in food preparation, storage, and distribution.

The people behind the invention:
Clarence Birdseye (1886-1956), a scientist and inventor
Donald K. Tressler (1894-1981), a researcher at Cornell University
Amanda Theodosia Jones (1835-1914), a food-preservation pioneer

Feeding the Family

In 1917, Clarence Birdseye developed a means of quick-freezing meat, fish, vegetables, and fruit without substantially changing their original taste. His system of freezing was called by *Fortune* magazine "one of the most exciting and revolutionary ideas in the history of food." Birdseye went on to refine and perfect his method and to promote the frozen foods industry until it became a commercial success nationwide.

It was during a trip to Labrador, where he worked as a fur trader, that Birdseye was inspired by this idea. Birdseye's new wife and five-week-old baby had accompanied him there. In order to keep his family well fed, he placed barrels of fresh cabbages in salt water and then exposed the vegetables to freezing winds. Successful at preserving vegetables, he went on to freeze a winter's supply of ducks, caribou, and rabbit meat.

In the following years, Birdseye experimented with many freezing techniques. His equipment was crude: an electric fan, ice, and salt water. His earliest experiments were on fish and rabbits, which he froze and packed in old candy boxes. By 1924, he had borrowed money against his life insurance and was lucky enough to find three partners willing to invest in his new General Seafoods Company (later renamed General Foods), located in Gloucester, Massachusetts.

Although it was Birdseye's genius that put the principles of quick-freezing to work, he did not actually invent quick-freezing. The scientific principles involved had been known for some time. As early as 1842, a patent for freezing fish had been issued in England. Nevertheless, the commercial exploitation of the freezing process could not have happened until the end of the 1800's, when mechanical refrigeration was invented. Even then, Birdseye had to overcome major obstacles.

Finding a Niche

By the 1920's, there still were few mechanical refrigerators in American homes. It would take years before adequate facilities for food freezing and retail distribution would be established across the United States. By the late 1930's, frozen foods had, indeed, found its role in commerce but still could not compete with canned or fresh foods. Birdseye had to work tirelessly to promote the industry, writing and delivering numerous lectures and articles to advance its popularity. His efforts were helped by scientific research conducted at Cornell University by Donald K. Tressler and by C. R. Fellers of what was then Massachusetts State College. Also, during World War II (1939-1945), more Americans began to accept the idea: Rationing, combined with a shortage of canned foods, contributed to the demand for frozen foods. The armed forces made large purchases of these items as well.

General Foods was the first to use a system of extremely rapid freezing of perishable foods in packages. Under the Birdseye system, fresh foods, such as berries or lobster, were packaged snugly in convenient square containers. Then, the packages were pressed between refrigerated metal plates under pressure at 50 degrees below zero. Two types of freezing machines were used. The "double belt" freezer consisted of two metal belts that moved through a 15-meter freezing tunnel, while a special salt solution was sprayed on the surfaces of the belts. This double-belt freezer was used only in permanent installations and was soon replaced by the "multiplate" freezer, which was portable and required only 11.5 square meters of floor space compared to the double belt's 152 square meters.

AMANDA THEODOSIA JONES

Amanda Theodosia Jones (1835-1914) was close to her brother. When he suddenly died while they were at school and she was left to contact relatives and make the necessary arrangements for his remains, she was devastated. She had a nervous breakdown at seventeen and could not believe he was entirely gone. She was sure that he remained an active presence in her life, and she became a spiritualist and medium so that they could talk during séances.

Jones always claimed she did not come up with the idea for the vacuum packing method for preserving food, an important technique before freezing foods became practicable. It was her brother who gave it to her. She did the actual experimental work herself, however, and with the aid of Leroy C. Cooley got the first of their seven patents for food processing. In 1873 she launched The Women's Canning and Preserving Company, and it was more than just a company. It was a mission. All the officers, stockholders, and employees were women. "This is a woman's industry," she insisted, and ran the company so that it was a training school for working women.

In the 1880's, the spirit of invention moved Jones again. Concerned about the high rate of accidents among oil drillers, she examined the problem. Simply add a safety valve to pipes to control the release of the crude oil, she told drillers in Pennsylvania. The idea had not occurred to them, but they tried it, and it so improved safety that Jones won wide praise.

The multiplate freezer also made it possible to apply the technique of quick-freezing to seasonal crops. People were able to transport these freezers easily from one harvesting field to another, where they were used to freeze crops such as peas fresh off the vine. The handy multiplate freezer consisted of an insulated cabinet equipped with refrigerated metal plates. Stacked one above the other, these plates were capable of being opened and closed to receive food products and to compress them with evenly distributed pressure. Each aluminum plate had internal passages through which ammonia flowed and expanded at a temperature of –3.8 degrees Celsius, thus causing the foods to freeze.

A major benefit of the new frozen foods was that their taste and vitamin content were not lost. Ordinarily, when food is frozen slowly, ice crystals form, which slowly rupture food cells, thus altering the taste of the food. With quick-freezing, however, the food looks, tastes, and smells like fresh food. Quick-freezing also cuts down on bacteria.

IMPACT

During the months between one food harvest and the next, humankind requires trillions of pounds of food to survive. In many parts of the world, an adequate supply of food is available; elsewhere, much food goes to waste and many go hungry. Methods of food preservation such as those developed by Birdseye have done much to help those who cannot obtain proper fresh foods. Preserving perishable foods also means that they will be available in greater quantity and variety all year-round. In all parts of the world, both tropical and arctic delicacies can be eaten in any season of the year.

With the rise in popularity of frozen "fast" foods, nutritionists began to study their effect on the human body. Research has shown that fresh is the most beneficial. In an industrial nation with many people, the distribution of fresh commodities is, however, difficult. It may be many decades before scientists know the long-term effects on generations raised primarily on frozen foods.

See also Electric refrigerator; Freeze-drying; Microwave cooking; Polystyrene; Refrigerant gas; Tupperware.

FURTHER READING

Altman, Linda Jacobs. *Women Inventors*. New York: Facts on File, 1997.
Tressler, Donald K. *The Memoirs of Donald K. Tressler*. Westport, Conn.: Avi Publishing, 1976.
_____, and Clifford F. Evers. *The Freezing Preservation of Foods*. New York: Avi Publishing, 1943.

FORTRAN PROGRAMMING LANGUAGE

THE INVENTION: The first major computer programming language, FORTRAN supported programming in a mathematical language that was natural to scientists and engineers and achieved unsurpassed success in scientific computation.

THE PEOPLE BEHIND THE INVENTION:

John Backus (1924-   ), an American software engineer and manager

John W. Mauchly (1907-1980), an American physicist and engineer

Herman Heine Goldstine (1913-   ), a mathematician and computer scientist

John von Neumann (1903-1957), a Hungarian American mathematician and physicist

Talking to Machines

Formula Translation, or FORTRAN—the first widely accepted high-level computer language—was completed by John Backus and his coworkers at the International Business Machines (IBM) Corporation in April, 1957. Designed to support programming in a mathematical language that was natural to scientists and engineers, FORTRAN achieved unsurpassed success in scientific computation.

Computer languages are means of specifying the instructions that a computer should execute and the order of those instructions. Computer languages can be divided into categories of progressively higher degrees of abstraction. At the lowest level is binary code, or machine code: Binary digits, or "bits," specify in complete detail every instruction that the machine will execute. This was the only language available in the early days of computers, when such machines as the ENIAC (Electronic Numerical Integrator and Calculator) required hand-operated switches and plugboard connections. All higher levels of language are implemented by having a program translate instructions written in the higher language into binary machine language (also called "object code"). High-level languages (also called "programming languages") are largely or entirely independent of the underlying machine structure. FORTRAN was the first language of this type to win widespread acceptance.

The emergence of machine-independent programming languages was a gradual process that spanned the first decade of electronic computation. One of the earliest developments was the invention of "flowcharts," or "flow diagrams," by Herman Heine Goldstine and John von Neumann in 1947. Flowcharting became the most influential software methodology during the first twenty years of computing.

Short Code was the first language to be implemented that contained some high-level features, such as the ability to use mathematical equations. The idea came from John W. Mauchly, and it was implemented on the BINAC (Binary Automatic Computer) in 1949 with an "interpreter"; later, it was carried over to the UNIVAC (Universal Automatic Computer) I. Interpreters are programs that do not translate commands into a series of object-code instructions; instead, they directly execute (interpret) those commands. Every time the interpreter encounters a command, that command must be interpreted again. "Compilers," however, convert the entire command into object code before it is executed.

Much early effort went into creating ways to handle commonly encountered problems—particularly scientific mathematical calculations. A number of interpretive languages arose to support these features. As long as such complex operations had to be performed by software (computer programs), however, scientific computation would be relatively slow. Therefore, Backus lobbied successfully for a direct hardware implementation of these operations on IBM's new scientific computer, the 704. Backus then started the Programming Research Group at IBM in order to develop a compiler that would allow programs to be written in a mathematically oriented language rather than a machine-oriented language. In November of 1954, the group defined an initial version of FORTRAN.

A MORE ACCESSIBLE LANGUAGE

Before FORTRAN was developed, a computer had to perform a whole series of tasks to make certain types of mathematical calculations. FORTRAN made it possible for the same calculations to be performed much more easily. In general, FORTRAN supported constructs with which scientists were already acquainted, such as functions and multidimensional arrays. In defining a powerful notation that was accessible to scientists and engineers, FORTRAN opened up programming to a much wider community.

Backus's success in getting the IBM 704's hardware to support scientific computation directly, however, posed a major challenge: Because such computation would be much faster, the object code produced by FORTRAN would also have to be much faster. The lower-level compilers preceding FORTRAN produced programs that were usually five to ten times slower than their hand-coded counterparts; therefore, efficiency became the primary design objective for Backus. The highly publicized claims for FORTRAN met with widespread skepticism among programmers. Much of the team's efforts, therefore, went into discovering ways to produce the most efficient object code.

The efficiency of the compiler produced by Backus, combined with its clarity and ease of use, guaranteed the system's success. By 1959, many IBM 704 users programmed exclusively in FORTRAN. By 1963, virtually every computer manufacturer either had delivered or had promised a version of FORTRAN.

Incompatibilities among manufacturers were minimized by the popularity of IBM's version of FORTRAN; every company wanted to be able to support IBM programs on its own equipment. Nevertheless, there was sufficient interest in obtaining a standard for FORTRAN that the American National Standards Institute adopted a formal standard for it in 1966. A revised standard was adopted in 1978, yielding FORTRAN 77.

CONSEQUENCES

In demonstrating the feasibility of efficient high-level languages, FORTRAN inaugurated a period of great proliferation of programming languages. Most of these languages attempted to provide similar or better high-level programming constructs oriented toward a different, nonscientific programming environment. COBOL, for example, stands for "Common Business Oriented Language."

FORTRAN, while remaining the dominant language for scientific programming, has not found general acceptance among nonscientists. An IBM project established in 1963 to extend FORTRAN found the task too unwieldy and instead ended up producing an entirely different language, PL/I, which was delivered in 1966. In the beginning, Backus and his coworkers believed that their revolutionary language would virtually eliminate the burdens of coding and debugging. Instead, FORTRAN launched software as a field of study and an industry in its own right.

In addition to stimulating the introduction of new languages, FORTRAN encouraged the development of operating systems. Programming languages had already grown into simple operating systems called "monitors." Operating systems since then have been greatly improved so that they support, for example, simultaneously active programs (multiprogramming) and the networking (combining) of multiple computers.

See also BASIC programming language; COBOL computer language; SAINT.

FURTHER READING

Goff, Leslie. "Born of Frustration." *Computerworld* 33, no. 6 (February 8, 1999).

Moreau, René. *The Computer Comes of Age: The People, the Hardware, and the Software*. Cambridge, Mass.: MIT Press, 1984.

Slater, Robert. *Portraits in Silicon*. Cambridge, Mass.: MIT Press, 1987.

Stern, Nancy B. *From ENIAC to UNIVAC: An Appraisal of the Eckert-Mauchly Computers*. Bedford, Mass.: Digital Press., 1981.

Freeze-Drying

THE INVENTION: Method for preserving foods and other organic matter by freezing them and using a vacuum to remove their water content without damaging their solid matter.

THE PEOPLE BEHIND THE INVENTION:

Earl W. Flosdorf (1904-   ), an American physician

Ronald I. N. Greaves (1908-   ), an English pathologist

Jacques Arsène d'Arsonval (1851-1940), a French physicist

FREEZE-DRYING FOR PRESERVATION

Drying, or desiccation, is known to preserve biomaterials, including foods. In freeze-drying, water is evaporated in a frozen state in a vacuum, by means of *sublimation* (the process of changing a solid to a vapor without first changing it to a liquid).

In 1811, John Leslie had first caused freezing by means of the evaporation and sublimation of ice. In 1813, William Wollaston demonstrated this process to the Royal Society of London. It does not seem to have occurred to either Leslie or Wollaston to use sublimation for drying. That distinction goes to Richard Altmann, a German histologist, who dried pieces of frozen tissue in 1890. Later, in 1903, Vansteenberghe freeze-dried the rabies virus. In 1906, Jacques Arsène d'Arsonval removed water at a low temperature for distillation.

Since water removal is the essence of drying, d'Arsonval is often credited with the discovery of freeze-drying, but the first clearly recorded use of sublimation for preservation was by Leon Shackell in 1909. His work was widely recognized, and he freeze-dried a variety of biological materials. The first patent for freeze-drying was issued to Henri Tival, a French inventor, in 1927. In 1934, William Elser received patents for a modern freeze-drying apparatus that supplied heat for sublimation.

In 1933, Earl W. Flosdorf had freeze-dried human blood serum and plasma for clinical use. The subsequent efforts of Flosdorf led to commercial freeze-drying applications in the United States.

FREEZE-DRYING OF FOODS

With the freeze-drying technique fairly well established for biological products, it was a natural extension for Flosdorf to apply the technique to the drying of foods. As early as 1935, Flosdorf experimented with the freeze-drying of fruit juices and milk. An early British patent was issued to Franklin Kidd, a British inventor, in 1941 for the freeze-drying of foods. An experimental program on the freeze-drying of food was also initiated at the Low Temperature Research Station at Cambridge University in England, but until World War II, freeze-drying was only an occasionally used scientific tool.

It was the desiccation of blood plasma from the frozen state, performed by the American Red Cross for the U.S. armed forces, that provided the first spectacular, extensive use of freeze-drying. This work demonstrated the vast potential of freeze-drying for commercial applications. In 1949, Flosdorf published the first book on freeze-drying, which laid the foundation for freeze-drying of foods and remains one of the most important contributions to large-scale operations in the field. In the book, Flosdorf described the freeze-drying of fruit juices, milk, meats, oysters, clams, fish fillets, coffee and tea extracts, fruits, vegetables, and other products. Flosdorf also devoted an entire chapter to describing the equipment used for both batch and continuous processing, and he discussed cost analysis. The holder of more than fifteen patents covering various aspects of freeze-drying, Flosdorf dominated the move toward commercialization in the United States.

Simultaneously, researchers in England were developing freeze-drying applications under the leadership of Ronald I. N. Greaves. The food crisis during World War II had led to the recognition that dried foods cut the costs of transporting, storing, and packaging foods in times of emergency. Thus, in 1951, the British Ministry of Food Research was established at Aberdeen, Scotland. Scientists at Aberdeen developed a vacuum contact plate freeze-dryer that improved product quality and reduced the time required for rehydration (replacement of the water removed in the freeze-drying process so that the food can be used).

In 1954, trials of initial freeze-drying, followed by the ordinary process of vacuum drying, were carried out. The abundance of membranes within plant and animal tissues was a major obstacle to the movement of water vapor, thus limiting the drying rate. In 1956, two Canadian scientists developed a new method of improving the freeze-drying rate for steaks by impaling the steaks on spiked heater plates. This idea was adapted in 1957 by interposing sheets of expanded metal, instead of spikes, between the drying surfaces of the frozen food and the heating platens. Because of the substantially higher freeze-drying rates that it achieved, the process was called "accelerated freeze-drying."

In 1960, Greaves described an ingenious method of freeze-drying liquids. It involved continuously scraping the dry layer during its formation. This led to a continuous process for freeze-drying liquids. During the remainder of the 1960's, freeze-drying applications proliferated with the advent of several techniques for controlling and improving the effectiveness of the freeze-drying process.

IMPACT

Flosdorf's vision and ingenuity in applying freeze-drying to foods has revolutionized food preservation. He was also responsible for making a laboratory technique a tremendous commercial success.

Freeze-drying is important because it stops the growth of microorganisms, inhibits deleterious chemical reactions, and facilitates distribution and storage. Freeze-dried foods are easily prepared for consumption by adding water (rehydration). When freeze-dried properly, most foods, either raw or cooked, can be rehydrated quickly to yield products that are equal in quality to their frozen counterparts. Freeze-dried products retain most of their nutritive qualities and have a long storage life, even at room temperature.

Freeze-drying is not, however, without disadvantages. The major disadvantage is the high cost of processing. Thus, to this day, the great potential of freeze-drying has not been fully realized. The drying of cell-free materials, such as coffee and tea extracts, has been extremely successful, but the obstacles imposed by the cell membranes in foods such as fruits, vegetables, and meats have limited the application to expensive specialty items such as freeze-dried soups and to foods for armies, campers, and astronauts. Future economic changes may create a situation in which the high cost of freeze-drying is more than offset by the cost of transportation and storage.

See also Electric refrigerator; Food freezing; Polystyrene; Tupperware.

FURTHER READING

Comello, Vic. "Improvements in Freeze Drying Expand Application Base." *Research and Development* 42, no. 5 (May, 2000).

Flosdorf, Earl William. *Freeze-Drying: Drying by Sublimation.* New York: Reinhold, 1949.

Noyes, Robert. *Freeze Drying of Foods and Biologicals, 1968.* Park Ridge, N.J.: Noyes Development Corporation, 1968.

Fuel cell

The invention: An electrochemical cell that directly converts energy from reactions between oxidants and fuels, such as liquid hydrogen, into electrical energy.

The people behind the invention:
Francis Thomas Bacon (1904-1992), an English engineer
Sir William Robert Grove (1811-1896), an English inventor
Georges Leclanché (1839-1882), a French engineer
Alessandro Volta (1745-1827), an Italian physicist

The Earth's Resources

Because of the earth's rapidly increasing population and the dwindling of fossil fuels (natural gas, coal, and petroleum), there is a need to design and develop new ways to obtain energy and to encourage its intelligent use. The burning of fossil fuels to create energy causes a slow buildup of carbon dioxide in the atmosphere, creating pollution that poses many problems for all forms of life on this planet. Chemical and electrical studies can be combined to create electrochemical processes that yield clean energy.

Because of their very high rate of efficiency and their nonpolluting nature, fuel cells may provide the solution to the problem of finding sufficient energy sources for humans. The simple reaction of hydrogen and oxygen to form water in such a cell can provide an enormous amount of clean (nonpolluting) energy. Moreover, hydrogen and oxygen are readily available.

Studies by Alessandro Volta, Georges Leclanché, and William Grove preceded the work of Bacon in the development of the fuel cell. Bacon became interested in the idea of a hydrogen-oxygen fuel cell in about 1932. His original intent was to develop a fuel cell that could be used in commercial applications.

The Fuel Cell Emerges

In 1800, the Italian physicist Alessandro Volta experimented with solutions of chemicals and metals that were able to conduct electricity. He found that two pieces of metal and such a solution could be arranged in such a way as to produce an electric current. His creation was the first electrochemical battery, a device that produced energy from a chemical reaction. Studies in this area were continued by various people, and in the late nineteenth century, Georges Leclanché invented the dry cell battery, which is now commonly used.

The work of William Grove followed that of Leclanché. His first significant contribution was the Grove cell, an improved form of the cells described above, which became very popular. Grove experimented with various forms of batteries and eventually invented the "gas battery," which was actually the earliest fuel cell. It is worth noting that his design incorporated separate test tubes of hydrogen and oxygen, which he placed over strips of platinum.

After studying the design of Grove's fuel cell, Bacon decided that, for practical purposes, the use of platinum and other precious metals should be avoided. By 1939, he had constructed a cell in which nickel replaced the platinum used.

The theory behind the fuel cell can be described in the following way. If a mixture of hydrogen and oxygen is ignited, energy is released in the form of a violent explosion. In a fuel cell, however, the reaction takes place in a controlled manner. Electrons lost by the hydrogen gas flow out of the fuel cell and return to be taken up by the oxygen in the cell. The electron flow provides electricity to any device that is connected to the fuel cell, and the water that the fuel cell produces can be purified and used for drinking.

Bacon's studies were interrupted by World War II. After the war was over, however, Bacon continued his work. Sir Eric Keightley Rideal of Cambridge University in England supported Bacon's studies; later, others followed suit. In January, 1954, Bacon wrote an article entitled "Research into the Properties of the Hydrogen/Oxygen Fuel Cell" for a British journal. He was surprised at the speed with which news of the article spread throughout the scientific world, particularly in the United States.

After a series of setbacks, Bacon demonstrated a forty-cell unit that had increased power. This advance showed that the fuel cell was not merely an interesting toy; it had the capacity to do useful work. At this point, the General Electric Company (GE), an American corporation, sent a representative to England to offer employment in the United States to senior members of Bacon's staff. Three scientists accepted the offer.

A high point in Bacon's career was the announcement that the American Pratt and Whitney Aircraft company had obtained an order to build fuel cells for the Apollo project, which ultimately put two men on the Moon in 1969. Toward the end of his career in 1978, Bacon hoped that commercial applications for his fuel cells would be found.

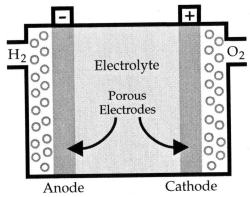

Parts of a basic fuel cell

IMPACT

Because they are lighter and more efficient than batteries, fuel cells have proved to be useful in the space program. Beginning with the Gemini 5 spacecraft, alkaline fuel cells (in which a water solution of potassium hydroxide, a basic, or alkaline, chemical, is placed) have been used for more than ten thousand hours in space. The fuel cells used aboard the space shuttle deliver the same amount of power as batteries weighing ten times as much. On a typical seven-day mission, the shuttle's fuel cells consume 680 kilograms (1,500 pounds) of hydrogen and generate 719 liters (190 gallons) of water that can be used for drinking.

Major technical and economic problems must be overcome in order to design fuel cells for practical applications, but some important advancements have been made. A few test vehicles that use fuel

FRANCIS BACON

Born in Billericay, England, in 1904, Francis Thomas Bacon completed secondary school at the prestigious Eton College and then attended Trinity College, Cambridge University. In 1932 he started his long search for a practical fuel cell based upon the oxygen-hydrogen (Hydrox) reaction with an alkaline electrolyte and inexpensive nickel electrodes. In 1940 the British Admiralty set him up in full-time experimental work at King's College, London, and then moved him to the Anti-Submarine Experimental Establishment because the Royal Navy wanted fuel cells for their submarines.

After World War II Cambridge University appointed him to the faculty at the Department of Chemical Engineering, and he worked intensively on his fuel cell research. In 1959 he proved the worth of his work by producing a fuel cell capable of powering a small truck. It was not until the 1990's, however, that fuel cells were taken seriously as the main power source for automobiles. In 1998, for instance, Iceland enlisted the help of Daimler-Chrysler, Shell Oil, and Norsk Hydro to convert all its transportation vehicles, including its fishing boats, to fuel cell power, part of its long-range plans for a completely "hydrogen economy." Meanwhile, Bacon had the satisfaction of seeing his invention become a power source for American space vehicles and stations. He died in 1992 in Cambridge.

cells as a source of power have been constructed. Fuel cells using hydrogen as a fuel and oxygen to burn the fuel have been used in a van built by General Motors Corporation. Thirty-two fuel cells are installed below the floorboards, and tanks of liquid oxygen are carried in the back of the van. A power plant built in New York City contains stacks of hydrogen-oxygen fuel cells, which can be put on line quickly in response to power needs. The Sanyo Electric Company has developed an electric car that is partially powered by a fuel cell.

These tremendous technical advances are the result of the single-minded dedication of Francis Thomas Bacon, who struggled all of his life with an experiment he was convinced would be successful.

See also Alkaline storage battery; Breeder reactor; Compressed-air-accumulating power plant; Fluorescent lighting; Geothermal power; Heat pump; Photoelectric cell; Photovoltaic cell; Solar thermal engine; Tidal power plant.

FURTHER READING

Eisenberg, Anne. "Fuel Cell May Be the Future 'Battery.'" *New York Times* (October 21, 1999).

Hoverstein, Paul. "Century-Old Invention Finding a Niche Today. *USA Today* (June 3, 1994).

Kufahl, Pam. "Electric: Lighting Up the Twentieth Century." *Unity Business* 3, no. 7 (June, 2000).

Stobart, Richard. *Fuel Cell Technology for Vehicles*. Warrendale, Pa.: Society of Automotive Engineers, 2001.

Gas-electric car

THE INVENTION: A hybrid automobile with both an internal combustion engine and an electric motor.

THE PEOPLE BEHIND THE INVENTION:
Victor Wouk (1919-   ), an American engineer
Tom Elliott, executive vice president of American Honda Motor Company
Hiroyuki Yoshino, president and chief executive officer of Honda Motor Company
Fujio Cho, president of Toyota Motor Corporation

Announcing Hybrid Vehicles

At the 2000 North American International Auto Show in Detroit, not only did the Honda Motor Company show off its new Insight model, it also announced expanded use of its new technology. Hiroyuki Yoshino, president and chief executive officer, said that Honda's integrated motor assist (IMA) system would be expanded to other mass-market models. The system basically fits a small electric motor directly on a one-liter, three-cylinder internal combustion engine. The two share the workload of powering the car, but the gasoline engine does not start up until it is needed. The electric motor is powered by a nickel-metal hydride (Ni-MH) battery pack, with the IMA system automatically recharging the energy pack during braking.

Tom Elliott, Honda's executive vice-president, said the vehicle was a continuation of the company's philosophy of making the latest environmental technology accessible to consumers. The $18,000 Insight was a two-seat sporty car that used many innovations to reduce its weight and improve its performance.

Fujio Cho, president of Toyota, also spoke at the Detroit show, where his company showed off its new $20,000 hybrid Prius. The Toyota Prius relied more on the electric motor and had more energy-storage capacity than the Insight, but was a four-door, five-seat model. The Toyota Hybrid System divided the power from its 1.5-liter gasoline engine and directed it to drive the wheels and a generator. The generator alternately powered the motor and recharged the batteries. The electric motor was coupled with the gasoline engine to power the wheels under normal driving. The gasoline engine supplied average power needs, with the electric motor helping the peaks; at low speeds, it was all electric. A variable transmission seamlessly switched back and forth between the gasoline engine and electric motor or applied both of them.

VARIATIONS ON AN IDEA

Automobiles generally use gasoline or diesel engines for driving, electric motors that start the main motors, and a means of recharging the batteries that power starter motors and other devices. In solely electric cars, gasoline engines are eliminated entirely, and the batteries that power the vehicles are recharged from stationary sources. In hybrid cars, the relationship between gasoline engines and electric motors is changed so that electric motors handle some or all of the driving. This is at the expense of an increased number of batteries or other energy-storage devices.

Possible in many combinations, "hybrids" couple the low-end torque and regenerative braking potential of electric motors with the range and efficient packaging of gasoline, natural gas, or even hydrogen fuel power plants. The return is greater energy efficiency and reduced pollution.

With sufficient energy-storage capacity, an electric motor can actually propel a car from a standing start to a moving speed. In hybrid vehicles, the gasoline engines—which are more energy-efficient at higher speeds, then kick in. However, the gasoline engines in these vehicles are smaller, lighter, and more efficient than ordinary gas engines. Designed for average—not peak—driving conditions, they reduce air pollution and considerably improve fuel economy.

Batteries in hybrid vehicles are recharged partly by the gas engines and partly by regenerative braking; a third of the energy from slowing the car is turned into electricity. What has finally made hybrids feasible at reasonable cost are the new developments in computer technology, allowing sophisticated controls to coordinate electrical and mechanical power.

Victor Wouk

H. Piper, an American engineer, filed the first patent for a hybrid gas-electric powered car in 1905, and from then until 1915 they were popular, although not common, because they could accelerate faster than plain gas-powered cars. Then the gas-only models became as swift. Their hybrid cousins fells by the wayside.

Interest in hybrids revived with the unheard-of gasoline prices during the 1973 oil crisis. The champion of their come-back—the father of the modern hybrid electric vehicle (HEV)—was Victor Wouk. Born in 1919 in New York City, Wouk earned a math and physics degree from Columbia University in 1939 and a doctorate in electrical engineering from the California Institute of Technology in 1942. In 1946 he founded Beta Electric Corporation, which he led until 1959, when he founded and was president of another company, Electronic Energy Conversion Corporation. After 1970, he became an independent consultant, hoping to build an HEV that people would prefer to gas-guzzlers.

With his partner, Charles Rosen, Wouk gutted the engine compartment of a Buick Skylark and installed batteries designed for police cars, a 20-watt direct-current electric motor, and an RX-2 Mazda rotary engine. Only a test vehicle, it still got better gas mileage (thirty miles per gallon) than the original Skylark and met the requirements for emissions control set by the Clean Air Act of 1970, unlike all American automobiles of the era. Moreover, Wouk designed an HEV that would get fifty miles per gallon and pollute one-eighth as much as gas-powered automobiles. However, the oil crisis ended, gas prices went down, and consumers and the government lost interest. Wouk continued to publish, lecture, and design; still, it was not until the 1990's that high gas prices and concerns over pollution made HEV's attractive yet again.

Wouk holds twelve patents, mostly for speed and braking controls in electric vehicles but also for air conditioning, high voltage direct-current power sources, and life extenders for incandescent lamps.

One way to describe hybrids is to separate them into two types: parallel, in which either of the two power plants can propel the vehicle, and series, in which the auxiliary power plant is used to charge the battery, rather than propel the vehicle.

Honda's Insight is a simplified parallel hybrid that uses a small but efficient gasoline engine. The electric motor assists the engine, providing extra power for acceleration or hill climbing, helps provide regenerative braking, and starts the engine. However, it cannot run the car by itself.

Toyota's Prius is a parallel hybrid whose power train allows some series features. Its engine runs only at an efficient speed and load and is combined with a unique power splitting device. It allows the car to operate like a parallel hybrid, motor alone, engine alone, or both. It can act as a series hybrid with the engine charging the batteries rather than powering the vehicle. It also provides a continually variable transmission using a planetary gear set that allows interaction between the engine, the motor, and the differential which drives the wheels.

Impact

In 2001 Honda and Toyota marketed gas-electric hybrids that offered better than 60-mile-per-gallon fuel economy and met California's stringent standards for "super ultra-low emissions" vehicles. Both comparnies achieved these standards without the inconvenience of fully electric cars which could go only about a hundred miles on a single battery charge and required such gimmicks as kerosene-powered heaters. As a result, other manufacturers were beginning to follow suit. Ford, for example, promised a hybrid sport utility vehicle (SUV) by 2003. Other automakers, including General Motors and DaimlerChrysler, also have announced development of alternative fuel and low emission vehicles. An example is the ESX3 concept car using a 1.5-liter, direct injection diesel combined with a electric motor and a lithium-ion battery

While American automakers were planning to offer some "full hybrids"—cars capable of running on battery power alone at low speeds—they were focusing more enthusiastically on electrically assisted gasoline engines called "mild hybrids." Full hybrids typically increase gas mileage by up to 60 percent; mild hybrids by only 10 or 20 percent. The "mild hybrid" approach uses regenerative braking with electrical systems of a much lower voltage and storage capacity than for full hybrids, a much cheaper approach. But there still is enough energy available to allow the gasoline engine to turn off automatically when a vehicle stops and turn on instantly when the accelerator is touched. Because the "mild hybrid" approach adds only $1000 to $1500 to a vehicle's price, it is likely to be used in many models. Full hybrids cost much more, but achieve more benefits.

See also Airplane; Diesel locomotive; Hovercraft; Internal combustion engine; Supersonic passenger plane; Turbojet.

FURTHER READING

Morton, Ian. "Honda Insight Hybrid Makes Heavy Use of Light Metal." *Automotive News* 74, no. 5853 (December 20, 1999).
Peters, Eric. "Hybrid Cars: The Hope, Hype, and Future." *Consumers' Research Magazine* 83, no. 6 (June, 2000).
Reynolds, Kim. "Burt Rutan Ponders the Hybrid Car." *Road and Track* 51, no. 11 (July, 2000).
Swoboda, Frank. "'Hybrid' Cars Draw Waiting List of Buyers." *Washington Post* (May 3, 2001).
Yamaguchi, Jack. "Toyota Prius IC/Electric Hybrid Update." *Automotive Engineering International* 108, no. 12 (December, 2000).

GEIGER COUNTER

THE INVENTION: the first electronic device able to detect and measure radioactivity in atomic particles.

THE PEOPLE BEHIND THE INVENTION:

Hans Geiger (1882-1945), a German physicist
Ernest Rutherford (1871-1937), a British physicist
Sir John Sealy Edward Townsend (1868-1957), an Irish physicist
Sir William Crookes (1832-1919), an English physicist
Wilhelm Conrad Röntgen (1845-1923), a German physicist
Antoine-Henri Becquerel (1852-1908), a French physicist

DISCOVERING NATURAL RADIATION

When radioactivity was discovered and first studied, the work was done with rather simple devices. In the 1870's, Sir William Crookes learned how to create a very good vacuum in a glass tube. He placed electrodes in each end of the tube and studied the passage of electricity through the tube. This simple device became known as the "Crookes tube." In 1895, Wilhelm Conrad Röntgen was experimenting with a Crookes tube. It was known that when electricity went through a Crookes tube, one end of the glass tube might glow. Certain mineral salts placed near the tube would also glow. In order to observe carefully the glowing salts, Röntgen had darkened the room and covered most of the Crookes tube with dark paper. Suddenly, a flash of light caught his eye. It came from a mineral sample placed some distance from the tube and shielded by the dark paper; yet when the tube was switched off, the mineral sample went dark. Experimenting further, Röntgen became convinced that some ray from the Crookes tube had penetrated the mineral and caused it to glow. Since light rays were blocked by the black paper, he called the mystery ray an "X ray," with "X" standing for unknown.

Antoine-Henri Becquerel heard of the discovery of X rays and, in February, 1886, set out to discover if glowing minerals themselves emitted X rays. Some minerals, called "phosphorescent," begin to glow when activated by sunlight. Becquerel's experiment involved wrapping photographic film in black paper and setting various phosphorescent minerals on top and leaving them in the sun. He soon learned that phosphorescent minerals containing uranium would expose the film.

A series of cloudy days, however, brought a great surprise. Anxious to continue his experiments, Becquerel decided to develop film that had not been exposed to sunlight. He was astonished to discover that the film was deeply exposed. Some emanations must be coming from the uranium, he realized, and they had nothing to do with sunlight. Thus, natural radioactivity was discovered by accident with a simple piece of photographic film.

RUTHERFORD AND GEIGER

Ernest Rutherford joined the world of international physics at about the same time that radioactivity was discovered. Studying the "Becquerel rays" emitted by uranium, Rutherford eventually distinguished three different types of radiation, which he named "alpha," "beta," and "gamma" after the first three letters of the Greek alphabet. He showed that alpha particles, the least penetrating of the three, are the nuclei of helium atoms (a group of two neutrons and a proton tightly bound together). It was later shown that beta particles are electrons. Gamma rays, which are far more penetrating than either alpha or beta particles, were shown to be similar to X rays, but with higher energies.

Rutherford became director of the associated research laboratory at Manchester University in 1907. Hans Geiger became an assistant. At this time, Rutherford was trying to prove that alpha particles carry a double positive charge. The best way to do this was to measure the electric charge that a stream of alpha particles would bring to a target. By dividing that charge by the total number of alpha particles that fell on the target, one could calculate the charge of a single alpha particle. The problem lay in counting the particles and in proving that every particle had been counted.

Basing their design upon work done by Sir John Sealy Edward Townsend, a former colleague of Rutherford, Geiger and Rutherford constructed an electronic counter. It consisted of a long brass tube sealed at both ends from which most of the air had been

HANS GEIGER

Atomic radiation was the first physical phenomenon that humans discovered that they could not detect with any of their five natural senses. Hans Geiger found a way to make radiation observable.

Born into a family with an academic tradition in 1882, Geiger became an academician himself. His father was a professor of linguistics at the University of Erlangen, where Geiger completed his own doctorate in physics in 1906. One of the world's centers for experimental physics at the time was England, and there Geiger went in 1907. He became an assistant to Ernest Rutherford at the University of Manchester and thereby began the first of a series of successful collaborations during his career—all devoted to detecting or explaining types of radiation.

Rutherford had distinguished three types of radiation. In 1908, he and Geiger built a device to sense the first alpha particles. It gave them evidence for Rutherford's conjecture that the atom was structured like a miniature solar system. Geiger also worked closely with Ernest Marsden, James Chadwick, and Walter Bothe on aspects of radiation physics.

Geiger's stay in England ended with the outbreak of World War I in 1914. He returned to Germany and served as an artillery officer. Immediately after the war he took up university posts again, first in Berlin, then in Kiel, Tubingen, and back to Berlin. With Walther Müller he perfected a compact version of the radiation detector in 1925, the Geiger- Müller counter. It became the standard radiation sensor for scientists thereafter, and, during the rush to locate uranium deposits during the 1950's, for prospectors.

Geiger used it to prove the existence of the Compton effect, which concerned the scattering of X rays, and his experiments further proved beyond doubt that light can take the form of quanta. He also discovered cosmic-ray showers with his detector.

Geiger remained in German during World War II, although he vigorously opposed the Nazi party's treatment of scientists. He died in Potsdam in 1945, after losing his home and possessions during the Allied occupation of Berlin.

pumped. A thin wire, insulated from the brass, was suspended down the middle of the tube. This wire was connected to batteries producing about thirteen hundred volts and to an electrometer, a device that could measure the voltage of the wire. This voltage could be increased until a spark jumped between the wire and the tube. If the voltage was turned down a little, the tube was ready to operate. An alpha particle entering the tube would ionize (knock some electrons away from) at least a few atoms. These electrons would be accelerated by the high voltage and, in turn, would ionize more atoms, freeing more electrons. This process would continue until an avalanche of electrons struck the central wire and the electrometer registered the voltage change. Since the tube was nearly ready to arc because of the high voltage, every alpha particle, even if it had very little energy, would initiate a discharge. The most complex of the early radiation detection devices—the forerunner of the Geiger counter—had just been developed. The two physicists reported their findings in February, 1908.

IMPACT

Their first measurements showed that one gram of radium emitted 34 thousand million alpha particles per second. Soon, the number was refined to 32.8 thousand million per second. Next, Geiger and Rutherford measured the amount of charge emitted by radium each second. Dividing this number by the previous number gave them the charge on a single alpha particle. Just as Rutherford had anticipated, the charge was double that of a hydrogen ion (a proton). This proved to be the most accurate determination of the fundamental charge until the American physicist Robert Andrews Millikan conducted his classic oil-drop experiment in 1911.

Another fundamental result came from a careful measurement of the volume of helium emitted by radium each second. Using that value, other properties of gases, and the number of helium nuclei emitted each second, they were able to calculate Avogadro's number more directly and accurately than had previously been possible. (Avogadro's number enables one to calculate the number of atoms in a given amount of material.)

The true Geiger counter evolved when Geiger replaced the central wire of the tube with a needle whose point lay just inside a thin entrance window. This counter was much more sensitive to alpha and beta particles and also to gamma rays. By 1928, with the assistance of Walther Müller, Geiger made his counter much more efficient, responsive, durable, and portable. There are probably few radiation facilities in the world that do not have at least one Geiger counter or one of its compact modern relatives.

See also Carbon dating; Gyrocompass; Radar; Sonar; Richter scale.

FURTHER READING

Campbell, John. *Rutherford: Scientist Supreme.* Christchurch, New Zealand: AAS Publications, 1999.

Halacy, D. S. *They Gave Their Names to Science.* New York: Putnam, 1967.

Krebs, A. T. "Hans Geiger: Fiftieth Anniversary of the Publication of His Doctoral Thesis, 23 July 1906." *Science* 124 (1956).

Weir, Fred. "Muscovites Check Radishes for Radiation; a $50 Personal Geiger Counter Gives Russians a Sense of Confidence at the Market." *Christian Science Monitor* (November 4, 1999).

Genetic "fingerprinting"

THE INVENTION: A technique for using the unique characteristics of each human being's DNA to identify individuals, establish connections among relatives, and identify criminals.

THE PEOPLE BEHIND THE INVENTION:
Alec Jeffreys (1950-   ), an English geneticist
Victoria Wilson (1950-   ), an English geneticist
Swee Lay Thein (1951-   ), a biochemical geneticist

Microscopic Fingerprints

In 1985, Alec Jeffreys, a geneticist at the University of Leicester in England, developed a method of deoxyribonucleic acid (DNA) analysis that provides a visual representation of the human genetic structure. Jeffreys's discovery had an immediate, revolutionary impact on problems of human identification, especially the identification of criminals. Whereas earlier techniques, such as conventional blood typing, provide evidence that is merely exclusionary (indicating only whether a suspect could or could not be the perpetrator of a crime), DNA fingerprinting provides positive identification.

For example, under favorable conditions, the technique can establish with virtual certainty whether a given individual is a murderer or rapist. The applications are not limited to forensic science; DNA fingerprinting can also establish definitive proof of parenthood (paternity or maternity), and it is invaluable in providing markers for mapping disease-causing genes on chromosomes. In addition, the technique is utilized by animal geneticists to establish paternity and to detect genetic relatedness between social groups.

DNA fingerprinting (also referred to as "genetic fingerprinting") is a sophisticated technique that must be executed carefully to produce valid results. The technical difficulties arise partly from the complex nature of DNA. DNA, the genetic material responsible for heredity in all higher forms of life, is an enormously long, double-stranded molecule composed of four different units called "bases." The bases on one strand of DNA pair with complementary bases on the other strand. A human being contains twenty-three pairs of chromosomes; one member of each chromosome pair is inherited from the mother, the other from the father. The order, or sequence, of bases forms the genetic message, which is called the "genome." Scientists did not know the sequence of bases in any sizable stretch of DNA prior to the 1970's because they lacked the molecular tools to split DNA into fragments that could be analyzed. This situation changed with the advent of biotechnology in the mid-1970's.

The door to DNA analysis was opened with the discovery of bacterial enzymes called "DNA restriction enzymes." A restriction enzyme binds to DNA whenever it finds a specific short sequence of base pairs (analogous to a code word), and it splits the DNA at a defined site within that sequence. A single enzyme finds millions of cutting sites in human DNA, and the resulting fragments range in size from tens of base pairs to hundreds or thousands. The fragments are exposed to a radioactive DNA probe, which can bind to specific complementary DNA sequences in the fragments. X-ray film detects the radioactive pattern. The developed film, called an "autoradiograph," shows a pattern of DNA fragments, which is similar to a bar code and can be compared with patterns from known subjects.

THE PRESENCE OF MINISATELLITES

The uniqueness of a DNA fingerprint depends on the fact that, with the exception of identical twins, no two human beings have identical DNA sequences. Of the three billion base pairs in human DNA, many will differ from one person to another.

In 1985, Jeffreys and his coworkers, Victoria Wilson at the University of Leicester and Swee Lay Thein at the John Radcliffe Hospital in Oxford, discovered a way to produce a DNA fingerprint. Jeffreys had found previously that human DNA contains many repeated minisequences called "minisatellites." Minisatellites consist of sequences of base pairs repeated in tandem, and the number of repeated units varies widely from one individual to another. Every person, with the exception of identical twins, has a different number of tandem repeats and, hence, different lengths of minisatellite DNA. By using two labeled DNA probes to detect two different minisatellite sequences, Jeffreys obtained a unique fragment band pattern that was completely specific for an individual.

The power of the technique derives from the law of chance, which indicates that the probability (chance) that two or more unrelated events will occur simultaneously is calculated as the multiplication product of the two separate probabilities. As Jeffreys discovered, the likelihood of two unrelated people having completely identical DNA fingerprints is extremely small—less than one in ten trillion. Given the population of the world, it is clear that the technique can distinguish any one person from everyone else. Jeffreys called his band patterns "DNA fingerprints" because of their ability to individualize. As he stated in his landmark research paper, published in the English scientific journal *Nature* in 1985, probes to minisatellite regions of human DNA produce "DNA 'fingerprints' which are completely specific to an individual (or to his or her identical twin) and can be applied directly to problems of human identification, including parenthood testing."

CONSEQUENCES

In addition to being used in human identification, DNA fingerprinting has found applications in medical genetics. In the search for a cause, a diagnostic test for, and ultimately the treatment of an inherited disease, it is necessary to locate the defective gene on a human chromosome. Gene location is accomplished by a technique called "linkage analysis," in which geneticists use marker sections of DNA as reference points to pinpoint the position of a defective gene on a chromosome. The minisatellite DNA probes developed by Jeffreys provide a potent and valuable set of markers that are of great value in locating disease-causing genes. Soon after its discovery, DNA fingerprinting was used to locate the defective genes responsible for several diseases, including fetal hemoglobin abnormality and Huntington's disease.

Genetic fingerprinting also has had a major impact on genetic studies of higher animals. Because DNA sequences are conserved in evolution, humans and other vertebrates have many sequences in common. This commonality enabled Jeffreys to use his probes to human minisatellites to bind to the DNA of many different vertebrates, ranging from mammals to birds, reptiles, amphibians, and fish; this made it possible for him to produce DNA fingerprints of these vertebrates. In addition, the technique has been used to discern the mating behavior of birds, to determine paternity in zoo primates, and to detect inbreeding in imperiled wildlife. DNA fingerprinting can also be applied to animal breeding problems, such as the identification of stolen animals, the verification of semen samples for artificial insemination, and the determination of pedigree.

The technique is not foolproof, however, and results may be far from ideal. Especially in the area of forensic science, there was a rush to use the tremendous power of DNA fingerprinting to identify a purported murderer or rapist, and the need for scientific standards was often neglected. Some problems arose because forensic DNA fingerprinting in the United States is generally conducted in private, unregulated laboratories. In the absence of rigorous scientific controls, the DNA fingerprint bands of two completely unknown samples cannot be matched precisely, and the results may be unreliable.

See also Amniocentesis; Artificial chromosome; Cloning; In vitro plant culture; Rice and wheat strains; Synthetic amino acid; Synthetic DNA; Synthetic RNA.

FURTHER READING

Bodmer, Walter, and Robin McKie. "Probing the Present." In *The Book of Man: The Human Genome Project*. New York: Scribner, 1985.
Caetano-Anolles, Gustavo, and Peter M. Gresshoff. *DNA Markers: Protocols, Applications, and Overviews*. New York: Wiley-VCH, 1997.
Krawezak, Michael, and Jorg Schmidtke. *DNA Fingerprinting*. 2d ed. New York: Springer-Verlag, 1998.
Schacter, Bernice Zeldin. *Issues and Dilemmas of Biotechnology: A Reference Guide*. Westport, Conn.: Greenwood Press, 1999.

Genetically Engineered Insulin

The invention: Artificially manufactured human insulin (Humulin) as a medication for people suffering from diabetes.

The people behind the invention:

Irving S. Johnson (1925-   ), an American zoologist who was vice president of research at Eli Lilly Research Laboratories

Ronald E. Chance (1934-   ), an American biochemist at Eli Lilly Research Laboratories

What Is Diabetes?

Carbohydrates (sugars and related chemicals) are the main food and energy source for humans. In wealthy countries such as the United States, more than 50 percent of the food people eat is made up of carbohydrates, while in poorer countries the carbohydrate content of diets is higher, from 70 to 90 percent.

Normally, most carbohydrates that a person eats are used (or metabolized) quickly to produce energy. Carbohydrates not needed for energy are either converted to fat or stored as a glucose polymer called "glycogen." Most adult humans carry about a pound of body glycogen; this substance is broken down to produce energy when it is needed.

Certain diseases prevent the proper metabolism and storage of carbohydrates. The most common of these diseases is diabetes mellitus, usually called simply "diabetes." It is found in more than seventy million people worldwide. Diabetic people cannot produce or use enough insulin, a hormone secreted by the pancreas. When their condition is not treated, the eyes may deteriorate to the point of blindness. The kidneys may stop working properly, blood vessels may be damaged, and the person may fall into a coma and die. In fact, diabetes is the third most common killer in the United States. Most of the problems surrounding diabetes are caused by high levels of glucose in the blood. Cataracts often form in diabetics, as excess glucose is deposited in the lens of the eye.

Important symptoms of diabetes include constant thirst, excessive urination, and large amounts of sugar in the blood and in the urine. The glucose tolerance test (GTT) is the best way to find out whether a person is suffering from diabetes. People given a GTT are first told to fast overnight. In the morning their blood glucose level is measured; then they are asked to drink about a fourth of a pound of glucose dissolved in water. During the next four to six hours, the blood glucose level is measured repeatedly. In nondiabetics, glucose levels do not rise above a certain amount during a GTT, and the level drops quickly as the glucose is assimilated by the body. In diabetics, the blood glucose levels rise much higher and do not drop as quickly. The extra glucose then shows up in the urine.

TREATING DIABETES

Until the 1920's, diabetes could be controlled only through a diet very low in carbohydrates, and this treatment was not always successful. Then Sir Frederick G. Banting and Charles H. Best found a way to prepare purified insulin from animal pancreases and gave it to patients. This gave diabetics their first chance to live a fairly normal life. Banting and his coworkers won the 1923 Nobel Prize in Physiology or Medicine for their work.

The usual treatment for diabetics became regular shots of insulin. Drug companies took the insulin from the pancreases of cattle and pigs slaughtered by the meat-packing industry. Unfortunately, animal insulin has two disadvantages. First, about 5 percent of diabetics are allergic to it and can have severe reactions. Second, the world supply of animal pancreases goes up and down depending on how much meat is being bought. Between 1970 and 1975, the supply of insulin fell sharply as people began to eat less red meat, yet the numbers of diabetics continued to increase. So researchers began to look for a better way to supply insulin.

Studying pancreases of people who had donated their bodies to science, researchers found that human insulin did not cause allergic reactions. Scientists realized that it would be best to find a chemical or biological way to prepare human insulin, and pharmaceutical companies worked hard toward this goal. Eli Lilly and Company was the first to succeed, and on May 14, 1982, it filed a new drug application with the Food and Drug Administration (FDA) for the human insulin preparation it named "Humulin."

Humulin is made by genetic engineering. Irving S. Johnson, who worked on the development of Humulin, described Eli Lilly's method for producing Humulin. The common bacterium *Escherichia coli* is used. Two strains of the bacterium are produced by genetic engineering: The first strain is used to make a protein called an "A chain," and the second strain is used to make a "B chain." After the bacteria are harvested, the A and B chains are removed and purified separately. Then the two chains are combined chemically. When they are purified once more, the result is Humulin, which has been proved by Ronald E. Chance and his Eli Lilly coworkers to be chemically, biologically, and physically identical to human insulin.

CONSEQUENCES

The FDA and other regulatory agencies around the world approved genetically engineered human insulin in 1982. Humulin does not trigger allergic reactions, and its supply does not fluctuate. It has brought an end to the fear that there would be a worldwide shortage of insulin.

Humulin is important as well in being the first genetically engineered industrial chemical. It began an era in which such advanced technology could be a source for medical drugs, chemicals used in farming, and other important industrial products. Researchers hope that genetic engineering will help in the understanding of cancer and other diseases, and that it will lead to ways to grow enough food for a world whose population continues to rise.

See also Artificial chromosome; Artificial insemination; Cloning; Genetic "fingerprinting"; Synthetic amino acid; Synthetic DNA; Synthetic RNA.

FURTHER READING

Berger, Abi. "Gut Cells Engineered to Produce Insulin." *British Medical Journal* 321, no. 7275 (December 16, 2000).
"Genetically Engineered Duckweed to Produce Insulin." *Resource* 6, no. 3 (March, 1999).

"Lilly Gets FDA Approval for New Insulin Formula." *Wall Street Journal* (October 3, 1985).

Williams, Linda. "UC Regents Sue Lilly in Dispute Over Biotech Patent for Insulin." *Los Angeles Times* (February 8, 1990).

Geothermal Power

THE INVENTION: Energy generated from the earth's natural hot springs.

THE PEOPLE BEHIND THE INVENTION:
Prince Piero Ginori Conti (1865-1939), an Italian nobleman and industrialist
Sir Charles Parsons (1854-1931), an English engineer
B. C. McCabe, an American businessman

Developing a Practical System

The first successful use of geothermal energy was at Larderello in northern Italy. The Larderello geothermal field, located near the city of Pisa about 240 kilometers northwest of Rome, contains many hot springs and fumaroles (steam vents). In 1777, these springs were found to be rich in boron, and in 1818, Francesco de Larderel began extracting the useful mineral borax from them. Shortly after 1900, Prince Piero Ginori Conti, director of the Larderello borax works, conceived the idea of using the steam for power production. An experimental electrical power plant was constructed at Larderello in 1904 to provide electric power to the borax plant. After this initial experiment proved successful, a 250-kilowatt generating station was installed in 1913 and commercial power production began.

As the Larderello field grew, additional geothermal sites throughout the region were prospected and tapped for power. Power production grew steadily until the 1940's, when production reached 130 megawatts; however, the Larderello power plants were destroyed late in World War II (1939-1945). After the war, the generating plants were rebuilt, and they were producing more than 400 megawatts by 1980.

The Larderello power plants encountered many of the technical problems that were later to concern other geothermal facilities. For example, hydrogen sulfide in the steam was highly corrosive to copper, so the Larderello power plant used aluminum for electrical connections much more than did conventional power plants of the time. Also, the low pressure of the steam in early wells at Larderello presented problems. The first generators simply used steam to drive a generator and vented the spent steam into the atmosphere. A system of this sort, called a "noncondensing system," is useful for small generators but not efficient to produce large amounts of power.

Most steam engines derive power not only from the pressure of the steam but also from the vacuum created when the steam is condensed back to water. Geothermal systems that generate power from condensation, as well as direct steam pressure, are called "condensing systems." Most large geothermal generators are of this type. Condensation of geothermal steam presents special problems not present in ordinary steam engines: There are other gases present that do not condense. Instead of a vacuum, condensation of steam contaminated with other gases would result in only a limited drop in pressure and, consequently, very low efficiency.

Initially, the operators of Larderello tried to use the steam to heat boilers that would, in turn, generate pure steam. Eventually, a device was developed that removed most of the contaminating gases from the steam. Although later wells at Larderello and other geothermal fields produced steam at greater pressure, these engineering innovations improved the efficiency of any geothermal power plant.

EXPANDING THE IDEA

In 1913, the English engineer Sir Charles Parsons proposed drilling an extremely deep (12-kilometer) hole to tap the earth's deep heat. Power from such a deep hole would not come from natural steam as at Larderello but would be generated by pumping fluid into the hole and generating steam (as hot as 500 degrees Celsius) at the bottom. In modern terms, Parsons proposed tapping "hot dry-rock" geothermal energy. (No such plant has been commercially operated yet, but research is being actively pursued in several countries.)

The first use of geothermal energy in the United States was for direct heating. In 1890, the municipal water company of Boise, Idaho, began supplying hot water from a geothermal well. Water was piped from the well to homes and businesses along appropriately named Warm Springs Avenue. At its peak, the system served more than four hundred customers, but as cheap natural gas became available, the number declined.

Although Larderello was the first successful geothermal electric power plant, the modern era of geothermal electric power began with the opening of the Geysers Geothermal Field in California. Early attempts began in the 1920's, but it was not until 1955 that B. C. McCabe, a Los Angeles businessman, leased 14.6 square kilometers in the Geysers area and founded the Magma Power Company. The first 12.5-megawatt generator was installed at the Geysers in 1960, and production increased steadily from then on. The Geysers surpassed Larderello as the largest producing geothermal field in the 1970's, and more than 1,000 megawatts were being generated by 1980. By the end of 1980, geothermal plants had been installed in thirteen countries, with a total capacity of almost 2,600 megawatts, and projects with a total capacity of more than 15,000 megawatts were being planned in more than twenty countries.

IMPACT

Geothermal power has many attractive features. Because the steam is naturally heated and under pressure, generating equipment can be simple, inexpensive, and quickly installed. Equipment and installation costs are offset by savings in fuel. It is economically practical to install small generators, a fact that makes geothermal plants attractive in remote or underdeveloped areas. Most important to a world faced with a variety of technical and environmental problems connected with fossil fuels, geothermal power does not deplete fossil fuel reserves, produces little pollution, and contributes little to the greenhouse effect.

Despite its attractive features, geothermal power has some limitations. Geologic settings suitable for easy geothermal power production are rare; there must be a hot rock or magma body close to the surface. Although it is technically possible to pump water from an external source into a geothermal well to generate steam, most geothermal sites require a plentiful supply of natural underground water that can be tapped as a source of steam. In contrast, fossil-fuel generating plants can be at any convenient location.

See also Breeder reactor; Compressed-air-accumulating power plant; Fuel cell; Heat pump; Nuclear power plant; Solar thermal engine; Thermal cracking process; Tidal power plant.

FURTHER READING

Appleyard, Rollo. *Charles Parsons: His Life and Work.* London: Constable, 1933.

Boyle, Godfrey. *Renewable Energy: Power for a Sustainable Future.* Oxford: Oxford University Press, 1998.

Cassedy, Edward S. *Prospects for Sustainable Energy: A Critical Assessment.* New York: Cambridge University Press, 2000.

Parsons, Robert Hodson. *The Steam Turbine and Other Inventions of Sir Charles Parsons, O.M.* New York: Longmans Green, 1946.

Gyrocompass

THE INVENTION: The first practical navigational device that enabled ships and submarines to stay on course without relying on the earth's unreliable magnetic poles.

THE PEOPLE BEHIND THE INVENTION:

Hermann Anschütz-Kaempfe (1872-1931), a German inventor and manufacturer

Jean-Bernard-Léon Foucault (1819-1868), a French experimental physicist and inventor

Elmer Ambrose Sperry (1860-1930), an American engineer and inventor

FROM TOYS TO TOOLS

A gyroscope consists of a rapidly spinning wheel mounted in a frame that enables the wheel to tilt freely in any direction. The amount of momentum allows the wheel to maintain its "attitude" even when the whole device is turned or rotated.

These devices have been used to solve problems arising in such areas as sailing and navigation. For example, a gyroscope aboard a ship maintains its orientation even while the ship is rolling. Among other things, this allows the extent of the roll to be measured accurately. Moreover, the spin axis of a free gyroscope can be adjusted to point toward true north. It will (with some exceptions) stay that way despite changes in the direction of a vehicle in which it is mounted. Gyroscopic effects were employed in the design of various objects long before the theory behind them was formally known. A classic example is a child's top, which balances, seemingly in defiance of gravity, as long as it continues to spin. Boomerangs and flying disks derive stability and accuracy from the spin imparted by the thrower. Likewise, the accuracy of rifles improved when barrels were manufactured with internal spiral grooves that caused the emerging bullet to spin.

In 1852, the French inventor Jean-Bernard-Léon Foucault built the first gyroscope, a measuring device consisting of a rapidly spinning wheel mounted within concentric rings that allowed the wheel to move freely about two axes. This device, like the Foucault pendulum, was used to demonstrate the rotation of the earth around its axis, since the spinning wheel, which is not fixed, retains its orientation in space while the earth turns under it. The gyroscope had a related interesting property: As it continued to spin, the force of the earth's rotation caused its axis to rotate gradually until it was oriented parallel to the earth's axis, that is, in a north-south direction. It is this property that enables the gyroscope to be used as a compass.

When Magnets Fail

In 1904, Hermann Anschütz-Kaempfe, a German manufacturer working in the Kiel shipyards, became interested in the navigation problems of submarines used in exploration under the polar ice cap. By 1905, efficient working submarines were a reality, and it was evident to all major naval powers that submarines would play an increasingly important role in naval strategy.

Submarine navigation posed problems, however, that could not be solved by instruments designed for surface vessels. A submarine needs to orient itself under water in three dimensions; it has no automatic horizon with respect to which it can level itself. Navigation by means of stars or landmarks is impossible when the submarine is submerged. Furthermore, in an enclosed metal hull containing machinery run by electricity, a magnetic compass is worthless. To a lesser extent, increasing use of metal, massive moving parts, and electrical equipment had also rendered the magnetic compass unreliable in conventional surface battleships.

It made sense for Anschütz-Kaempfe to use the gyroscopic effect to design an instrument that would enable a ship to maintain its course while under water. Yet producing such a device would not be easy. First, it needed to be suspended in such a way that it was free to turn in any direction with as little mechanical resistance as possible. At the same time, it had to be able to resist the inevitable pitching and rolling of a vessel at sea. Finally, a continuous power supply was required to keep the gyroscopic wheels spinning at high speed.

The original Anschütz-Kaempfe gyrocompass consisted of a pair of spinning wheels driven by an electric motor. The device was connected to a compass card visible to the ship's navigator. Motor, gyro-

---

ELMER SPERRY

Although Elmer Ambrose Sperry, born in 1860, had only a grade school education as a child in rural New York, the equipment used on local farms piqued his interest in machinery and he learned about technology on his own. He attended a local teachers' college, and graduating in 1880, he was determined to become an inventor.

He was especially interested in the application of electricity. He designed his own arc lighting system and opened the Sperry Electric Light, Motor, and Car Brake Company to sell it, changing its name to Sperry Electric Company in 1887. He made such progress in devising electric mining equipment, electric brakes for automobiles and streetcars, and his own electric car that General Electric bought him out.

In 1900 Sperry opened a laboratory in Washington, D.C., and continued research on a gyroscope that he began in 1896. After more than a decade he patented his device, and after successful trials aboard the USS *Worden*, he established the Sperry Gyroscope Company in 1910, later supplying the American, British, and Russian navies as well as commercial ships. In 1914 he successfully demonstrated a gyrostabilizer for aircraft and expanded his company to manufacture aeronautical technology. Before he sold the company in 1926 he had registered more than four hundred patents. Sperry died in Brooklyn in 1930.

---

scope, and suspension system were mounted in a frame that allowed the apparatus to remain stable despite the pitch and roll of the ship.

In 1906, the German navy installed a prototype of the Anschütz-Kaempfe gyrocompass on the battleship *Undine* and subjected it to exhaustive tests under simulated battle conditions, sailing the ship under forced draft and suddenly reversing the engines, changing the position of heavy turrets and other mechanisms, and firing heavy guns. In conditions under which a magnetic compass would have been worthless, the gyrocompass proved a satisfactory navigational tool, and the results were impressive enough to convince the German navy to undertake installation of gyrocompasses in submarines and heavy battleships, including the battleship *Deutschland*.

Elmer Ambrose Sperry, a New York inventor intimately associated with pioneer electrical development, was independently working on a design for a gyroscopic compass at about the same time. In 1907, he patented a gyrocompass consisting of a single rotor mounted within two concentric shells, suspended by fine piano wire from a frame mounted on gimbals. The rotor of the Sperry compass operated in a vacuum, which enabled it to rotate more rapidly. The Sperry gyrocompass was in use on larger American battleships and submarines on the eve of World War I (1914-1918).

IMPACT

The ability to navigate submerged submarines was of critical strategic importance in World War I. Initially, the German navy had an advantage both in the number of submarines at its disposal and in their design and maneuverability. The German U-boat fleet declared all-out war on Allied shipping, and, although their efforts to blockade England and France were ultimately unsuccessful, the tremendous toll they inflicted helped maintain the German position and prolong the war. To a submarine fleet operating throughout the Atlantic and in the Caribbean, as well as in near-shore European waters, effective long-distance navigation was critical.

Gyrocompasses were standard equipment on submarines and battleships and, increasingly, on larger commercial vessels during World War I, World War II (1939-1945), and the period between the wars. The devices also found their way into aircraft, rockets, and guided missiles. Although the compasses were made more accurate and easier to use, the fundamental design differed little from that invented by Anschütz-Kaempfe.

See also Atomic-powered ship; Dirigible; Hovercraft; Radar; Sonar.

FURTHER READING

Hughes, Thomas Parke. *Elmer Sperry: Inventor and Engineer.* Baltimore: Johns Hopkins University Press, 1993.
_____. *Science and the Instrument-Maker: Michelson, Sperry, and the Speed of Light.* Washington: Smithsonian Institution Press, 1976.
Sorg, H. W. "From Serson to Draper: Two Centuries of Gyroscopic Development." *Journal of the Institute of Navigation* 23, no. 4 (Winter, 1976-1977).

Hard disk

The invention: A large-capacity, permanent magnetic storage device built into most personal computers.

The people behind the invention:
Alan Shugart (1930-   ), an engineer who first developed the floppy disk
Philip D. Estridge (1938?-1985), the director of IBM's product development facility
Thomas J. Watson, Jr. (1914-1993), the chief executive officer of IBM

The Personal Oddity

When the International Business Machines (IBM) Corporation introduced its first microcomputer, called simply the IBM PC (for "personal computer"), the occasion was less a dramatic invention than the confirmation of a trend begun some years before. A number of companies had introduced microcomputers before IBM; one of the best known at that time was Apple Corporation's Apple II, for which software for business and scientific use was quickly developed. Nevertheless, the microcomputer was quite expensive and was often looked upon as an oddity, not as a useful tool.

Under the leadership of Thomas J. Watson, Jr., IBM, which had previously focused on giant mainframe computers, decided to develop the PC. A design team headed by Philip D. Estridge was assembled in Boca Raton, Florida, and it quickly developed its first, pacesetting product. It is an irony of history that IBM anticipated selling only one hundred thousand or so of these machines, mostly to scientists and technically inclined hobbyists. Instead, IBM's product sold exceedingly well, and its design parameters, as well as its operating system, became standards.

The earliest microcomputers used a cassette recorder as a means of mass storage; a floppy disk drive capable of storing approximately 160 kilobytes of data was initially offered only as an option. While home hobbyists were accustomed to using a cassette recorder for storage purposes, such a system was far too slow and awkward for use in business and science. As a result, virtually every IBM PC sold was equipped with at least one 5.25-inch floppy disk drive.

MEMORY REQUIREMENTS

All computers require memory of two sorts in order to carry out their tasks. One type of memory is main memory, or random access memory (RAM), which is used by the computer's central processor to store data it is using while operating. The type of memory used for this function is built typically of silicon-based integrated circuits that have the advantage of speed (to allow the processor to fetch or store the data quickly), but the disadvantage of possibly losing or "forgetting" data when the electric current is turned off. Further, such memory generally is relatively expensive.

To reduce costs, another type of memory—long-term storage memory, known also as "mass storage"—was developed. Mass storage devices include magnetic media (tape or disk drives) and optical media (such as the compact disc, read-only memory, or CD-ROM). While the speed with which data may be retrieved from or stored in such devices is rather slow compared to the central processor's speed, a disk drive—the most common form of mass storage used in PCs—can store relatively large amounts of data quite inexpensively.

Early floppy disk drives (so called because the magnetically treated material on which data are recorded is made of a very flexible plastic) held 160 kilobytes of data using only one side of the magnetically coated disk (about eighty pages of normal, double-spaced, typewritten information). Later developments increased storage capacities to 360 kilobytes by using both sides of the disk and later, with increasing technological ability, 1.44 megabytes (millions of bytes). In contrast, mainframe computers, which are typically connected to large and expensive tape drive storage systems, could store gigabytes (millions of megabytes) of information.

While such capacities seem large, the needs of business and scientific users soon outstripped available space. Since even the mailing list of a small business or a scientist's mathematical model of a chemical reaction easily could require greater storage potential than early PCs allowed, the need arose for a mass storage device that could accommodate very large files of data.

The answer was the hard disk drive, also known as a "fixed disk drive," reflecting the fact that the disk itself is not only rigid but also permanently installed inside the machine. In 1955, IBM had envisioned the notion of a fixed, hard magnetic disk as a means of storing computer data, and, under the direction of Alan Shugart in the 1960's, the floppy disk was developed as well.

As the engineers of IBM's facility in Boca Raton refined the idea of the original PC to design the new IBM PC XT, it became clear that chief among the needs of users was the availability of large-capability storage devices. The decision was made to add a 10-megabyte hard disk drive to the PC. On March 8, 1983, less than two years after the introduction of its first PC, IBM introduced the PC XT. Like the original, it was an evolutionary design, not a revolutionary one. The inclusion of a hard disk drive, however, signaled that mass storage devices in personal computers had arrived.

CONSEQUENCES

Above all else, any computer provides a means for storing, ordering, analyzing, and presenting information. If the personal computer is to become the information appliance some have suggested it will be, the ability to manipulate very large amounts of data will be of paramount concern. Hard disk technology was greeted enthusiastically in the marketplace, and the demand for hard drives has seen their numbers increase as their quality increases and their prices drop.

It is easy to understand one reason for such eager acceptance: convenience. Floppy-bound computer users find themselves frequently changing (or "swapping") their disks in order to allow programs to find the data they need. Moreover, there is a limit to how much data a single floppy disk can hold. The advantage of a hard drive is that it allows users to keep seemingly unlimited amounts of data and programs stored in their machines and readily available.

Also, hard disk drives are capable of finding files and transferring their contents to the processor much more quickly than a floppy drive. A user may thus create exceedingly large files, keep them on hand at all times, and manipulate data more quickly than with a floppy. Finally, while a hard drive is a slow substitute for main memory, it allows users to enjoy the benefits of larger memories at significantly lower cost.

The introduction of the PC XT with its 10-megabyte hard drive was a milestone in the development of the PC. Over the next two decades, the size of computer hard drives increased dramatically. By 2001, few personal computers were sold with hard drives with less than three *giga*bytes of storage capacity, and hard drives with more than thirty gigabytes were becoming the standard. Indeed, for less money than a PC XT cost in the mid-1980's, one could buy a fully equipped computer with a hard drive holding sixty gigabytes—a storage capacity equivalent to six thousand 10-megabyte hard drives.

See also Bubble memory; Compact disc; Computer chips; Floppy disk; Optical disk; Personal computer.

FURTHER READING

Chposky, James, and Ted Leonsis. *Blue Magic: The People, Power, and Politics Behind the IBM Personal Computer.* New York: Facts on File, 1988.
Freiberger, Paul, and Michael Swaine. *Fire in the Valley: The Making of the Personal Computer.* 2d ed. New York: McGraw-Hill, 2000.
Grossman, Wendy. *Remembering the Future: Interviews from Personal Computer World.* New York: Springer, 1997.
Watson, Thomas J., and Peter Petre. *Father, Son and Co.: My Life at IBM and Beyond.* New York: Bantam Books, 2000.

Hearing aid

The invention: Miniaturized electronic amplifier worn inside the ears of hearing-impaired persons.

The organization behind the invention:
Bell Labs, the research and development arm of the American Telephone and Telegraph Company

Trapped in Silence

Until the middle of the twentieth century, people who experienced hearing loss had little hope of being able to hear sounds without the use of large, awkward, heavy appliances. For many years, the only hearing aids available were devices known as ear trumpets. The ear trumpet tried to compensate for hearing loss by increasing the number of sound waves funneled into the ear canal. A wide, bell-like mouth similar to the bell of a musical trumpet narrowed to a tube that the user placed in his or her ear. Ear trumpets helped a little, but they could not truly increase the volume of the sounds heard.

Beginning in the nineteenth century, inventors tried to develop electrical devices that would serve as hearing aids. The telephone was actually a by-product of Alexander Graham Bell's efforts to make a hearing aid. Following the invention of the telephone, electrical engineers designed hearing aids that employed telephone technology, but those hearing aids were only a slight improvement over the old ear trumpets. They required large, heavy battery packs and used a carbon microphone similar to the receiver in a telephone. More sensitive than purely physical devices such as the ear trumpet, they could transmit a wider range of sounds but could not amplify them as effectively as electronic hearing aids now do.

Transistors Make Miniaturization Possible

Two types of hearing aids exist: body-worn and head-worn. Body-worn hearing aids permit the widest range of sounds to be heard, but because of the devices' larger size, many hearing-impaired persons do not like to wear them. Head-worn hearing aids, especially those worn completely in the ear, are much less conspicuous. In addition to in-ear aids, the category of head-worn hearing aids includes both hearing aids mounted in eyeglass frames and those worn behind the ear.

All hearing aids, whether head-worn or body-worn, consist of four parts: a microphone to pick up sounds, an amplifier, a receiver, and a power source. The microphone gathers sound waves and converts them to electrical signals; the amplifier boosts, or increases, those signals; and the receiver then converts the signals back into sound waves. In effect, the hearing aid is a miniature radio. After the receiver converts the signals back to sound waves, those waves are directed into the ear canal through an earpiece or ear mold. The ear mold generally is made of plastic and is custom fitted from an impression taken from the prospective user's ear.

Effective head-worn hearing aids could not be built until the electronic circuit was developed in the early 1950's. The same invention—the transistor—that led to small portable radios and tape players allowed engineers to create miniaturized, inconspicuous hearing aids. Depending on the degree of amplification required, the amplifier in a hearing aid contains three or more transistors. Transistors first replaced vacuum tubes in devices such as radios and phonographs, and then engineers realized that they could be used in devices for the hearing-impaired.

The research at Bell Labs that led to the invention of the transistor rose out of military research during World War II. The vacuum tubes used in, for example, radar installations to amplify the strength of electronic signals were big, were fragile because they were made of blown glass, and gave off high levels of heat when they were used. Transistors, however, made it possible to build solid-state, integrated circuits. These are made from crystals of metals such as germanium or arsenic alloys and therefore are much less fragile than glass. They are also extremely small (in fact, some integrated circuits are barely visible to the naked eye) and give off no heat during use.

The number of transistors in a hearing aid varies depending upon the amount of amplification required. The first transistor is the most important for the listener in terms of the quality of sound heard. If the frequency response is set too high—that is, if the device is too sensitive—the listener will be bothered by distracting background noise. Theoretically, there is no limit on the amount of amplification that a hearing aid can be designed to provide, but there are practical limits. The higher the amplification, the more power is required to operate the hearing aid. This is why body-worn hearing aids can convey a wider range of sounds than head-worn devices can. It is the power source—not the electronic components—that is the limiting factor. A body-worn hearing aid includes a larger battery pack than can be used with a head-worn device. Indeed, despite advances in battery technology, the power requirements of a head-worn hearing aid are such that a 1.4-volt battery that could power a wristwatch for several years will last only a few days in a hearing aid.

Consequences

The invention of the electronic hearing aid made it possible for many hearing-impaired persons to participate in a hearing world. Prior to the invention of the hearing aid, hearing-impaired children often were unable to participate in routine school activities or function effectively in mainstream society. Instead of being able to live at home with their families and enjoy the same experiences that were available to other children their age, often they were forced to attend special schools operated by the state or by charities.

Hearing-impaired people were singled out as being different and were limited in their choice of occupations. Although not every hearing-impaired person can be helped to hear with a hearing aid—particularly in cases of total hearing loss—the electronic hearing aid has ended restrictions for many hearing-impaired people. Hearing-impaired children are now included in public school classes, and hearing-impaired adults can now pursue occupations from which they were once excluded.

Today, many deaf and hearing-impaired persons have chosen to live without the help of a hearing aid. They believe that they are not disabled but simply different, and they point out that their "disability" often allows them to appreciate and participate in life in unique and positive ways. For them, the use of hearing aids is a choice, not a necessity. For those who choose, hearing aids make it possible to participate in the hearing world.

See also Artificial heart; Artificial kidney; Cell phone; Contact lenses; Heart-lung machine; Pacemaker.

FURTHER READING

Alexander, Howard. "Hearing Aids: Smaller and Smarter." *New York Times* (November 26, 1998).
Fong, Petti. "Guess What's the New Buzz in Hearing Aids." *Business Week*, no. 3730 (April 30, 2001).
Levitt, Harry. "Noise Reduction in Hearing Aids: A Review." *Journal of Rehabilitation Research and Development* 38, no. 1 (January/February, 2001).

HEART-LUNG MACHINE

THE INVENTION: The first artificial device to oxygenate and circulate blood during surgery, the heart-lung machine began the era of open-heart surgery.

THE PEOPLE BEHIND THE INVENTION:

John H. Gibbon, Jr. (1903-1974), a cardiovascular surgeon

Mary Hopkinson Gibbon (1905-   ), a research technician

Thomas J. Watson (1874-1956), chairman of the board of IBM

T. L. Stokes and J. B. Flick, researchers in Gibbon's laboratory

Bernard J. Miller (1918-   ), a cardiovascular surgeon and research associate

Cecelia Bavolek, the first human to undergo open-heart surgery successfully using the heart-lung machine

A YOUNG WOMAN'S DEATH

In the first half of the twentieth century, cardiovascular medicine had many triumphs. Effective anesthesia, antiseptic conditions, and antibiotics made surgery safer. Blood-typing, anti-clotting agents, and blood preservatives made blood transfusion practical. Cardiac catheterization (feeding a tube into the heart), electrocardiography, and fluoroscopy (visualizing living tissues with an X-ray machine) made the nonsurgical diagnosis of cardiovascular problems possible.

As of 1950, however, there was no safe way to treat damage or defects within the heart. To make such a correction, this vital organ's function had to be interrupted. The problem was to keep the body's tissues alive while working on the heart. While some surgeons practiced so-called blind surgery, in which they inserted a finger into the heart through a small incision without observing what they were attempting to correct, others tried to reduce the body's need for circulation by slowly chilling the patient until the heart stopped. Still other surgeons used "cross-circulation," in which the patient's circulation was connected to a donor's circulation. All these approaches carried profound risks of hemorrhage, tissue damage, and death.

In February of 1931, Gibbon witnessed the death of a young

woman whose lung circulation was blocked by a blood clot. Because her blood could not pass through her lungs, she slowly lost consciousness from lack of oxygen. As he monitored her pulse and breathing, Gibbon thought about ways to circumvent the obstructed lungs and straining heart and provide the oxygen required. Because surgery to remove such a blood clot was often fatal, the woman's surgeons operated only as a last resort. Though the surgery took only six and one-half minutes, she never regained consciousness. This experience prompted Gibbon to pursue what few people then considered a practical line of research: a way to circulate and oxygenate blood outside the body.

A WOMAN'S LIFE RESTORED

Gibbon began the project in earnest in 1934, when he returned to the laboratory of Edward D. Churchill at Massachusetts General Hospital for his second surgical research fellowship. He was assisted by Mary Hopkinson Gibbon. Together, they developed, using cats, a surgical technique for removing blood from a vein, supplying the blood with oxygen, and returning it to an artery using tubes inserted into the blood vessels. Their objective was to create a device that would keep the blood moving, spread it over a very thin layer to pick up oxygen efficiently and remove carbon dioxide, and avoid both clotting and damaging blood cells. In 1939, they reported that prolonged survival after heart-lung bypass was possible in experimental animals.

World War II (1939-1945) interrupted the progress of this work; it was resumed by Gibbon at Jefferson Medical College in 1944. Shortly thereafter, he attracted the interest of Thomas J. Watson, chairman of the board of the International Business Machines (IBM) Corporation, who provided the services of IBM's experimental physics laboratory and model machine shop as well as the assistance of staff engineers. IBM constructed and modified two experimental machines over the next seven years, and IBM engineers contributed significantly to the evolution of a machine that would be practical in humans.

Gibbon's first attempt to use the pump-oxygenator in a human being was in a fifteen-month-old baby. This attempt failed, not because of a malfunction or a surgical mistake but because of a misdiagnosis. The child died following surgery because the real problem had not been corrected by the surgery.

On May 6, 1953, the heart-lung machine was first used successfully on Cecelia Bavolek. In the six months before surgery, Bavolek had been hospitalized three times for symptoms of heart failure when she tried to engage in normal activity. While her circulation was connected to the heart-lung machine for forty-five minutes, the surgical team headed by Gibbon was able to close an opening between her atria and establish normal heart function. Two months later, an examination of the defect revealed that it was fully closed; Bavolek resumed a normal life. The age of open-heart surgery had begun.

CONSEQUENCES

The heart-lung bypass technique alone could not make openheart surgery truly practical. When it was possible to keep tissues alive by diverting blood around the heart and oxygenating it, other questions already under investigation became even more critical: how to prolong the survival of bloodless organs, how to measure oxygen and carbon dioxide levels in the blood, and how to prolong anesthesia during complicated surgery. Thus, following the first successful use of the heart-lung machine, surgeons continued to refine the methods of open-heart surgery.

The heart-lung apparatus set the stage for the advent of "replacement parts" for many types of cardiovascular problems. Cardiac valve replacement was first successfully accomplished in 1960 by placing an artificial ball valve between the left atrium and ventricle. In 1957, doctors performed the first coronary bypass surgery, grafting sections of a leg vein into the heart's circulation system to divert blood around clogged coronary arteries. Likewise, the first successful heart transplant (1967) and the controversial Jarvik-7 artificial heart implantation (1982) required the ability to stop the heart and keep the body's tissues alive during time-consuming and delicate surgical procedures. Gibbon's heart-lung machine paved the way for all these developments.

See also Artificial heart; Blood transfusion; CAT scanner; Coronary artery bypass surgery; Electrocardiogram; Iron lung; Mammography; Nuclear magnetic resonance; Pacemaker; X-ray image intensifier.

FURTHER READING

DeJauregui, Ruth. *One Hundred Medical Milestones That Shaped World History.* San Mateo, Calif.: Bluewood Books, 1998.

Romaine-Davis, Ada. *John Gibbon and his Heart-Lung Machine.* Philadelphia: University of Pennsylvania Press, 1991.

Shumacker, Harris B. *A Dream of the Heart: The Life of John H. Gibbon, Jr., Father of the Heart-Lung Machine.* Santa Barbara, Calif.: Fithian Press, 1999.

Watson, Thomas J., and Peter Petre. *Father, Son and Co.: My Life at IBM and Beyond.* New York: Bantam Books, 2000.

HEAT PUMP

THE INVENTION: A device that warms and cools buildings efficiently and cheaply by moving heat from one area to another.

THE PEOPLE BEHIND THE INVENTION:

T. G. N. Haldane, a British engineer

Lord Kelvin (William Thomson, 1824-1907), a British mathematician, scientist, and engineer

Sadi Carnot (1796-1832), a French physicist and thermodynamicist

THE HEAT PUMP

A heat pump is a device that takes in heat at one temperature and releases it at a higher temperature. When operated to provide heat (for example, for space heating), the heat pump is said to operate in the heating mode; when operated to remove heat (for example, for air conditioning), it is said to operate in the cooling mode. Some type of work must be done to drive the pump, no matter which mode is being used.

There are two general types of heat pumps: vapor compression pumps and absorption pumps. The basic principle of vapor compression cycle heat pumps is derived from the work of Sadi Carnot in the early nineteenth century. Carnot's work was published in 1824. It was William Thomson (later to become known as Lord Kelvin), however, who first proposed a practical heat pump system, or "heat multiplier," as it was known then, and he also indicated that a refrigerating machine could be used for heating.

Thomson's heat pump used air as its working fluid. Thomson claimed that his heat pump was able to produce heat by using only 3 percent of the energy that would be required for direct heating. Absorption cycle machines have an even longer history. Refrigerators based on the use of sulfuric acid and water date back to 1777. Systems using this fluid combination, improved and modified by Edmond Carré, were used extensively in Paris cafés in the late 1800's. In 1849, a patent was filed by Ferdinand Carré for the working-fluid pair of ammonia and water in absorption cycle machines.

REFRIGERATOR OR HEATER

In the early nineteenth century, many people (including some electrical engineers) believed that electrical energy could never be used economically to produce large quantities of heat under ordinary conditions. A few researchers, however, believed that it was possible to produce heat by using electrical energy if that energy was first converted to mechanical energy and if the Carnot principle was then used to pump heat from a lower to a higher temperature.

In 1927, T. G. N. Haldane carried out detailed experiments showing that the heat pump can be made to operate in either the heating mode or the cooling mode. A heat pump in the cooling mode works like a refrigerator; a heat pump in the heating mode supplies heat for heating. Haldane demonstrated that a refrigerator could be modified to work as a heating unit. He used a vapor compression cycle refrigerator for his demonstration.

In the design of a refrigerating device, the primary objective is the production of cold rather than heat, but the two operations are complementary. The process of producing cold is simply that of pumping heat from a relatively cold to a relatively hot source, but in the refrigeration process particular attention is paid to the prevention of the leakage of heat into the cold source, whereas no attempt is made to prevent the escape of heat from the hot source. If a refrigerating device were treated as a heat pump in which the primary product is the heat rejected to the hot source, the order of importance would be reversed, and every opportunity would be taken to allow heat to leak into the cold source and every precaution would be taken against allowing heat to leak out of the hot source.

The components of a heat pump that operates on the principle of vapor compression include an electric motor, a compressor, an evaporator, and a condenser. The compressor sucks in gas from the evaporator and compresses it to a pressure that corresponds to a saturation temperature that is slightly higher than that of the required heat. From the compressor, the compressed gas passes to the condenser, where it is cooled and condensed, thereby giving up a large quantity of heat to the water or other substance that it is intended to heat. The condensed gas then passes through the expansion valve, where a sudden reduction of pressure takes place. This reduction of pressure lowers the boiling point of the liquid, which therefore vaporizes and takes in heat from the medium surrounding the evaporator. After evaporation, the gas passes on to the compressor, and the cycle is complete.

Haldane was the first person in the United Kingdom to install a heat pump. He was also the first person to install a domestic heat pump to provide hot water and space heating.

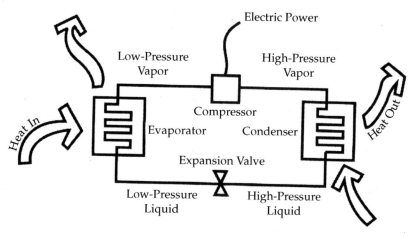

Components of a heat pump.

IMPACT

Since Haldane's demonstration of the use of the heat pump, the device has been highly successful in people's homes, especially in those regions where both heating and cooling are required for single- and multifamily residences (for example, Australia, Japan, and the United States). This is the case because the heat pump can provide both heating and cooling; therefore, the cost of a heat pump system can be spread over both heating and cooling seasons. Total annual sales of heat pumps worldwide have risen to the millions, with most sales being made in Japan and the United States.

The use of heat pumps can save energy. In addition, because they are electric, they can save significant quantities of oil, especially in the residential retrofit and replacement markets and when used as add-on devices for existing heating systems. Some heat pumps are now available that may compete cost-effectively with other heating systems in meeting the heating demands of cooler regions.

Technological developments by heat pump manufacturers are continually improving the performance and cost-effectiveness of heat pumps. The electric heat pump will continue to dominate the residential market, although engine-driven systems are likely to have a greater impact on the multifamily market.

See also Breeder reactor; Compressed-air-accumulating power plant; Fuel cell; Geothermal power; Nuclear power plant; Solar thermal engine; Tidal power plant.

FURTHER READING

Kavanaugh, Stephen P., and Kevin D. Rafferty. *Ground-Source Heat Pumps: Design of Geothermal Systems for Commercial and Institutional Buildings.* Atlanta: American Society of Heating, Refrigerating and Air-Conditioning Engineers, 1997.
Nisson, Ned. "Efficient and Affordable." *Popular Science* 247, no. 2 (August, 1995).
Using the Earth to Heat and Cool Homes. Washington, D.C.: U.S. Department of Energy, 1983.

HOLOGRAPHY

THE INVENTION: A lensless system of three-dimensional photography that was one of the most important developments in twentieth century optical science.

THE PEOPLE BEHIND THE INVENTION:
Dennis Gabor (1900-1979), a Hungarian-born inventor and
    physicist who was awarded the 1971 Nobel Prize in Physics
Emmett Leith (1927-   ), a radar researcher who, with Juris
    Upatnieks, produced the first laser holograms
Juris Upatnieks (1936-   ), a radar researcher who, with
    Emmett Leith, produced the first laser holograms

EASTER INSPIRATION

The development of photography in the early 1900's made possible the recording of events and information in ways unknown before the twentieth century: the photographing of star clusters, the recording of the emission spectra of heated elements, the storing of data in the form of small recorded images (for example, microfilm), and the photographing of microscopic specimens, among other things. Because of its vast importance to the scientist, the science of photography has developed steadily.

An understanding of the photographic and holographic processes requires some knowledge of the wave behavior of light. Light is an electromagnetic wave that, like a water wave, has an amplitude and a phase. The amplitude corresponds to the wave height, while the phase indicates which part of the wave is passing a given point at a given time. A cork floating in a pond bobs up and down as waves pass under it. The position of the cork at any time depends on both amplitude and phase: The phase determines on which part of the wave the cork is floating at any given time, and the amplitude determines how high or low the cork can be moved. Waves from more than one source arriving at the cork combine in ways that depend on their relative phases. If the waves meet in the same phase, they add and produce a large amplitude; if they arrive out of phase, they subtract and produce a small amplitude. The total amplitude, or intensity, depends on the phases of the combining waves.

Dennis Gabor, the inventor of holography, was intrigued by the way in which the photographic image of an object was stored by a photographic plate but was unable to devote any consistent research effort to the question until the 1940's. At that time, Gabor was involved in the development of the electron microscope. On Easter morning in 1947, as Gabor was pondering the problem of how to improve the electron microscope, the solution came to him. He would attempt to take a poor electron picture and then correct it optically. The process would require coherent electron beams—that is, electron waves with a definite phase.

This two-stage method was inspired by the work of Lawrence Bragg. Bragg had formed the image of a crystal lattice by diffracting the photographic X-ray diffraction pattern of the original lattice. This double diffraction process is the basis of the holographic process. Bragg's method was limited because of his inability to record the phase information of the X-ray photograph. Therefore, he could study only those crystals for which the phase relationship of the reflected waves could be predicted.

WAITING FOR THE LASER

Gabor devised a way of capturing the phase information after he realized that adding coherent background to the wave reflected from an object would make it possible to produce an interference pattern on the photographic plate. When the phases of the two waves are identical, a maximum intensity will be recorded; when they are out of phase, a minimum intensity is recorded. Therefore, what is recorded in a hologram is not an image of the object but rather the interference pattern of the two coherent waves. This pattern looks like a collection of swirls and blank spots. The hologram (or photograph) is then illuminated by the reference beam, and part of the transmitted light is a replica of the original object wave. When viewing this object wave, one sees an exact replica of the original object.

The major impediment at the time in making holograms using any form of radiation was a lack of coherent sources. For example, the coherence of the mercury lamp used by Gabor and his assistant

DENNIS GABOR

The eldest son of a mine director, Dennis Gabor was born in 1900 in Budapest, Hungary. At fifteen, suddenly developing an intense interest in optics and photography, Gabor and his brother sent up their own home laboratory and experimented in those fields as well as with X rays and radioactivity. The love of physics never left him.

Gabor graduated from the Berlin Technische Hochschule in 1924 and earned a doctorate of engineering in 1927 after developing a high-speed cathode ray oscillograph and a new kind of magnetic lens for controlling electrons. After graduate school he joined Siemens and Halske Limited and invented a high-pressure mercury lamp, which was later used widely in street lamps. In 1933, Gabor left Germany because of the rise of Nazism and moved to England. He worked in industrial research until 1948, improving gas-discharge tubes and stereoscopic cinematography, but he also published scientific papers on his own, including the first of many on communications theory. At the beginning of 1949, Gabor became a faculty member of the Imperial College of Science and Technology in London, first as a reader in electronics and later as a professor of applied physics.

During his academic years came more inventions, including the hologram, an electron-velocity spectroscope, an analog computer, a flat color television tube, and a new type of thermionic converter. He also build a cloud chamber for detecting subatomic particles and used it to study electron interactions. As interested in theory as he was in applied physics, Gabor published papers on theoretical aspects of communications, plasma, magnetrons, and fusion. In his later years he worried deeply about the modern tendency for technology to advance out of step with social institutions and wrote popular books outlining his belief that social reform should be given priority.

Gabor became a member of Britain's Royal Society in 1956 and was awarded its Rumsford Medal in 1968. In 1971 he received the Nobel Prize in Physics for inventing holography. He died in London in 1979.

Ivor Williams was so short that they were able to make holograms of only about a centimeter in diameter. The early results were rather poor in terms of image quality and also had a double image. For this reason, there was little interest in holography, and the subject lay almost untouched for more than ten years.

Interest in the field was rekindled after the laser (*l*ight *a*mplification by *s*timulated *e*mission of *r*adiation) was developed in 1962. Emmett Leith and Juris Upatnieks, who were conducting radar research at the University of Michigan, published the first laser holographs in 1963. The laser was an intense light source with a very long coherence length. Its monochromatic nature improved the resolution of the images greatly. Also, there was no longer any restriction on the size of the object to be photographed.

The availability of the laser allowed Leith and Upatnieks to propose another improvement in holographic technique. Before 1964, holograms were made of only thin transparent objects. A small region of the hologram bore a one-to-one correspondence to a region of the object. Only a small portion of the image could be viewed at one time without the aid of additional optical components. Illuminating the transparency diffusely allowed the whole image to be seen at one time. This development also made it possible to record holograms of diffusely reflected three-dimensional objects. Gabor had seen from the beginning that this should make it possible to create three-dimensional images.

After the early 1960's, the field of holography developed very quickly. Because holography is different from conventional photography, the two techniques often complement each other. Gabor saw his idea blossom into a very important technique in optical science.

IMPACT

The development of the laser and the publication of the first laser holograms in 1963 caused a blossoming of the new technique in many fields. Soon, techniques were developed that allowed holograms to be viewed with white light. It also became possible for holograms to reconstruct multicolored images. Holographic methods have been used to map terrain with radar waves and to conduct surveillance in the fields of forestry, agriculture, and meteorology.

By the 1990's, holography had become a multimillion-dollar industry, finding applications in advertising, as an art form, and in security devices on credit cards, as well as in scientific fields. An alternate form of holography, also suggested by Gabor, uses sound waves. Acoustical imaging is useful whenever the medium around the object to be viewed is opaque to light rays—for example, in medical diagnosis. Holography has affected many areas of science, technology, and culture.

See also Color film; Electron microscope; Infrared photography; Laser; Mammography; Mass spectrograph; X-ray crystallography.

FURTHER READING

Greguss, Pál, Tung H. Jeong, and Dennis Gabor. *Holography: Commemorating the Ninetieth Anniversary of the Birth of Dennis Gabor.* Bellingham, Wash.: SPIE Optical Engineering Press, 1991.
Kasper, Joseph Emil, and Steven A. Feller. *The Complete Book of Holograms: How They Work and How to Make Them.* Mineola, N.Y.: Dover, 2001.
McNair, Don. *How to Make Holograms.* Blue Ridge Summit, Pa.: Tab Books, 1983.
Saxby, Graham. *Holograms: How to Make and Display Them.* New York: Focal Press, 1980.

Hovercraft

THE INVENTION: A vehicle requiring no surface contact for traction that moves freely over a variety of surfaces—particularly water—while supported on a self-generated cushion of air.

THE PEOPLE BEHIND THE INVENTION:

Christopher Sydney Cockerell (1910-   ), a British engineer who built the first hovercraft

Ronald A. Shaw (1910-   ), an early pioneer in aerodynamics who experimented with hovercraft

Sir John Isaac Thornycroft (1843-1928), a Royal Navy architect who was the first to experiment with air-cushion theory

AIR-CUSHION TRAVEL

The air-cushion vehicle was first conceived by Sir John Isaac Thornycroft of Great Britain in the 1870's. He theorized that if a ship had a plenum chamber (a box open at the bottom) for a hull and it were pumped full of air, the ship would rise out of the water and move faster, because there would be less drag. The main problem was keeping the air from escaping from under the craft.

In the early 1950's, Christopher Sydney Cockerell was experimenting with ways to reduce both the wave-making and frictional resistance that craft had to water. In 1953, he constructed a punt with a fan that supplied air to the bottom of the craft, which could thus glide over the surface with very little friction. The air was contained under the craft by specially constructed side walls. In 1955, the first true "hovercraft," as Cockerell called it, was constructed of balsa wood. It weighed only 127 grams and traveled over water at a speed of 13 kilometers per hour.

On November 16, 1956, Cockerell successfully demonstrated his model hovercraft at the patent agent's office in London. It was immediately placed on the "secret" list, and Saunders-Roe Ltd. was given the first contract to build hovercraft in 1957. The first experimental piloted hovercraft, the SR.N1, which had a weight of 3,400 kilograms and could carry three people at the speed of 25 knots, was completed on May 28, 1959, and publicly demonstrated on June 11, 1959.

GROUND EFFECT PHENOMENON

In a hovercraft, a jet airstream is directed downward through a hole in a metal disk, which forces the disk to rise. The jet of air has a reverse effect of its own that forces the disk away from the surface. Some of the air hitting the ground bounces back against the disk to add further lift. This is called the "ground effect." The ground effect is such that the greater the under-surface area of the hovercraft, the greater the reverse thrust of the air that bounces back. This makes the hovercraft a mechanically efficient machine because it provides three functions.

First, the ground effect reduces friction between the craft and the earth's surface. Second, it acts as a spring suspension to reduce some of the vertical acceleration effects that arise from travel over an uneven surface. Third, it provides a safe and comfortable ride at high speed, whatever the operating environment. The air cushion can distribute the weight of the hovercraft over almost its entire area so that the cushion pressure is low.

The basic elements of the air-cushion vehicle are a hull, a propulsion system, and a lift system. The hull, which accommodates the crew, passengers, and freight, contains both the propulsion and lift systems. The propulsion and lift systems can be driven by the same power plant or by separate power plants. Early designs used only one unit, but this proved to be a problem when adequate power was not achieved for movement and lift. Better results are achieved when two units are used, since far more power is used to lift the vehicle than to propel it.

For lift, high-speed centrifugal fans are used to drive the air through jets that are located under the craft. A redesigned aircraft propeller is used for propulsion. Rudderlike fins and an air fan that can be swiveled to provide direction are placed at the rear of the craft.

Several different air systems can be used, depending on whether a skirt system is used in the lift process. The plenum chamber system, the peripheral jet system, and several types of recirculating air systems have all been successfully tried without skirting. A variety of rigid and flexible skirts have also proved to be satisfactory, depending on the use of the vehicle.

Skirts are used to hold the air for lift. Skirts were once hung like cur-

---

Sir John Isaac Thornycroft

To be truly ahead of one's time as an inventor, one must simply know everything there is to know about a specialty and then imagine something useful that contemporary technology is not quite ready for.

John Isaac Thornycroft was such an inventor. Born in 1843 in what were then the Papal States (Rome, Italy), he trained as an engineer and became a naval architect. He opened a boat-building and engineering company at Chiswick in London in 1866 and began looking for ways to improve the performance of small seacraft. In 1877 he delivered the HMS *Lightning*, England's first torpedo boat, to the Royal Navy. He continued to make torpedo boats for coastal waters, nicknamed "scooters," and made himself a leading expert on boat design. He introduced stabilizers and modified hull and propeller shapes in order to reduce drag from the hull's contact with water and thereby increase a boat's speed.

One of his best ideas was to have the boat ride on a cushion of air, so that air acted as a lubricant between the hull and water. He even filed patents for the concept and built models, but the power-source technology of the day was simply too inefficient. Engines were too heavy for the amount of power they put out. None could lift a full-size boat off the water and keep it on an air cushion. So the hovercraft had to wait until the 1950's and incorporation of sophisticated internal combustion engines into the design.

Meanwhile, Thornycroft and the company named after him continued to make innovative transports and engines: a steam-powered van in 1896, a gas engine in 1902, and heavy trucks in 1912 that the British government used during World War I. By the time Thornycroft died in 1928, on the Isle of Wight, he had been knighted by a grateful government, which would benefit from his company's products and his advanced ideas for the rest of the twentieth century.

tains around hovercraft. Instead of simple curtains to contain the air, there are now complicated designs that contain the cushion, duct the air, and even provide a secondary suspension. The materials used in the skirting have also changed from a rubberized fabric to pure rubber and nylon and, finally, to neoprene, a lamination of nylon and plastic.

The three basic types of hovercraft are the amphibious, non-amphibious, and semiamphibious models. The amphibious type can travel over water and land, whereas the nonamphibious type is restricted to water travel. The semiamphibious model is also restricted to water travel but may terminate travel by nosing up on a prepared ramp or beach. All hovercraft contain built-in buoyancy tanks in the side skirting as a safety measure in the event that a hovercraft must settle on the water. Most hovercraft are equipped with gas turbines and use either propellers or water-jet propulsion.

IMPACT

Hovercraft are used primarily for short passenger ferry services. Great Britain was the only nation to produce a large number of hovercraft. The British built larger and faster craft and pioneered their successful use as ferries across the English Channel, where they could reach speeds of 111 kilometers per hour (160 knots) and carry more than four hundred passengers and almost one hundred vehicles. France and the former Soviet Union have also effectively demonstrated hovercraft river travel, and the Soviets have experimented with military applications as well.

The military adaptations of hovercraft have been more diversified. Beach landings have been performed effectively, and the United States used hovercraft for river patrols during the Vietnam War.

Other uses also exist for hovercraft. They can be used as harbor pilot vessels and for patrolling shores in a variety of police-and customs-related duties. Hovercraft can also serve as flood-rescue craft and fire-fighting vehicles. Even a hoverfreighter is being considered.

The air-cushion theory in transport systems is rapidly developing. It has spread to trains and smaller people movers in many countries. Their smooth, rapid, clean, and efficient operation makes hovercraft attractive to transportation designers around the world.

See also Airplane; Atomic-powered ship; Bullet train; Gyro-compass.

FURTHER READING

Amyot, Joseph R. *Hovercraft Technology, Economics, and Applications.* Amsterdam: Elsevier, 1989.

Croome, Angela. *Hover Craft*. 4th ed. London: Hodder and Stoughton, 1984.

Gromer, Cliff. "Flying Low." *Popular Mechanics* 176, no. 9 (September, 1999).

McLeavy, Roy. *Hovercraft and Hydrofoils*. London: Jane's Publishing, 1980.

Pengelley, Rupert. "Hovercraft Cushion the Blow of Amphibious Operations." *Jane's Navy International* 104, no. 008 (October 1, 1999).

Robertson, Don. *A Restless Spirit*. New Port, Isle of Wight: Cross Publishing, 1994.

Hydrogen Bomb

THE INVENTION: Popularly known as the "H-Bomb," the hydrogen bomb differs from the original atomic bomb in using fusion, rather than fission, to create a thermonuclear explosion almost a thousand times more powerful.

THE PEOPLE BEHIND THE INVENTION:
Edward Teller (1908-   ), a Hungarian-born theoretical physicist
Stanislaw Ulam (1909-1984), a Polish-born mathematician

CRASH DEVELOPMENT

A few months before the 1942 creation of the Manhattan Project, the United States-led effort to build the atomic (fission) bomb, physicist Enrico Fermi suggested to Edward Teller that such a bomb could release more energy by the process of heating a mass of the hydrogen isotope deuterium and igniting the fusion of hydrogen into helium. *Fusion* is the process whereby two atoms come together to form a larger atom, and this process usually occurs only in stars, such as the Sun. Physicists Hans Bethe, George Gamow, and Teller had been studying fusion since 1934 and knew of the tremendous energy than could be released by this process—even more energy than the *fission* (atom-splitting) process that would create the atomic bomb. Initially, Teller dismissed Fermi's idea, but later in 1942, in collaboration with Emil Konopinski, he concluded that a hydrogen bomb, or superbomb, could be made.

For practical considerations, it was decided that the design of the superbomb would have to wait until after the war. In 1946, a secret conference on the superbomb was held in Los Alamos, New Mexico, that was attended by, among other Manhattan Project veterans, Stanislaw Ulam and Klaus Emil Julius Fuchs. Supporting the investigation of Teller's concept, the conferees requested a more complete mathematical analysis of his own admittedly crude calculations on the dynamics of the fusion reaction. In 1947, Teller believed that these calculations might take years. Two years later, however, the Soviet explosion of an atomic bomb convinced Teller that America's Cold War adversary was hard at work on its own superbomb. Even when new calculations cast further doubt on his designs, Teller began a vigorous campaign for crash development of the hydrogen bomb, or H-bomb.

THE SUPERBOMB

Scientists knew that fusion reactions could be induced by the explosion of an atomic bomb. The basic problem was simple and formidable: How could fusion fuel be heated and compressed long enough to achieve significant thermonuclear burning before the atomic fission explosion blew the assembly apart? A major part of the solution came from Ulam in 1951. He proposed using the energy from an exploding atomic bomb to induce significant thermonuclear reactions in adjacent fusion fuel components.

This arrangement, in which the A-bomb (the primary) is physically separated from the H-bomb's (the secondary's) fusion fuel, became known as the "Teller-Ulam configuration." All H-bombs are cylindrical, with an atomic device at one end and the other components filling the remaining space. Energy from the exploding primary could be transported by X rays and would therefore affect the fusion fuel at near light speed—before the arrival of the explosion.

Frederick de Hoffman's work verified and enriched the new concept. In the revised method, moderated X rays from the primary irradiate a reactive plastic medium surrounding concentric and generally cylindrical layers of fusion and fission fuel in the secondary. Instantly, the plastic becomes a hot plasma that compresses and heats the inner layer of fusion fuel, which in turn compresses a central core of fissile plutonium to supercriticality. Thus compressed, and bombarded by fusion-produced, high-energy neutrons, the fission element expands rapidly in a chain reaction from the inside out, further compressing and heating the surrounding fusion fuel, releasing more energy and more neutrons that induce fission in a fuel casing-tamper made of normally stable uranium 238.

With its equipment to refrigerate the hydrogen isotopes, the device created to test Teller's new concept weighed more than sixty tons. During Operation Ivy, it was tested at Elugelab in the Marshall

EDWARD TELLER

To call Edward Teller "controversial" is equivalent to saying that the hydrogen bomb is "destructive"—an enormous understatement. His forceful support for nuclear arms prompted some to label him a war criminal while others consider him to be one of the most thoughtful statesmen among scientists.

Teller was born into a Jewish family in Budapest, Hungary, in 1908. He left his homeland to flee the anti-Semitic fascist government of the late 1920's and attended the University of Leipzig in Germany. In 1930 he completed his doctorate and hoped to settle into an academic career there, but he fled Germany when Adolf Hitler came to power. Teller migrated to the United States in 1935 and taught at George Washington University, where with George Gamow he studied aspects of quantum mechanics and nuclear physics. He became a U.S. citizen in 1941.

Teller was among the first physicists to realize the possibility of an atomic (fission) bomb, and he became a central figure in the Manhattan Project that built it during World War II. However, he was already exploring the idea of a "superbomb" that explodes because of a fusion reaction. He helped persuade President Harry Truman to finance a project to build it and continued to influence the politics of nuclear weapons and power afterward. Teller developed the theoretical basis for the hydrogen bomb and its rough design—and so is know as its father. However, controversy later erupted over credit. Mathematician Stanislaw Ulam claimed he contributed key insights and calculations, a claim Teller vehemently denied. Teller, however, did credit a young physicist, Richard L. Garwin, with creating the successful working design for the first bomb.

Fiercely anticommunist, Teller argued for a strong nuclear arsenal to make the Soviet Union afraid of attacking the United States and supported space-based missile defense systems. He served as director of the Lawrence Livermore National Laboratory, professor at the University of California at Berkeley, and senior fellow at the nearby Hoover Institution. In his nineties he outraged environmentalists by suggesting that the atmosphere could be manipulated with technology to offset the effects of global warming.

Islands on November 1, 1952. Exceeding the expectations of all concerned and vaporizing the island, the explosion equaled 10.4 million tons of trinitrotoluene (TNT), which meant that it was about seven hundred times more powerful than the atomic bomb dropped on Hiroshima, Japan, in 1945. A version of this device weighing about 20 tons was prepared for delivery by specially modified Air Force B-36 bombers in the event of an emergency during wartime.

In development at Los Alamos before the 1952 test was a device weighing only about 4 tons, a "dry bomb" that did not require refrigeration equipment or liquid fusion fuel; when sufficiently compressed and heated in its molded-powder form, the new fusion fuel component, lithium-6 deutride, instantly produced tritium, an isotope of hydrogen. This concept was tested during Operation Castle at Bikini atoll in 1954 and produced a yield of 15 million tons of TNT, the largest-ever nuclear explosion created by the United States.

CONSEQUENCES

Teller was not alone in believing that the world could produce thermonuclear devices capable of causing great destruction. Months before Fermi suggested to Teller the possibility of explosive thermonuclear reactions on Earth, Japanese physicist Tokutaro Hagiwara had proposed that a uranium 235 bomb could ignite significant fusion reactions in hydrogen. The Soviet Union successfully tested an H-bomb dropped from an airplane in 1955, one year before the United States did so.

Teller became the scientific adviser on nuclear affairs of many presidents, from Dwight D. Eisenhower to Ronald Reagan. The widespread blast and fallout effects of H-bombs assured the mutual destruction of the users of such weapons. During the Cold War (from about 1947 to 1981), both the United States and the Soviet Union possessed H-bombs. "Testing" these bombs made each side aware of how powerful the other side was. Everyone wanted to avoid nuclear war. It was thought that no one would try to start a war that would end in the world's destruction. This theory was called *deterrence*: The United States wanted to let the Soviet Union know that it had just as many bombs, or more, than it did, so that the leaders of the Sovet Union would be *deterred* from starting a war.

Teller knew that the availability of H-bombs on both sides was not enough to guarantee that such weapons would never be used. It was also necessary to make the Soviet Union aware of the existence of the bombs through testing. He consistently advised against U.S. participation with the Soviet Union in a moratorium (period of waiting) on nuclear weapons testing. Largely based on Teller's urging that underground testing be continued, the United States rejected a total moratorium in favor of the 1963 Atmospheric Test Ban Treaty.

During the 1980's, Teller, among others, convinced President Reagan to embrace the Strategic Defense Initiative (SDI). Teller argued that SDI components, such as the space-based "Excalibur," a nuclear bomb-powered X-ray laser weapon proposed by the Lawrence-Livermore National Laboratory, would make thermonuclear war not unimaginable, but theoretically impossible.

See also Airplane; Atomic bomb; Cruise missile; Rocket; Stealth aircraft; V-2 rocket.

FURTHER READING

Blumberg, Stanley A., and Louis G. Panos. *Edward Teller, Giant of the Golden Age of Physics: A Biography.* New York: Scribner's, 1990.

Clash, James M. "Teller Tells It." *Forbes* (May 17, 1999).

Teller, Edward, Wendy Teller, and Wilson Talley. *Conversations on the Dark Secrets of Physics.* New York: Plenum Press, 1991.

York, Herbert E. *The Advisors: Oppenheimer, Teller, and the Superbomb.* Stanford, Calif.: Stanford University Press, 1989.

IBM MODEL 1401 COMPUTER

THE INVENTION: A relatively small, simple, and inexpensive computer that is often credited with having launched the personal computer age.

THE PEOPLE BEHIND THE INVENTION:

Howard H. Aiken (1900-1973), an American mathematician
Charles Babbage (1792-1871), an English mathematician and inventor
Herman Hollerith (1860-1929), an American inventor

COMPUTERS: FROM THE BEGINNING

Computers evolved into their modern form over a period of thousands of years as a result of humanity's efforts to simplify the process of counting. Two counting devices that are considered to be very simple, early computers are the abacus and the slide rule. These calculating devices are representative of digital and analog computers, respectively, because an abacus counts numbers of things, while the slide rule calculates length measurements.

The first modern computer, which was planned by Charles Babbage in 1833, was never built. It was intended to perform complex calculations with a data processing/memory unit that was controlled by punched cards. In 1944, Harvard University's Howard H. Aiken and the International Business Machines (IBM) Corporation built such a computer—the huge, punched-tape-controlled Automatic Sequence Controlled Calculator, or Mark I ASCC, which could perform complex mathematical operations in seconds. During the next fifteen years, computer advances produced digital computers that used binary arithmetic for calculation, incorporated simplified components that decreased the sizes of computers, had much faster calculating speeds, and were transistorized.

Although practical computers had become much faster than they had been only a few years earlier, they were still huge and extremely expensive. In 1959, however, IBM introduced the Model 1401 computer. Smaller, simpler, and much cheaper than the multimillion-dollar computers that were available, the IBM Model 1401 computer was also relatively easy to program and use. Its low cost, simplicity of operation, and very wide use have led many experts to view the IBM Model 1401 computer as beginning the age of the personal computer.

COMPUTER OPERATION AND IBM'S MODEL 1401

Modern computers are essentially very fast calculating machines that are capable of sorting, comparing, analyzing, and outputting information, as well as storing it for future use. Many sources credit Aiken's Mark I ASCC as being the first modern computer to be built. This huge, five-ton machine used thousands of relays to perform complex mathematical calculations in seconds. Soon after its introduction, other companies produced computers that were faster and more versatile than the Mark I. The computer development race was on.

All these early computers utilized the decimal system for calculations until it was found that binary arithmetic, whose numbers are combinations of the binary digits 1 and 0, was much more suitable for the purpose. The advantage of the binary system is that the electronic switches that make up a computer (tubes, transistors, or chips) can be either on or off; in the binary system, the on state can be represented by the digit 1, the off state by the digit 0. Strung together correctly, binary numbers, or digits, can be inputted rapidly and used for high-speed computations. In fact, the computer term bit is a contraction of the phrase "binary digit."

A computer consists of input and output devices, a storage device (memory), arithmetic and logic units, and a control unit. In most cases, a central processing unit (CPU) combines the logic, arithmetic, memory, and control aspects. Instructions are loaded into the memory via an input device, processed, and stored. Then, the CPU issues commands to the other parts of the system to carry out computations or other functions and output the data as needed. Most output is printed as hard copy or displayed on cathode-ray tube monitors, or screens.

The early modern computers—such as the Mark I ASCC—were huge because their information circuits were large relays or tubes. Computers became smaller and smaller as the tubes were replaced—

first with transistors, then with simple integrated circuits, and then with silicon chips. Each technological changeover also produced more powerful, more cost-effective computers.

In the 1950's, with reliable transistors available, IBM began the development of two types of computers that were completed by about 1959. The larger version was the Stretch computer, which was advertised as the most powerful computer of its day. Customized for each individual purchaser (for example, the Atomic Energy Commission), a Stretch computer cost $10 million or more. Some innovations in Stretch computers included semiconductor circuits, new switching systems that quickly converted various kinds of data into one language that was understood by the CPU, rapid data readers, and devices that seemed to anticipate future operations.

CONSEQUENCES

The IBM Model 1401 was the first computer sold in very large numbers. It led IBM and other companies to seek to develop less expensive, more versatile, smaller computers that would be sold to small businesses and to individuals. Six years after the development of the Model 1401, other IBM models—and those made by other companies—became available that were more compact and had larger memories. The search for compactness and versatility continued. A major development was the invention of integrated circuits by Jack S. Kilby of Texas Instruments; these integrated circuits became available by the mid-1960's. They were followed by even smaller "microprocessors" (computer chips) that became available in the 1970's. Computers continued to become smaller and more powerful.

Input and storage devices also decreased rapidly in size. At first, the punched cards invented by Herman Hollerith, founder of the Tabulation Machine Company (which later became IBM), were read by bulky readers. In time, less bulky magnetic tapes and more compact readers were developed, after which magnetic disks and compact disc drives were introduced.

Many other advances have been made. Modern computers can talk, create art and graphics, compose music, play games, and operate robots. Further advancement is expected as societal needs change. Many experts believe that it was the sale of large numbers of IBM Model 1401 computers that began the trend.

See also Apple II computer; BINAC computer; Colossus computer; ENIAC computer; Personal computer; Supercomputer; UNIVAC computer.

FURTHER READING

Carroll, Paul. *Big Blues: The Unmaking of IBM.* New York: Crown, 1993.
Chposky, James, and Ted Leonsis. *Blue Magic: The People, Power, and Politics Behind the IBM Personal Computer.* New York: Facts on File, 1988.
Manes, Stephen, and Paul Andrews. *Gates: How Microsoft's Mogul Reinvented an Industry.* New York: Doubleday, 1993.

In Vitro Plant Culture

The invention: Method for propagating plants in artificial media that has revolutionized agriculture.

The people behind the invention:
Georges Michel Morel (1916-1973), a French physiologist
Philip Cleaver White (1913-   ), an American chemist

Plant Tissue Grows "In Glass"

In the mid-1800's, biologists began pondering whether a cell isolated from a multicellular organism could live separately if it were provided with the proper environment. In 1902, with this question in mind, the German plant physiologist Gottlieb Haberlandt attempted to culture (grow) isolated plant cells under sterile conditions on an artificial growth medium. Although his cultured cells never underwent cell division under these "in vitro" (in glass) conditions, Haberlandt is credited with originating the concept of cell culture.

Subsequently, scientists attempted to culture plant tissues and organs rather than individual cells and tried to determine the medium components necessary for the growth of plant tissue in vitro. In 1934, Philip White grew the first organ culture, using tomato roots. The discovery of plant hormones, which are compounds that regulate growth and development, was crucial to the successful culture of plant tissues; in 1939, Roger Gautheret, P. Nobécourt, and White independently reported the successful culture of plant callus tissue. "Callus" is an irregular mass of dividing cells that often results from the wounding of plant tissue. Plant scientists were fascinated by the perpetual growth of such tissue in culture and spent years establishing optimal growth conditions and exploring the nutritional and hormonal requirements of plant tissue.

Plants by the Millions

A lull in botanical research occurred during World War II, but immediately afterward there was a resurgence of interest in applying tissue culture techniques to plant research. Georges Morel, a plant physiologist at the National Institute for Agronomic Research in France, was one of many scientists during this time who had become interested in the formation of tumors in plants as well as in studying various pathogens such as fungi and viruses that cause plant disease.

To further these studies, Morel adapted existing techniques in order to grow tissue from a wider variety of plant types in culture, and he continued to try to identify factors that affected the normal growth and development of plants. Morel was successful in culturing tissue from ferns and was the first to culture monocot plants. Monocots have certain features that distinguish them from the other classes of seed-bearing plants, especially with respect to seed structure. More important, the monocots include the economically important species of grasses (the major plants of range and pasture) and cereals.

For these cultures, Morel utilized a small piece of the growing tip of a plant shoot (the shoot apex) as the starting tissue material. This tissue was placed in a glass tube, supplied with a medium containing specific nutrients, vitamins, and plant hormones, and allowed to grow in the light. Under these conditions, the apex tissue grew roots and buds and eventually developed into a complete plant. Morel was able to generate whole plants from pieces of the shoot apex that were only 100 to 250 micrometers in length.

Morel also investigated the growth of parasites such as fungi and viruses in dual culture with host-plant tissue. Using results from these studies and culture techniques that he had mastered, Morel and his colleague Claude Martin regenerated virus-free plants from tissue that had been taken from virally infected plants. Tissues from certain tropical species, dahlias, and potato plants were used for the original experiments, but after Morel adapted the methods for the generation of virus-free orchids, plants that had previously been difficult to propagate by any means, the true significance of his work was recognized.

Morel was the first to recognize the potential of the in vitro culture methods for the mass propagation of plants. He estimated that several million plants could be obtained in one year from a single small piece of shoot-apex tissue. Plants generated in this manner were clonal (genetically identical organisms prepared from a single plant).

In vitro plant culture has been especially useful for species such as palm trees that cannot be propagated by other methods, such as by sowing seeds or grafting. (PhotoDisc)

With other methods of plant propagation, there is often a great variation in the traits of the plants produced, but as a result of Morel's ideas, breeders could select for some desirable trait in a particular plant and then produce multiple clonal plants, all of which expressed the desired trait. The methodology also allowed for the production of virus-free plant material, which minimized both the spread of potential pathogens during shipping and losses caused by disease.

Consequences

Variations on Morel's methods are used to propagate plants used for human food consumption; plants that are sources of fiber, oil, and livestock feed; forest trees; and plants used in landscaping and in the floral industry. In vitro stocks are preserved under deepfreeze conditions, and disease-free plants can be proliferated quickly at any time of the year after shipping or storage.

The in vitro multiplication of plants has been especially useful for species such as coconut and certain palms that cannot be propagated by other methods, such as by sowing seeds or grafting, and has also become important in the preservation and propagation of rare plant species that might otherwise have become extinct. Many of these plants are sources of pharmaceuticals, oils, fragrances, and other valuable products.

The capability of regenerating plants from tissue culture has also been crucial in basic scientific research. Plant cells grown in culture can be studied more easily than can intact plants, and scientists have gained an in-depth understanding of plant physiology and biochemistry by using this method. This information and the methods of Morel and others have made possible the genetic engineering and propagation of crop plants that are resistant to disease or disastrous environmental conditions such as drought and freezing. In vitro techniques have truly revolutionized agriculture.

See also Artificial insemination; Cloning; Genetically engineered insulin; Rice and wheat strains.

FURTHER READING

Arbury, Jim, Richard Bird, Mike Honour, Clive Innes, and Mike Salmon. *The Complete Book of Plant Propagation.* Newtown, Conn.: Taunton Press, 1997.

Clarke, Graham. *The Complete Book of Plant Propagation.* London: Seven Dials, 2001.

Hartmann, Hudson T. *Plant Propagation: Principles and Practices.* 6th ed. London: Prentice-Hall, 1997.

Heuser, Charles. *The Complete Book of Plant Propagation.* Newtown, Conn.: Taunton Press, 1997.

INFRARED PHOTOGRAPHY

THE INVENTION: The first application of color to infrared photography, which performs tasks not possible for ordinary photography.

THE PERSON BEHIND THE INVENTION:
Sir William Herschel (1738-1822), a pioneering English astronomer

INVISIBLE LIGHT

Photography developed rapidly in the nineteenth century when it became possible to record the colors and shades of visible light on sensitive materials. Visible light is a form of radiation that consists of electromagnetic waves, which also make up other forms of radiation such as X rays and radio waves. Visible light occupies the range of wavelengths from about 400 nanometers (1 nanometer is 1 billionth of a meter) to about 700 nanometers in the electromagnetic spectrum.

Infrared radiation occupies the range from about 700 nanometers to about 1,350 nanometers in the electromagnetic spectrum. Infrared rays cannot be seen by the human eye, but they behave in the same way that rays of visible light behave; they can be reflected, diffracted (broken), and refracted (bent).

Sir William Herschel, a British astronomer, discovered infrared rays in 1800 by calculating the temperature of the heat that they produced. The term "infrared," which was probably first used in 1800, was used to indicate rays that had wavelengths that were longer than those on the red end (the high end) of the spectrum of visible light but shorter than those of the microwaves, which appear higher on the electromagnetic spectrum. Infrared film is therefore sensitive to the infrared radiation that the human eye cannot see or record. Dyes that were sensitive to infrared radiation were discovered early in the twentieth century, but they were not widely used until the 1930's. Because these dyes produced only black-and-white images, their usefulness to artists and researchers was limited. After 1930, however, a tidal wave of infrared photographic applications appeared.

THE DEVELOPMENT OF COLOR-SENSITIVE INFRARED FILM

In the early 1940's, military intelligence used infrared viewers for night operations and for gathering information about the enemy. One device that was commonly used for such purposes was called a "snooper scope." Aerial photography with black-and-white infrared film was used to locate enemy hiding places and equipment. The images that were produced, however, often lacked clear definition.

The development in 1942 of the first color-sensitive infrared film, Ektachrome Aero Film, became possible when researchers at the Eastman Kodak Company's laboratories solved some complex chemical and physical problems that had hampered the development of color infrared film up to that point. Regular color film is sensitive to all visible colors of the spectrum; infrared color film is sensitive to violet, blue, and red light as well as to infrared radiation. Typical color film has three layers of emulsion, which are sensitized to blue, green, and red. Infrared color film, however, has its three emulsion layers sensitized to green, red, and infrared. Infrared wavelengths are recorded as reds of varying densities, depending on the intensity of the infrared radiation. The more infrared radiation there is, the darker the color of the red that is recorded.

In infrared photography, a filter is placed over the camera lens to block the unwanted rays of visible light. The filter blocks visible and ultraviolet rays but allows infrared radiation to pass. All three layers of infrared film are sensitive to blue, so a yellow filter is used. All blue radiation is absorbed by this filter.

In regular photography, color film consists of three basic layers: the top layer is sensitive to blue light, the middle layer is sensitive to green, and the third layer is sensitive to red. Exposing the film to light causes a latent image to be formed in the silver halide crystals that make up each of the three layers. In infrared photography, color film consists of a top layer that is sensitive to infrared radiation, a middle layer sensitive to green, and a bottom layer sensitive to red. "Reversal processing" produces blue in the infrared-sensitive layer, yellow in the green-sensitive layer, and magenta in the red-sensitive layer. The blue, yellow, and magenta layers of the film produce the "false colors" that accentuate the various levels of infrared radiation shown as red in a color transparency, slide, or print.

SIR WILLIAM HERSCHEL

During his long career Sir William Herschel passed from human music to the music of the spheres, and in doing so revealed the invisible unlike any astronomer before him.

He was born Friedrich Wilhelm Herschel in Hannover, Germany, in 1738. Like his brothers, he trained to be a musician in a local regimental band. In 1757 he had to flee to England because his regiment was on the losing side of a war. Settling in the town of Bath, he supported himself with music, eventually becoming the organist for the city's celebrated Octagon Chapel. He studied the music theory in Robert Smith's book on harmonics and, discovering another book by Smith about optics and astronomy, read that too. He was immediately hooked. By 1773 he was assembling his own telescopes, and within ten years he had built the most powerful instruments in the land. He interested King George III in astronomy and was rewarded with a royal pension that gave him the leisure to survey the heavens.

(Library of Congress)

Herschel looked deeper into space than anyone before him. He discovered thousands of double stars and nebulae that had been invisible to astronomers with less powerful telescopes than his. He was the first person in recorded history to discover a planet—Uranus. While trying to learn the construction of the sun, he conducted hundreds of experiments with light. He found, unexpectedly, that he could feel heat from the sun even when visible light was filtered out, and concluded that some solar radiation—in this case infrared—was invisible to human eyes.

Late in his career Herschel addressed the grandest of all invisible aspects of the nature: the structure of the universe. His investigations led him to conclude that the nebulae he had so often observed were in themselves vast clouds of stars, very far away—they were galaxies. It was a key conceptual step in the development of modern cosmology.

By the time Herschel died in 1822, he had trained his sister Caroline and his son John to carry on his work. Both became celebrated astronomers in their own right.

The color of the dye that is formed in a particular layer bears no relationship to the color of light to which the layer is sensitive. If the relationship is not complementary, the resulting colors will be false. This means that objects whose colors appear to be similar to the human eye will not necessarily be recorded as similar colors on infrared film. A red rose with healthy green leaves will appear on infrared color film as being yellow with red leaves, because the chlorophyll contained in the plant leaf reflects infrared radiation and causes the green leaves to be recorded as red. Infrared radiation from about 700 nanometers to about 900 nanometers on the electromagnetic spectrum can be recorded by infrared color film. Above 900 nanometers, infrared radiation exists as heat patterns that must be recorded by nonphotographic means.

Impact

Infrared photography has proved to be valuable in many of the sciences and the arts. It has been used to create artistic images that are often unexpected visual explosions of everyday views. Because infrared radiation penetrates haze easily, infrared films are often used in mapping areas or determining vegetation types. Many cloud-covered tropical areas would be impossible to map without infrared photography. False-color infrared film can differentiate between healthy and unhealthy plants, so it is widely used to study insect and disease problems in plants. Medical research uses infrared photography to trace blood flow, detect and monitor tumor growth, and to study many other physiological functions that are invisible to the human eye.

Some forms of cancer can be detected by infrared analysis before any other tests are able to perceive them. Infrared film is used in criminology to photograph illegal activities in the dark and to study evidence at crime scenes. Powder burns around a bullet hole, which are often invisible to the eye, show clearly on infrared film. In addition, forgeries in documents and works of art can often be seen clearly when photographed on infrared film. Archaeologists have used infrared film to locate ancient sites that are invisible in daylight. Wildlife biologists also document the behavior of animals at night with infrared equipment.

See also Autochrome plate; Color film; Fax machine; Instant photography.

FURTHER READING

Collins, Douglas. *The Story of Kodak*. New York: Harry N. Abrams, 1990.
Cummins, Richard. "Infrared Revisited." *Petersen's Photographic Magazine* 23 (February, 1995).
Paduano, Joseph. *The Art of Infrared Photography*. 4th ed. Buffalo, N.Y: Amherst Media, 1998.
Richards, Dan. "The Strange Otherworld of Infrared." *Popular Photography* 62, no. 6 (June, 1998).
White, Laurie. *Infrared Photography Handbook*. Amherst, N.Y.: Amherst Media, 1995.

Instant photography

The invention: Popularly known by its Polaroid tradename, a camera capable of producing finished photographs immediately after its film was exposed.

The people behind the invention:
Edwin Herbert Land (1909-1991), an American physicist and chemist
Howard G. Rogers (1915-　　), a senior researcher at Polaroid and Land's collaborator
William J. McCune (1915-　　), an engineer and head of the Polaroid team
Ansel Adams (1902-1984), an American photographer and Land's technical consultant

The Daughter of Invention

Because he was a chemist and physicist interested primarily in research relating to light and vision, and to the materials that affect them, it was inevitable that Edwin Herbert Land should be drawn into the field of photography. Land founded the Polaroid Corporation in 1929. During the summer of 1943, while Land and his wife were vacationing in Santa Fe, New Mexico, with their three-year-old daughter, Land stopped to take a picture of the child. After the picture was taken, his daughter asked to see it. When she was told she could not see the picture immediately, she asked how long it would be. Within an hour after his daughter's question, Land had conceived a preliminary plan for designing the camera, the film, and the physical chemistry of what would become the instant camera. Such a device would, he hoped, produce a picture immediately after exposure.

Within six months, Land had solved most of the essential problems of the instant photography system. He and a small group of associates at Polaroid secretly worked on the project. Howard G. Rogers was Land's collaborator in the laboratory. Land conferred the responsibility for the engineering and mechanical phase of the project on William J. McCune, who led the team that eventually designed the original camera and the machinery that produced both the camera and Land's new film.

The first Polaroid Land camera—the Model 95—produced photographs measuring 8.25 by 10.8 centimeters; there were eight pictures to a roll. Rather than being black-and-white, the original Polaroid prints were sepia-toned (producing a warm, reddish-brown color). The reasons for the sepia coloration were chemical rather than aesthetic; as soon as Land's researchers could devise a workable formula for sharp black-and-white prints (about ten months after the camera was introduced commercially), they replaced the sepia film.

A SOPHISTICATED CHEMICAL REACTION

Although the mechanical process involved in the first demonstration camera was relatively simple, this process was merely the means by which a highly sophisticated chemical reaction— the diffusion transfer process—was produced.

In the basic diffusion transfer process, when an exposed negative image is developed, the undeveloped portion corresponds to the opposite aspect of the image, the positive. Almost all self-processing instant photography materials operate according to three phases—negative development, diffusion transfer, and positive development. These occur simultaneously, so that positive image formation begins instantly. With black-and-white materials, the positive was originally completed in about sixty seconds; with color materials (introduced later), the process took somewhat longer.

The basic phenomenon of silver in solution diffusing from one emulsion to another was first observed in the 1850's, but no practical use of this action was made until 1939. The photographic use of diffusion transfer for producing normal-continuous-tone images was investigated actively from the early 1940's by Land and his associates. The instant camera using this method was demonstrated in 1947 and marketed in 1948.

The fundamentals of photographic diffusion transfer are simplest in a black-and-white peel-apart film. The negative sheet is exposed in the camera in the normal way. It is then pulled out of the camera, or film pack holder, by a paper tab. Next, it passes through a set of rollers, which press it face-to-face with a sheet of receiving ma-

Edwin H. Land

Born in Bridgeport, Connecticut in 1909, Edwin Herbert Land developed an obsession with color vision. As a boy, he slept with a copy of an optics textbook under his pillow. When he went to Harvard to study physics, he found the instruction too elementary and spent much of the time educating himself at the New York Public Library. While there, he thought of the first of his many sight-related inventions.

He realized that by lining up tiny crystals and embedding them in clear plastic he could make a large, inexpensive light polarizer. He patented the idea for this "Polaroid" lens in 1929 (the first of more than five hundred patents) and in 1932 set up a commercial laboratory with his Harvard physics professor, George Wheelwright III. Five years later he opened the Polaroid Corporation in Boston to exploit the commercial potential of the lenses. They were to be used most famously as sunglasses, camera filters, eyeglasses for producing three-dimensional effects in movies, and glare-reduction screens for visual display terminals.

In 1937, with Joseph Mallory, Land invented the vectograph—a device that superimposed two photographs in order to create a three-dimensional image. The invention dramatically improved the aerial photography during World War II and the Cold War. In fact, Land had a hand in designing both the camera carried aboard Lockheed's U2 spyplane and the plane itself.

While not busy running the Polaroid Corporation and overseeing development of its cameras, Land pursued his passion for experimenting with color and developed a widely respected theory of color vision. When he retired in 1982, he launched the Rowland Institute for Science in Boston, once described as a cross between a private laboratory and a private art gallery. (Land had a deep interest in modern art.) He and other scientists there conducted research on artificial intelligence, genetics, microscopy, holography, protein dynamics, and color vision. Land died in 1991 in Cambridge, Massachusetts, but the institute carries forward his legacy of scientific curiosity and practical application.

terial included in the film pack. Simultaneously, the rollers rupture a pod of reagent chemicals that are spread evenly by the rollers between the two layers. The reagent contains a strong alkali and a silver halide solvent, both of which diffuse into the negative emulsion. There the alkali activates the developing agent, which immediately reduces the exposed halides to a negative image. At the same time, the solvent dissolves the unexposed halides. The silver in the dissolved halides forms the positive image.

IMPACT

The Polaroid Land camera had a tremendous impact on the photographic industry as well as on the amateur and professional photographer. Ansel Adams, who was known for his monumental, ultrasharp black-and-white panoramas of the American West, suggested to Land ways in which the tonal value of Polaroid film could be enhanced, as well as new applications for Polaroid photographic technology.

Soon after it was introduced, Polaroid photography became part of the American way of life and changed the face of amateur photography forever. By the 1950's, Americans had become accustomed to the world of recorded visual information through films, magazines, and newspapers; they also had become enthusiastic picture-takers as a result of the growing trend for simpler and more convenient cameras. By allowing these photographers not only to record their perceptions but also to see the results almost immediately, Polaroid brought people closer to the creative process.

See also Autochrome plate; Brownie camera; Color film; Fax machine; Xerography.

FURTHER READING

Adams, Ansel. *Polaroid Land Photography Manual*. New York: Morgan & Morgan, 1963.

Innovation/Imagination: Fifty Years of Polaroid Photography. New York: H. N. Abrams in association with the Friends of Photography, 1999.

McElheny, Victor K. *Insisting on the Impossible: The Life of Edwin Land*. Cambridge, Mass.: Perseus Books, 1998.

Olshaker, Mark. *The Instant Image*. New York: Stein & Day, 1978.

Wensberg, Peter C. *Land's Polaroid*. Boston: Houghton Mifflin, 1987.

Interchangeable Parts

THE INVENTION: A key idea in the late Industrial Revolution, the interchangeability of parts made possible mass production of identical products.

THE PEOPLE BEHIND THE INVENTION:

Henry M. Leland (1843-1932), president of Cadillac Motor Car Company in 1908, known as a master of precision

Frederick Bennett, the British agent for Cadillac Motor Car Company who convinced the Royal Automobile Club to run the standardization test at Brooklands, England

Henry Ford (1863-1947), founder of Ford Motor Company who introduced the moving assembly line into the automobile industry in 1913

AN AMERICAN IDEA

Mass production is a twentieth century methodology that for the most part is a result of nineteenth century ideas. It is a phenomenon that, although its origins were mostly American, has consequently changed the entire world. The use of interchangeable parts, the feasibility of which was demonstrated by the Cadillac Motor Car Company in 1908, was instrumental in making mass production possible.

The British phase of the Industrial Revolution saw the application of division of labor, the first principle of industrialization, to capitalist-directed manufacturing processes. Centralized power sources were connected through shafts, pulleys, and belts to machines housed in factories. Even after these dramatic changes, the British preferred to produce unique, handcrafted products formed one step at a time using general-purpose machine tools. Seldom did they make separate components to be assembled into standardized products.

Stories about American products that were assembled from fully interchangeable parts began to reach Great Britain. In 1851, the British public saw a few of these products on display at an exhibition in London's Crystal Palace. In 1854, they were informed by one of their own investigative commissions that American manufacturers were building military weapons and a number of consumer products with separately made parts that could be easily assembled, with little filing and fitting, by semiskilled workers.

English industrialists had probably heard as much as they ever wanted to about this so-called "American system of manufacturing" by the first decade of the twentieth century, when word came that American companies were building automobiles with parts manufactured so precisely that they were interchangeable.

THE CADILLAC

During the fall of 1907, Frederick Bennett, an Englishman who served as the British agent for the Cadillac Motor Car Company, paid a visit to the company's Detroit, Michigan, factory and was amazed at what he saw. He later described the assembling of the relatively inexpensive Cadillac vehicles as a demonstration of the beauty and practicality of precision. He was convinced that if his countrymen could see what he had seen they would also be impressed.

Most automobile builders at the time claimed that their vehicles were built with handcrafted quality, yet at the same time they advertised that they could supply repair parts that would fit perfectly. In actuality, machining and filing were almost always required when parts were replaced, and only shops with proper equipment could do the job.

Upon his return to London, Bennett convinced the Royal Automobile Club to sponsor a test of the precision of automobile parts. A standardization test was set to begin on February 29, 1908, and all of the companies then selling automobiles were invited to participate. Only the company that Bennett represented, Cadillac, was willing to enter the contest.

Three one-cylinder Cadillacs, each painted a different color, were taken from stock at the company's warehouse in London to a garage near the Brooklands race track. The cars were first driven around the track ten times to prove that they were operable. British mechanics then dismantled the vehicles, placing their parts in piles in the center of the garage, making sure that there was no way of identifying from which car each internal piece came. Then, as a further test, eighty-nine randomly selected parts were removed from the piles and replaced with new ones straight from Cadillac's storeroom in London. The mechanics then proceeded to reassemble the automobiles, using only screwdrivers and wrenches.

After the reconstruction, which took two weeks, the cars were driven from the garage. They were a motley looking trio, with fenders, doors, hoods, and wheels of mixed colors. All three were then driven five hundred miles around the Brooklands track. The British were amazed. Cadillac was awarded the club's prestigious Dewar Trophy, considered in the young automobile industry to be almost the equivalent of a Nobel Prize. A number of European and American automobile manufacturers began to consider the promise of interchangeable parts and the assembly line system.

HENRY M. LELAND

Cadillac's precision-built automobiles were the result of a lifetime of experience of Henry M. Leland, an American engineer. Known in Detroit at the turn of the century as a master of precision, Leland became the primary connection between a series of nineteenth century attempts to make interchangeable parts and the large-scale use of precision parts in mass production manufacturing during the twentieth century.

The first American use of truly interchangeable parts had occurred in the military, nearly three-quarters of a century before the test at Brooklands. Thomas Jefferson had written from France about a demonstration of uniform parts for musket locks in 1785. A few years later, Eli Whitney attempted to make muskets for the American military by producing separate parts for assembly using specialized machines. He was never able to produce the precision necessary for truly interchangeable parts, but he promoted the idea intensely. It was in 1822 at the Harpers Ferry Armory in Virginia, and then a few years later at the Springfield Armory in Massachusetts, that the necessary accuracy in machining was finally achieved on a relatively large scale.

Leland began his career at the Springfield Armory in 1863, at the age of nineteen. He worked as a tool builder during the Civil War years and soon became an advocate of precision manufacturing. In 1890, Leland moved to Detroit, where he began a firm, Leland &

HENRY MARTYN LELAND

Henry Martyn Leland (1843-1932) is the unsung giant of early automobile manufacturers, launching two of the best-known American car companies, Cadillac and Lincoln, and influenced the success of General Motors, as well as introducing the use of interchangeable parts. Had he allowed a model to be named after him, as did Henry Ford and Ransom Olds, he might have become a household name too, but he refused any such suggestion.

Leland worked in factories during his youth. During the Civil War he honed his skills as a machinist at the U.S. Armory in Springfield, Massachusetts, helping build rifles with interchangeable parts. After the war, he learned how to machine parts to within one-thousandth of an inch, fabricated the first mechanical barber's clippers, and refined the workings of air brakes for locomotives.

This was all warm-up. In 1890 he moved to Detroit and opened his own business, Leland and Faulconer Manufacturing Company, specializing in automobile engines. The 10.25-horsepower engine he built for Olds in 1901 was rejected, but the single-cylinder ("one-lunger") design that powered the first Cadillacs set him on the high road in the automotive industry. More innovations followed. He developed the electric starter, electric lights, and dimmable headlights. During World War I he built airplane engines for the U.S. government, and afterward converted the design for use in his new creation, the Lincoln.

Throughout, he demanded precision from himself and those working for him. Once, for example, he complained to Alfred P. Sloan that a lot of ball bearings that Sloan had sold him varied from the required engineering tolerances and showed Sloan a few misshapen bearings to prove the claim. "Even though you make thousands," Leland admonished Sloan, "the first and last should be precisely the same." Sloan took the lesson very seriously. When he later led General Motors to the top of the industry, he credited Leland with teaching him what mass production was all about.

Faulconer, that would become internationally known for precision machining. His company did well supplying parts to the bicycle industry and internal combustion engines and transmissions to early automobile makers. In 1899, Leland & Faulconer became the primary supplier of engines to the first of the major automobile producers, the Olds Motor Works.

In 1902, the directors of another Detroit firm, the Henry Ford Company, found themselves in a desperate situation. Henry Ford, the company founder and chief engineer, had resigned after a disagreement with the firm's key owner, William Murphy. Leland was asked to take over the reorganization of the company. Because it could no longer use Ford's name, the business was renamed in memory of the French explorer who had founded Detroit two hundred years earlier, Antoine de la Mothe Cadillac.

Leland was appointed president of the Cadillac Motor Car Company. The company, under his influence, soon became known for its precision manufacturing. He disciplined its suppliers, rejecting anything that did not meet his specifications, and insisted on precision machining for all parts. By 1906, Cadillac was outselling all of its competitors, including Oldsmobile and Ford's new venture, the Ford Motor Company. After the Brooklands demonstration in 1908, Cadillac became recognized worldwide for quality and interchangeability at a reasonable price.

IMPACT

The Brooklands demonstration went a long way in proving that mass-produced goods could be durable and of relatively high quality. It showed that standardized products, although often less costly to make, were not necessarily cheap substitutes for handcrafted and painstakingly fitted products. It also demonstrated that, through the use of interchangeable parts, the job of repairing such complex machines as automobiles could be made comparatively simple, moving maintenance and repair work from the well-equipped machine shop to the neighborhood garage or even to the home.

Because of the international publicity Cadillac received, Leland's methods began to be emulated by others in the automobile industry. His precision manufacturing, as his daughter-in-law would later write in his biography, "laid the foundation for the future American [automobile] industry." The successes of automobile manufacturers quickly led to the introduction of mass production methods, and strategies designed to promote their necessary corollary mass consumption, in many other American businesses.

In 1909, Cadillac was acquired by William Crapo Durant as the flagship company of his new holding company, which he labeled General Motors. Leland continued to improve his production methods, while also influencing his colleagues in the other General Motors companies to implement many of his techniques. By the mid-1920's, General Motors had become the world's largest manufacturer of automobiles. Much of its success resulted from extensions of Leland's ideas. The company began offering a number of brand name vehicles in a variety of price ranges for marketing purposes, while still keeping the costs of production down by including in each design a large number of commonly used, highly standardized components.

Henry Leland resigned from Cadillac during World War I after trying to convince Durant that General Motors should play an important part in the war effort by contracting to build Liberty aircraft engines for the military. He formed his own firm, named after his favorite president, Abraham Lincoln, and went on to build about four thousand aircraft engines in 1917 and 1918. In 1919, ready to make automobiles again, Leland converted the Lincoln Motor Company into a car manufacturer. Again he influenced the industry by setting high standards for precision, but in 1921 an economic recession forced his new venture into receivership. Ironically, Lincoln was purchased at auction by Henry Ford. Leland retired, his name overshadowed by those of individuals to whom he had taught the importance of precision and interchangeable parts. Ford, as one example, went on to become one of America's industrial legends by applying the standardized parts concept.

Ford and the Assembly Line

In 1913, Henry Ford, relying on the ease of fit made possible through the use of machined and stamped interchangeable parts, introduced the moving assembly line to the automobile industry. He had begun production of the Model T in 1908 using stationary assembly methods, bringing parts to assemblers. After having learned how to increase component production significantly, through experiments with interchangeable parts and moving assembly methods in the magneto department, he began to apply this same concept to final assembly. In the spring of 1913, Ford workers began dragging car frames past stockpiles of parts for assembly. Soon a power source was attached to the cars through a chain drive, and the vehicles were pulled past the stockpiles at a constant rate.

From this time on, the pace of tasks performed by assemblers would be controlled by the rhythm of the moving line. As demand for the Model T increased, the number of employees along the line was increased and the jobs were broken into smaller and simpler tasks. With stationary assembly methods, the time required to assemble a Model T had averaged twelve and one-half person-hours. Dragging the chassis to the parts cut the time to six hours per vehicle, and the power-driven, constant-rate line produced a Model T with only ninety-three minutes of labor time. Because of these amazing increases in productivity, Ford was able to lower the selling price of the basic model from $900 in 1910 to $260 in 1925. He had revolutionized automobile manufacturing: The average family could now afford an automobile.

Soon the average family would also be able to afford many of the other new products they had seen in magazines and newspapers. At the turn of the century, there were many new household appliances, farm machines, ready-made fashions, and prepackaged food products on the market, but only the wealthier class could afford most of these items. Major consumer goods retailers such as Sears, Roebuck and Company, Montgomery Ward, and the Great Atlantic and Pacific Tea Company were anxious to find lower-priced versions of these products to sell to a growing middle-class constituency. The methods of mass production that Henry Ford had popularized seemed to carry promise for these products as well. During the 1920's, by working with such key manufacturers as Whirlpool, Hoover, General Electric, and Westinghouse, these large distributors helped introduce mass production methods into a large number of consumer product industries. They changed class markets into mass markets.

The movement toward precision also led to the birth of a separate industry based on the manufacture of machine tools. A general purpose lathe, milling machine, or grinder could be used for a number of operations, but mass production industries called for narrow-purpose machines designed for high-speed use in performing one specialized step in the production process. Many more machines were now required, one at each step in the production process. Each machine had to be simpler to operate, with more automatic features, because of an increased dependence on unskilled workers. The machine tool industry became the foundation of modern production.

The miracle of mass production that followed, in products as diverse as airplanes, communication systems, and hamburgers, would not have been possible without the precision insisted upon by Henry Leland in the first decade of the twentieth century. It would not have come about without the lessons learned by Henry Ford in the use of specialized machines and assembly methods, and it would not have occurred without the growth of the machine tool industry. Cadillac's demonstration at Brooklands in 1908 proved the practicality of precision manufacturing and interchangeable parts to the world. It inspired American manufacturers to continue to develop these ideas; it convinced Europeans that such production was possible; and, for better or for worse, it played a major part in changing the world.

See also CAD/CAM; Assembly line; Internal combustion engine.

FURTHER READING

Hill, Frank Ernest. *The Automobile: How It Came, Grew, and Has Changed Our Lives.* New York: Dodd, Mead, 1967.

Hounshell, David A. *From the American System to Mass Production, 1800-1932.* Baltimore: Johns Hopkins University Press, 1984.

Leland, Ottilie M., and Minnie Dubbs Millbrook. *Master of Precision: Henry M. Leland.* 1966. Reprint. Detroit: Wayne State University Press, 1996.

Marcus, Alan I., and Howard P. Segal. *Technology in America: A Brief History.* Fort Worth, Texas: Harcourt Brace College, 1999.

Nevins, Allan, and Frank Ernest Hill. *The Times, the Man, the Company.* Vol. 1 in *Ford.* New York: Charles Scribner's Sons, 1954.

INTERNAL COMBUSTION ENGINE

THE INVENTION: The most common type of engine in automobiles and many other vehicles, the internal combusion engine is characterized by the fact that it burns its liquid fuelly *internally*—in contrast to engines, such as the steam engine, that burn fuel in external furnaces.

THE PEOPLE BEHIND THE INVENTION:
Sir Harry Ralph Ricardo (1885-1974), an English engineer
Oliver Thornycroft (1885-1956), an engineer and works manager
Sir David Randall Pye (1886-1960), an engineer and
    administrator
Sir Robert Waley Cohen (1877-1952), a scientist and industrialist

THE INTERNAL COMBUSTION ENGINE: 1900-1916

By the beginning of the twentieth century, internal combustion engines were almost everywhere. City streets in Berlin, London, and New York were filled with automobile and truck traffic; gasoline- and diesel-powered boat engines were replacing sails; stationary steam engines for electrical generation were being edged out by internal combustion engines. Even aircraft use was at hand: To progress from the Wright brothers' first manned flight in 1903 to the fighting planes of World War I took only a little more than a decade.

The internal combustion engines of the time, however, were primitive in design. They were heavy (10 to 15 pounds per output horsepower, as opposed to 1 to 2 pounds today), slow (typically 1,000 or fewer revolutions per minute or less, as opposed to 2,000 to 5,000 today), and extremely inefficient in extracting the energy content of their fuel. These were not major drawbacks for stationary applications, or even for road traffic that rarely went faster than 30 or 40 miles per hour, but the advent of military aircraft and tanks demanded that engines be made more efficient.

ENGINE AND FUEL DESIGN

Harry Ricardo, son of an architect and grandson (on his mother's side) of an engineer, was a central figure in the necessary redesign of internal combustion engines. As a schoolboy, he built a coal-fired steam engine for his bicycle, and at Cambridge University he produced a single-cylinder gasoline motorcycle, incorporating many of his own ideas, which won a fuel-economy competition when it traveled almost 40 miles on a quart of gasoline. He also began development of a two-cycle engine called the "Dolphin," which later was produced for use in fishing boats and automobiles. In fact, in 1911, Ricardo took his new bride on their honeymoon trip in a Dolphin-powered car.

The impetus that led to major engine research came in 1916 when Ricardo was an engineer in his family's firm. The British government asked for newly designed tank engines, which had to operate in the dirt and mud of battle, at a tilt of up to 35 degrees, and could not give off telltale clouds of blue oil smoke. Ricardo solved the problem with a special piston design and with air circulation around the carburetor and within the engine to keep the oil cool.

Design work on the tank engines turned Ricardo into a full-fledged research engineer. In 1917, he founded his own company, and a remarkable series of discoveries quickly followed. He investigated the problem of detonation of the fuel-air mixture in the internal combustion cylinder. The mixture is supposed to be ignited by the spark plug at the top of the compression stroke, with a controlled flame front spreading at a rate about equal to the speed of the piston head as it moves downward in the power stroke. Some fuels, however, detonated (ignited spontaneously throughout the entire fuel-air mixture) as a result of the compression itself, causing loss of fuel efficiency and damage to the engine.

With the cooperation of Robert Waley Cohen of Shell Petroleum, Ricardo evaluated chemical mixtures of fuels and found that paraffins (such as *n*-heptane, the current low-octane standard) detonated readily, but aromatics such as toluene were nearly immune to detonation. He established a "toluene number" rating to describe the tendency of various fuels to detonate; this number was replaced in

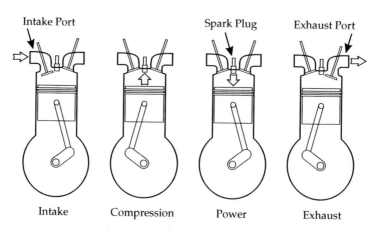

Intake Port | Spark Plug | Exhaust Port

Intake   Compression   Power   Exhaust

The four cycles of a standard internal combustion engine (left to right): (1) intake, when air enters the cylinder and mixes with gasoline vapor; (2) compression, when the cylinder is sealed and the piston moves up to compress the air-fuel mixture; (3) power, when the spark plug ignites the mixture, creating more pressure that propels the piston downward; and (4) exhaust, when the burned gases exit the cylinder through the exhaust port.

the 1920's by the "octane number" devised by Thomas Midgley at the Delco laboratories in Dayton, Ohio.

The fuel work was carried out in an experimental engine designed by Ricardo that allowed direct observation of the flame front as it spread and permitted changes in compression ratio while the engine was running. Three principles emerged from the investigation: the fuel-air mixture should be admitted with as much turbulence as possible, for thorough mixing and efficient combustion; the spark plug should be centrally located to prevent distant pockets of the mixture from detonating before the flame front reaches them; and the mixture should be kept as cool as possible to prevent detonation.

These principles were then applied in the first truly efficient side-valve ("L-head") engine—that is, an engine with the valves in a chamber at the side of the cylinder, in the engine block, rather than overhead, in the engine head. Ricardo patented this design, and after winning a patent dispute in court in 1932, he received royalties or consulting fees for it from engine manufacturers all over the world.

IMPACT

The side-valve engine was the workhorse design for automobile and marine engines until after World War II. With its valves actuated directly by a camshaft in the crankcase, it is simple, rugged, and easy to manufacture. Overhead valves with overhead camshafts are the standard in automobile engines today, but the side-valve engine is still found in marine applications and in small engines for lawn mowers, home generator systems, and the like. In its widespread use and its decades of employment, the side-valve engine represents a scientific and technological breakthrough in the twentieth century.

Ricardo and his colleagues, Oliver Thornycroft and D. R. Pye, went on to create other engine designs—notably, the sleeve-valve aircraft engine that was the basic pattern for most of the great British planes of World War II and early versions of the aircraft jet engine. For his technical advances and service to the government, Ricardo was elected a Fellow of the Royal Society in 1929, and he was knighted in 1948.

See also Alkaline storage battery; Assembly line; Diesel locomotive; Dirigible; Gas-electric car; Interchangeable parts; Thermal cracking process.

FURTHER READING

A History of the Automotive Internal Combustion Engine. Warrendale, Pa.: Society of Automotive Engineers, 1976.

Mowery, David C., and Nathan Rosenberg. *Paths of Innovation: Technological Change in Twentieth Century America.* New York: Cambridge University Press, 1999.

Ricardo, Harry R. *Memories and Machines: The Pattern of My Life.* London: Constable, 1968.

THE INTERNET

THE INVENTION: A worldwide network of interlocking computer systems, developed out of a U.S. government project to improve military preparedness.

THE PEOPLE BEHIND THE INVENTION:
  Paul Baran, a researcher for the RAND corporation
  Vinton G. Cerf (1943-   ), an American computer scientist regarded as the "father of the Internet"

COLD WAR COMPUTER SYSTEMS

In 1957, the world was stunned by the launching of the satellite *Sputnik I* by the Soviet Union. The international image of the United States as the world's technology superpower and its perceived edge in the Cold War were instantly brought into question. As part of the U.S. response, the Defense Department quickly created the Advanced Research Projects Agency (ARPA) to conduct research into "command, control, and communications" systems. Military planners in the Pentagon ordered ARPA to develop a communications network that would remain usable in the wake of a nuclear attack. The solution, proposed by Paul Baran, a scientist at the RAND Corporation, was the creation of a network of linked computers that could route communications around damage to any part of the system. Because the centralized control of data flow by major "hub" computers would make such a system vulnerable, the system could not have any central command, and all surviving points had to be able to reestablish contact following an attack on any single point. This redundancy of connectivity (later known as "packet switching") would not monopolize a single circuit for communications, as telephones do, but would automatically break up computer messages into smaller packets, each of which could reach a destination by rerouting along different paths.

ARPA then began attempting to link university computers over telephone lines. The historic connecting of four sites conducting ARPA research was accomplished in 1969 at a computer laboratory at the University of California at Los Angeles (UCLA), which was connected to computers at the University of California at Santa Barbara, the Stanford Research Institute, and the University of Utah. UCLA graduate student Vinton Cerf played a major role in establishing the connection, which was first known as "ARPAnet." By 1971, more than twenty sites had been connected to the network, including supercomputers at the Massachusetts Institute of Technology and Harvard University; by 1981, there were more than two hundred computers on the system.

The Development of the Internet

Because factors such as equipment failure, overtaxed telecommunications lines, and power outages can quickly reduce or abort ("crash") computer network performance, the ARPAnet managers and others quickly sought to build still larger "internetting" projects. In the late 1980's, the National Science Foundation built its own network of five supercomputer centers to give academic researchers access to high-power computers that had previously been available only to military contractors. The "NSFnet" connected university networks by linking them to the closest regional center; its development put ARPAnet out of commission in 1990. The economic savings that could be gained from the use of electronic mail ("e-mail"), which reduced postage and telephone costs, were motivation enough for many businesses and institutions to invest in hardware and network connections.

The evolution of ARPAnet and NSFnet eventually led to the creation of the "Internet," an international web of interconnected government, education, and business computer networks that has been called "the largest machine ever constructed." Using appropriate software, a computer terminal or personal computer can send and receive data via an "Internet Protocol" packet (an electronic envelope with an address). Communications programs on the intervening networks "read" the addresses on packets moving through the Internet and forward the packets toward their destinations. From approximately one thousand networks in the mid-1980's, the Internet grew to an estimated thirty thousand connected networks by 1994, with an estimated 25 million users accessing it regularly. The

Vinton Cerf

Although Vinton Cerf is widely hailed as the "father of the Internet," he himself disavows that honor. He has repeatedly emphasized that the Internet was built on the work of countless others, and that he and his partner merely happened to make a crucial contribution at a turning point in Internet development.

The path leading Cerf to the Internet began early. He was born in New Haven, Connecticut, in 1943. He read widely, devouring L. Frank Baum's Oz books and science fiction novels—especially those dealing with real-science themes. When he was ten, a book called *The Boy Scientist* fired his interest in science. After starting high school in Los Angeles in 1958, he got his first glimpse of computers, which were very different devices in those days. During a visit to a Santa Monica lab, he inspected a computer filling three rooms with wires and vacuum tubes that analyzed data from a Canadian radar system built to detect sneak missile attacks from the Soviet Union. Two years later he and a friend began programming a paper-tape computer at UCLA while they were still in high school.

After graduating from Stanford University in 1965 with a degree in computer science, Cerf worked for IBM for two years, then entered graduate school at UCLA. His work on multi-processing computer systems got sidetracked when a Defense Department request came in asking for help on a packet-switching project. This new project drew him into the brand-new field of computer networking on a system that became known as the ARPAnet. In 1972 Cerf returned to Stanford as an assistant professor. There he and a colleague, Robert Kahn, developed the concepts and protocols that became the basis of the modern Internet—a term they coined in a paper they delivered in 1974.

Afterward Cerf made development of the Internet the focus of his distinguished career, and he later moved back into the business world. In 1994 he returned to MCI as senior vice president of Internet architecture. Meanwhile, he founded the Internet Society in 1992 and the Internet Societal Task Force in 1999.

majority of Internet users live in the United States and Europe, but the Internet has continued to expand internationally as telecommunications lines are improved in other countries.

IMPACT

Most individual users access the Internet through modems attached to their home personal computers by subscribing to local area networks. These services make information sources available such as on-line encyclopedias and magazines and embrace electronic discussion groups and bulletin boards on nearly every specialized interest area imaginable. Many universities converted large libraries to electronic form for Internet distribution, with an ambitious example being Cornell University's conversion to electronic form of more than 100,000 books on the development of America's infrastructure.

Numerous corporations and small businesses soon began to market their products and services over the Internet. Problems soon became apparent with the commercial use of the new medium, however, as the protection of copyrighted material proved to be difficult; data and other text available on the system can be "downloaded," or electronically copied. To protect their resources from unauthorized use via the Internet, therefore, most companies set up a "firewall" computer to screen incoming communications.

The economic policies of the Bill Clinton administration highlighted the development of the "information superhighway" for improving the delivery of social services and encouraging new businesses; however, many governmental agencies and offices, including the U.S. Senate and House of Representative, have been slow to install high-speed fiber-optic network links. Nevertheless, the Internet soon came to contain numerous information sites to improve public access to the institutions of government.

See also Cell phone; Communications satellite; Fax machine; Personal computer.

FURTHER READING

Abbate, Janet. *Inventing the Internet.* Cambridge, Mass.: MIT Press, 2000.

Brody, Herb. "Net Cerfing." *Technology Review (Cambridge, Mass.)* 101, no. 3 (May-June, 1998).

Bryant, Stephen. *The Story of the Internet.* London: Peason Education, 2000.

Rodriguez, Karen. "Plenty Deserve Credit as 'Father' of the Internet." *Business Journal* 17, no. 27 (October 22, 1999).

Stefik, Mark J., and Vinton Cerf. *Internet Dreams: Archetypes, Myths, and Metaphors.* Cambridge, Mass.: MIT Press, 1997.

"Vint Cerf." *Forbes* 160, no. 7 (October 6, 1997).

Wollinsky, Art. *The History of the Internet and the World Wide Web.* Berkeley Heights, N.J.: Enslow, 1999.

Iron lung

The invention: A mechanical respirator that saved the lives of victims of poliomyelitis.

The people behind the invention:
Philip Drinker (1894-1972), an engineer who made many contributions to medicine
Louis Shaw (1886-1940), a respiratory physiologist who assisted Drinker
Charles F. McKhann III (1898-1988), a pediatrician and founding member of the American Board of Pediatrics

A Terrifying Disease

Poliomyelitis (polio, or infantile paralysis) is an infectious viral disease that damages the central nervous system, causing paralysis in many cases. Its effect results from the destruction of neurons (nerve cells) in the spinal cord. In many cases, the disease produces crippled limbs and the wasting away of muscles. In others, polio results in the fatal paralysis of the respiratory muscles. It is fortunate that use of the Salk and Sabin vaccines beginning in the 1950's has virtually eradicated the disease.

In the 1920's, poliomyelitis was a terrifying disease. Paralysis of the respiratory muscles caused rapid death by suffocation, often within only a few hours after the first signs of respiratory distress had appeared. In 1929, Philip Drinker and Louis Shaw, both of Harvard University, reported the development of a mechanical respirator that would keep those afflicted with the disease alive for indefinite periods of time. This device, soon nicknamed the "iron lung," helped thousands of people who suffered from respiratory paralysis as a result of poliomyelitis or other diseases.

Development of the iron lung arose after Drinker, then an assistant professor in Harvard's Department of Industrial Hygiene, was appointed to a Rockefeller Institute commission formed to improve methods for resuscitating victims of electric shock. The best-known use of the iron lung—treatment of poliomyelitis—was a result of numerous epidemics of the disease that occurred from 1898 until the 1920's, each leaving thousands of Americans paralyzed.

The concept of the iron lung reportedly arose from Drinker's observation of physiological experiments carried out by Shaw and Drinker's brother, Cecil. The experiments involved the placement of a cat inside an airtight box—a body plethysmograph—with the cat's head protruding from an airtight collar. Shaw and Cecil Drinker then measured the volume changes in the plethysmograph to identify normal breathing patterns. Philip Drinker then placed cats paralyzed by curare inside plethysmographies and showed that they could be kept breathing artificially by use of air from a hypodermic syringe connected to the device.

Next, they proceeded to build a human-sized plethysmograph-like machine, with a five-hundred-dollar grant from the New York Consolidated Gas Company. This was done by a tinsmith and the Harvard Medical School machine shop.

BREATH FOR PARALYZED LUNGS

The first machine was tested on Drinker and Shaw, and after several modifications were made, a workable iron lung was made available for clinical use. This machine consisted of a metal cylinder large enough to hold a human being. One end of the cylinder, which contained a rubber collar, slid out on casters along with a stretcher on which the patient was placed. Once the patient was in position and the collar was fitted around the patient's neck, the stretcher was pushed back into the cylinder and the iron lung was made airtight.

The iron lung then "breathed" for the patient by using an electric blower to remove and replace air alternatively inside the machine. In the human chest, inhalation occurs when the diaphragm contracts and powerful muscles (which are paralyzed in poliomyelitis sufferers) expand the rib cage. This lowers the air pressure in the lungs and allows inhalation to occur. In exhalation, the diaphragm and chest muscles relax, and air is expelled as the chest cavity returns to its normal size. In cases of respiratory paralysis treated with an iron lung, the air coming into or leaving the iron lung alternately compressed the patient's chest, producing artificial exhalation, and the allowed it to expand to so that the chest could fill with air. In this way, iron lungs "breathed" for the patients using them.

Careful examination of each patient was required to allow technicians to adjust the rate of operation of the machine. A cooling system and ports for drainage lines, intravenous lines, and the other apparatus needed to maintain a wide variety of patients were included in the machine.

The first person treated in an iron lung was an eight-year-old girl afflicted with respiratory paralysis resulting from poliomyelitis. The iron lung kept her alive for five days. Unfortunately, she died from heart failure as a result of pneumonia. The next iron lung patient, a Harvard University student, was confined to the machine for several weeks and later recovered enough to resume a normal life.

Impact

The Drinker respirator, or iron lung, came into use in 1929 and soon was considered indispensable, saving lives of poliomyelitis victims until the development of the Salk vaccine in the 1950's.

Although the iron lung is no longer used, it played a critical role in the development of modern respiratory care, proving that large numbers of patients could be kept alive with mechanical support. The iron lung and polio treatment began an entirely new era in treatment of respiratory conditions.

In addition to receiving a number of awards and honorary degrees for his work, Drinker was elected president of the American Industrial Hygiene Association in 1942 and became chairman of Harvard's Department of Industrial Hygiene.

See also Electrocardiogram; Electroencephalogram; Heart-lung machine; Pacemaker; Polio vaccine (Sabin); Polio vaccine (Salk).

Further Reading

DeJauregui, Ruth. *One Hundred Medical Milestones That Shaped World History*. San Mateo, Calif.: Bluewood Books, 1998.

Hawkins, Leonard C. *The Man in the Iron Lung: The Frederick B. Snite, Jr., Story*. Garden City, N.Y.: Doubleday, 1956.

Rudulph, Mimi. *Inside the Iron Lung*. Buckinghamshire: Kensal Press, 1984.

Laminated Glass

THE INVENTION: Double sheets of glass separated by a thin layer of plastic sandwiched between them.

THE PEOPLE BEHIND THE INVENTION:
Edouard Benedictus (1879-1930), a French artist
Katherine Burr Blodgett (1898-1979), an American physicist

THE QUEST FOR UNBREAKABLE GLASS

People have been fascinated for centuries by the delicate transparency of glass and the glitter of crystals. They have also been frustrated by the brittleness and fragility of glass. When glass breaks, it forms sharp pieces that can cut people severely. During the 1800's and early 1900's, a number of people demonstrated ways to make "unbreakable" glass. In 1855 in England, the first "unbreakable" glass panes were made by embedding thin wires in the glass. The embedded wire grid held the glass together when it was struck or subjected to the intense heat of a fire. Wire glass is still used in windows that must be fire resistant. The concept of embedding the wire within a glass sheet so that the glass would not shatter was a predecessor of the concept of laminated glass.

A series of inventors in Europe and the United States worked on the idea of using a durable, transparent inner layer of plastic between two sheets of glass to prevent the glass from shattering when it was dropped or struck by an impact. In 1899, Charles E. Wade of Scranton, Pennsylvania, obtained a patent for a kind of glass that had a sheet or netting of mica fused within it to bind it. In 1902, Earnest E. G. Street of Paris, France, proposed coating glass battery jars with pyroxylin plastic (celluloid) so that they would hold together if they cracked. In Swindon, England, in 1905, John Crewe Wood applied for a patent for a material that would prevent automobile windshields from shattering and injuring people when they broke. He proposed cementing a sheet of material such as celluloid between two sheets of glass. When the window was broken, the inner material would hold the glass splinters together so that they would not cut anyone.

KATHARINE BURR BLODGETT

Besides the danger of shattering, glass poses another problem. It reflects light, as much as 10 percent of the rays hitting it, and that is bad for many precision instruments. Katharine Burr Blodgett cleared away that problem.

Blodgett was born in 1898 in Schenectady, New York, just months after her father died. Her widowed mother, intent upon giving her and her brother the best upbringing possible, devoted herself to their education and took them abroad to live for extended periods. She succeeded. Blodgett attended Bryn Mawr and then earned a master's degree in physics from the University of Chicago. With the help of a family friend, Irving Langmuir, who later won a Nobel Prize in Chemistry, she was promised a job at the General Electric (GE) research laboratory. However, Langmuir first wanted her to study more physics. Blodgett went to Cambridge University and under the guidance of Ernest Rutherford became the first women to receive a doctorate in physics there. Then she went to work at GE.

Collaborating with Langmuir, Blodgett found that she could coat glass with a film one layer of molecules at a time, a feat never accomplished before. Moreover, the color of light reflected differed with the number of layers of film. She discovered that by adjusting the number of layers she could cancel out the light reflected by the glass beneath, so as much as 99 percent of natural light would pass through the glass. Producing almost no reflection, this treated glass was "invisible." It was perfect for lenses, such as those in cameras and microscopes. Blodgett also devised a way to measure the thickness of films based on the wavelengths of light they reflect—a color gauge—that became a standard laboratory technique.

Blodgett died in the town of her birth in 1979.

REMEMBERING A FORTUITOUS FALL

In his patent application, Edouard Benedictus described himself as an artist and painter. He was also a poet, musician, and philosopher who was descended from the philosopher Baruch Benedictus Spinoza; he seemed an unlikely contributor to the progress of glass manufacture. In 1903, Benedictus was cleaning his laboratory when he dropped a glass bottle that held a nitro-cellulose solution. The solvents, which had evaporated during the years that the bottle had sat on a shelf, had left a strong celluloid coating on the glass. When Benedictus picked up the bottle, he was surprised to see that it had not shattered: It was starred, but all the glass fragments had been held together by the internal celluloid coating. He looked at the bottle closely, labeled it with the date (November, 1903) and the height from which it had fallen, and put it back on the shelf.

One day some years later (the date is uncertain), Benedictus became aware of vehicular collisions in which two young women received serious lacerations from broken glass. He wrote a poetic account of a daydream he had while he was thinking intently about the two women. He described a vision in which the faintly illuminated bottle that had fallen some years before but had not shattered appeared to float down to him from the shelf. He got up, went into his laboratory, and began to work on an idea that originated with his thoughts of the bottle that would not splinter.

Benedictus found the old bottle and devised a series of experiments that he carried out until the next evening. By the time he had finished, he had made the first sheet of Triplex glass, for which he applied for a patent in 1909. He also founded the Société du Verre Triplex (The Triplex Glass Society) in that year. In 1912, the Triplex Safety Glass Company was established in England. The company sold its products for military equipment in World War I, which began two years later.

Triplex glass was the predecessor of laminated glass. Laminated glass is composed of two or more sheets of glass with a thin layer of plastic (usually polyvinyl butyral, although Benedictus used pyroxylin) laminated between the glass sheets using pressure and heat. The plastic layer will yield rather than rupture when subjected to loads and stresses. This prevents the glass from shattering into sharp pieces. Because of this property, laminated glass is also known as "safety glass."

IMPACT

Even after the protective value of laminated glass was known, the product was not widely used for some years. There were a number of technical difficulties that had to be solved, such as the discoloring of the plastic layer when it was exposed to sunlight; the relatively high cost; and the cloudiness of the plastic layer, which obscured vision—especially at night. Nevertheless, the expanding automobile industry and the corresponding increase in the number of accidents provided the impetus for improving the qualities and manufacturing processes of laminated glass. In the early part of the century, almost two-thirds of all injuries suffered in automobile accidents involved broken glass.

Laminated glass is used in many applications in which safety is important. It is typically used in all windows in cars, trucks, ships, and aircraft. Thick sheets of bullet-resistant laminated glass are used in banks, jewelry displays, and military installations. Thinner sheets of laminated glass are used as security glass in museums, libraries, and other areas where resistance to break-in attempts is needed. Many buildings have large ceiling skylights that are made of laminated glass; if the glass is damaged, it will not shatter, fall, and hurt people below. Laminated glass is used in airports, hotels, and apartments in noisy areas and in recording studios to reduce the amount of noise that is transmitted. It is also used in safety goggles and in viewing ports at industrial plants and test chambers. Edouard Benedictus's recollection of the bottle that fell but did not shatter has thus helped make many situations in which glass is used safer for everyone.

See also Buna rubber; Contact lenses; Neoprene; Plastic; Pyrex glass; Silicones.

Further Reading

Eastman, Joel W. *Styling vs. Safety: The American Automobile Industry and the Development of Automotive Safety, 1900-1966.* Lanham: University Press of America, 1984.
Fariss, Robert H. "Fifty Years of Safer Windshields." *CHEMTECH* 23, no. 9 (September, 1993).
Miel, Rhoda. "New Process Promises Safer Glass." *Automotive News* 74, no. 5863 (February 28, 2000).

Polak, James L. "Eighty Years Plus of Automotive Glass Development: Windshields Were Once an Option, Today They Are an Integral Part of the Automobile." *Automotive Engineering* 98, no. 6 (June, 1990).